BIOTECHNOLOGY PROCESSES

Scale-up and Mixing

BIOTECHNOLOGY PROCESSES
Scale-up and Mixing

Edited by

CHESTER S. HO
State University of New York at Buffalo
Buffalo, N.Y.

JAMES Y. OLDSHUE
Mixing Equipment Company
Rochester, New York

American Institute of Chemical Engineers
New York, New York

Library of Congress Cataloging-in-Publication Data

Biotechnology processes.

Includes index.
1. Biochemical engineering—Congresses. 2. Biotechnology—Methodology—Congresses. I. Ho, Chester S., 1950– . II. Oldshue, James Y. III. American Institute of Chemical Engineers.
TP248.3.B62 1987 660'.6 87-14393
ISBN 0–8169–0410–3

MANUFACTURED IN THE UNITED STATES OF AMERICA

Contents

BIOPROCESS SCALE-UP

BIOPROCESS APPLICATIONS

Preface

The American Institute of Chemical Engineers held a major symposium on Scale-up and Mixing of Bioprocesses at the Miami annual meeting in November, 1986. The sessions attracted twenty presentations. Another seven, which were not included in the oral presentations, are included in the volume. This represents a major collection of twenty-seven papers that relate to many different aspects of chemical engineering in bioprocessing.

The papers encompass experimental and design work on existing fermentors, which include the typical aerobic process well known in the industry. Results of studies of the mixing characteristics include a variety of experimental observations with different kinds of impellers, particularly the high pumping capacity, low shear rate impellers introduced in the last three or four years. These are usually patterned after airfoil designs, and are typically called either hydrofoil, liquidfoil, or laserfoil impeller.

Scale-up is a key item in bioprocessing, and several papers discuss the principles of scale-up and give examples of various types of scale-up calculations.

Some of the latest bioprocesses involve mammalian cells or other biological organisms that are much more sensitive to fluid shear stress than previously well-known free-suspending organisms. This may well indicate a different kind of impeller system, particularly where aerobic requirements are either low or absent, and the mixer is to provide primarily a blending and cell suspension function. The characteristics of airlift, airpulse, and bubble columns are included in the contributions presented in this volume. Of additional interest are methods of handling the non-Newtonian character of many fluids in bioprocessing. A variety of papers in this volume look at some of the basic concepts of mass transfer in impeller mixed fermentors as well as airlift types.

Particularly encouraging is a large number of papers devoted to actual bioprocess operations. These include descriptions of the requirements for several different products. Also very encouraging is the large number of papers from industrial sources, and the data presented which will be of help to all concerned in the pursuit of information in the whole spectrum of bioprocessing.

The editors would like to thank the authors for their fine contributions, and their diligent attention to time schedules and proper form for the paper preparation which made the production of this book a reality. The helpful suggestions and assistance of the AIChE staff and Stanley Barnett of the University of Rhode Island are also acknowledged.

Chester S. Ho
State University of New York at Buffalo
Buffalo, N.Y.

James Y. Oldshue
Mixing Equipment Company
Rochester, N.Y.

BIOPROCESS MIXING

Stirred Bioreactors

Current Situation in Fluid Mixing for Fermentation Processes

JAMES Y. OLDSHUE
Mixing Equipment Company
Rochester, New York

Fermentation involves anaerobic and aerobic processes. In anaerobic processes there is no need for high oxygen uptake rates and the main mixer design parameters are involved with mass transfer across the liquid-solid boundary, possible effects of shear rate on the organisms, and the overall blending and mixing of nutrients in the tank.

The main consideration with looking at mixer designs is to examine the role of fluid shear, power and pumping capacity.

Looking first at turbulent flow situations, in which the Reynolds number around the impeller is 1,000 or higher, we can take a measurement of the velocity at a point and we will normally get a fluctuating velocity as shown in Figure 1. From this fluctuating velocity profile with time, we can calculate the average velocity as well as the fluctuating component. The fluctuating component is normally expressed as a root mean square velocity fluctuation.

It turns out that large particles (on the order of 1,000 microns or larger) see only the average velocity and the velocity gradients between adjacent layers of fluid. These are shown in Figure 2 and it is seen that around a radial flow impeller, that there is a maximum shear rate, related to tip speed, and an average shear rate related to impeller operating speed.

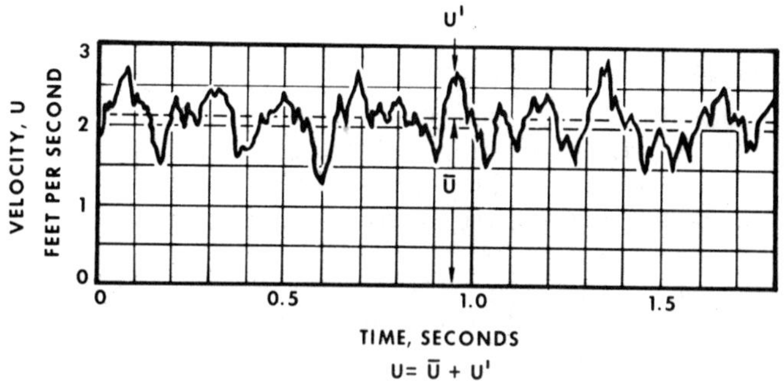

FIGURE 1. Typical velocity fluctuation in turbulent flow.

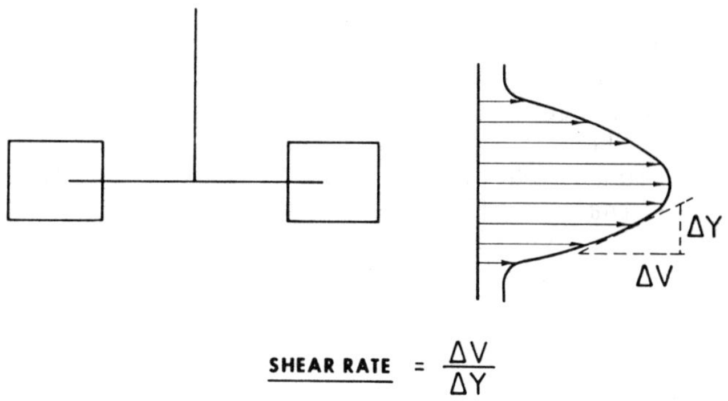

FIGURE 2. Method of calculation of average and maximum macro scale impeller zone shear rates.

All the power dissipates as heat. The only way to go from heat to power is through viscous shear. Therefore, the velocity gradients that affect the large scale particles, the macro scale, do not absorb a large part of the energy. The energy, which is represented by power per unit of volume, is largely dissipated through the action of the turbulent small scale viscous shear rates, which are related to the root mean square velocity fluctuation.

Micro scale shear rates affect particles largely 100 microns or smaller and are related in a major way to the power per unit of volume, as well as the root mean square velocity fluctuations.

The large particles respond to the macro scale shear rates based on average velocities, and don't know in general that the micro scale shear rates exist. On the other hand, small scale particles go along for the ride on the macro scale velocities, but respond directly to the high frequency fluctuations that are in their size regime.

The old saying that you must match the shear stresses and the responses to the particles is involved here; that you don't use a rivet hammer drill tooth--you don't use an ultrasonic drill to drive a rivet.

Every mixing problem has a micro scale mixing floor which is necessary to carry out the proper combination of fluid particles and materials at micro scale levels. It can also be very responsive to the micro scale level, and so process results can markedly change by power per unit of volume. On the other hand, many of the other combinations of functions are related to shear rates on a macro scale basis.

In the pilot plant the purpose is usually to determine which variables are the most important. This means that the scale of the pilot plant is very important. For macro scale mixing processes, the impeller blades should physically be at least three times bigger than the biggest particle, bubble, droplet or fluid clump that is of interest to the process. Otherwise, scaling will not be relative to what will happen in the full scale system.

A small mixing tank has a very short blend time and a much lower maximum shear rate impeller zone relative to large tanks. It may be necessary to use a non-geometric impeller and mixing tank on a small scale to duplicate the performance of proper variables on full scale.

Sufficient oxygen mass transfer is usually no problem in aerobic systems in the pilot plant. Without regard to scale, comparable mass transfer rates per unit of volume can be achieved on small and large systems. The main problem comes into the blending phenomena. The large tank has a much longer blend time than the small tank and this can lead to oxygen deficiencies in remote parts of the tank. Mass transfer zones should be made a part of the overall mixing flow pattern in the entire system.

In terms of minimum fluid shear rates required for some mammalian cells, and other types of shear sensitive cells, there will need to be a different approach to mixing requirements.

There is a whole family of fluid foil impellers, Figure 3, which are able to produce motion with the very least amount of shear rate--both macro and micro scale. However, if these impellers are required on small scale, then the same questions that existed previously, "how do we keep the blend time at a high enough and reasonable level on full scale and the shear rates down to a reasonable maximum on full scale?" apply to these impellers as well.

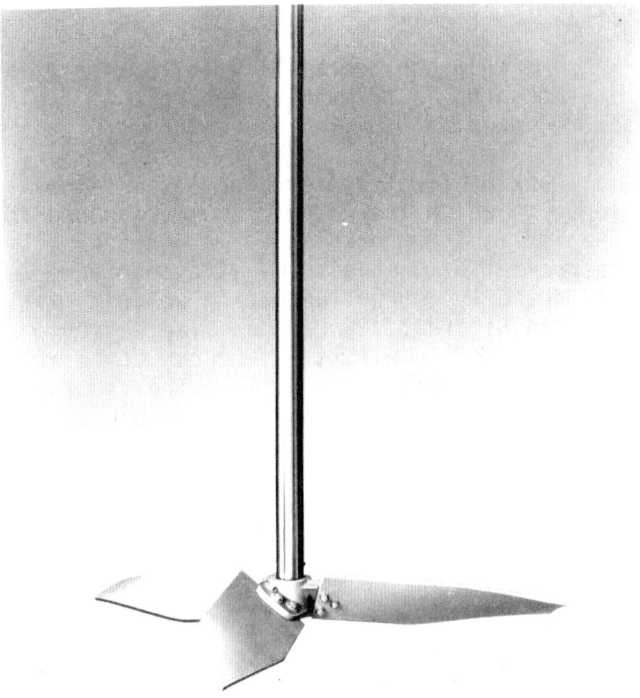

FIGURE 3. Typical fluid foil impeller.

There are other kinds of impellers such as anchors, Figure 4, and helicals. Figure 5, that have a much different kind of shear rate and flow pattern. These impellers may well end up to be the impellers of choice in cultures that are very shear sensitive.

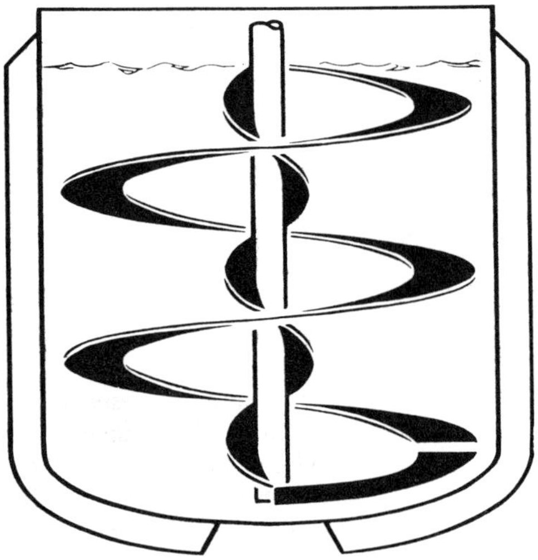

FIGURE 5. Typical helical impeller.

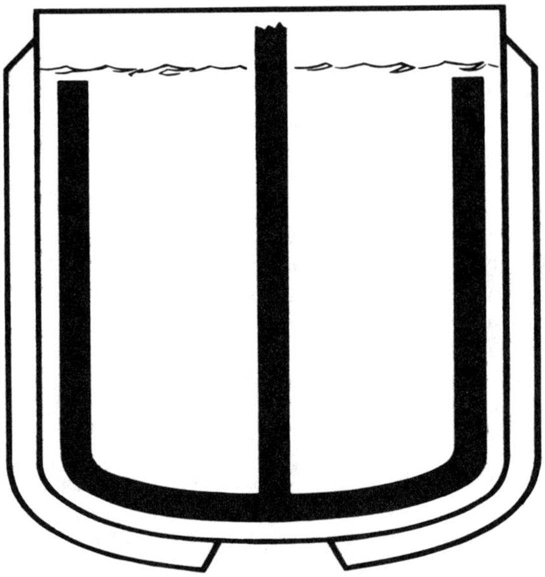

FIGURE 4. Typical anchor impeller.

A Variable Volume Two-Zone Mixing Model

S.J. GIBBS
D.F. LOY
K.A. DEBELAK
R.D. TANNER
Department of Chemical Engineering
Vanderbilt University
Nashville, Tennessee 37235

An analysis is presented to determine the dilution rates in a two-zone mixing model when the volumes in each zone are unequal and constant. Also, a model is developed to account for the effect of variable volume mixing zones. The model describes the effect of imperfect mixing on the progress of a first-order irreversible reaction in a batch reactor. The model is described by a system of two nonlinear differential equations. The equations are solved numerically for various values of the mixing parameters in order to study the qualitative effects on the behavior of the model. The variable volume model was qualitatively confirmed by the shape of experimental crossplots of the concentration of reactant in one zone versus the concentration of reactant in the other zone. In addition, the model was tested by comparison with data from an imperfectly mixed batch reactor in which the progress of a pseudo-first-order irreversible reaction was studied.

The two zone mixing model describing the effect of imperfect mixing on a first order irreversible reaction in a batch reactor was developed in two earlier papers (1,2). The model system consisted of a two-zone reactor with equal constant volumes and flow rates, as seen in Figure 1. In another previous paper (3), we studied the effect of constant unequal volumes on the progress of the irreversible A-->B reaction occurring in each zone. In this paper we will present a procedure to determine the dilution rates when the volumes are unequal and constant, and derive a model for the case of a variable volume in the mixing zones. This two zone, variable volume mixing model may have application to both chemical and biomedical models. For example, in many bioprocesses, the effect of poor mixing can be more of a problem at large scale than at small scale because of local buildups, of either product or substrate, leading to product or substrate inhibition. Toxicity caused by excessive concentrations of either products or substrates causes a similar set of problems (4)

In addition, since the model is based on positional differentiation by zones, those bioprocesses which produce thickening agents like gum xanthan or microbial cellulose (biopolymers) can lead to process control problems which arise from the dilemma that some parts of bioreactor are quite fluid while others are viscous (4). When sensors are placed in the bioreactor in different locations they give drastically different information (4). Simple mixing models such as those described in this paper may help deal with such difficulties.

CONSTANT AND UNEQUAL VOLUMES

The equations relating the concentrations in each mixing zone for the irreversible A-- > B reaction resulting from the solution of the material balances on zones one and two are (3):

$$A_1 = \frac{A^*}{D_1+D_2}\left[D_1 e^{-(k+D_1+D_2)t} + D_2 e^{-kt}\right] \quad (1)$$

$$A_2 = \frac{A^* D_2}{D_1+D_2}\left[e^{-kt} - e^{-(k+D_1+D_2)}\right] \quad (2)$$

Subtracting Equation (2) from Equation (1) gives the equation relating the concentrations of A in each zone.

$$A_1 = A_2 + A^* e^{(-k+D_1+D_2)t} \quad (3)$$

where,

A_1 = Concentration of A in zone 1 (mol/l)

A_2 = Concentration of A in zone 2 (mol/l)

A* = Initial Concentration of A in zone 1 (mol/l)

D_i = Interzone flow rate, F, divided by zone volume, V_i, F/V_i, (min^{-1}), the mixing parameter

k = first-order reaction rate constant, (min^{-1})

F = flow rate, (l/min)

V_i = Volume in each zone, i=1,2

For the case where no reaction takes place, k equals zero and

$$A_1 = A^* - \frac{D_1}{D_2} A_2 \qquad (4)$$

This is a line with slope $-D_1/D_2$ and intercept A^*. Equations (3) and (4) can be used to estimate D_1 and D_2 for this ideal case. If the first-order reaction rate constant, k, is known, then the sum $D_1 + D_2$ can be evaluated in this simplified model by taking the natural log of Equation (3) and rearranging

$$\ln\left[\frac{1}{A_1-A_2}\right] = (k+D_1+D_2)t + \ln\frac{1}{A^*} \qquad (5)$$

From the slope of the $\ln(1/(A_1 - A_2))$ versus time plot, we can find $(k+D_1+D_2)$. The initial concentration of reactant in zone 1 can be found or verified from the intercept. By subtracting the known value of k from $(k+D_1+D_2)$ we can find (D_1+D_2). Obtaining D_1/D_2 for the non-reacting case allows us to calculate the individual dilution rates D_1 and D_2.

Figure 2 shows a crossplot of concentrations for experimental data taken from the crystal violet-sodium hydroxide system. Both the reaction and no reaction cases are shown. The experimental system used to study this pseudo-first order system was previously described (2). From the slope of the no reaction case, the ratio D_1/D_2 is determined to be 1.22, if the initial transient data are neglected ($A_2 < 0.3$). The data for the reaction case are replotted in Figure 3 according to Equation (5) to determine $(k+D_1+D_2)$. The value of k, the pseudo-first order rate constant, at 20°C and a sodium hydroxide concentration of 0.01M was reported to be 3.42×10^{-3} sec^{-1}, from Cosaro (5). The values of D_1 and D_2 are therefore calculated to be 0.247 sec^{-1} and 0.203 sec^{-1}, respectively. The value of A* can be estimated from either the extrapolated intercept of Equation (4) or from the intercept of Equation (5). The value of A^* determined from Equation (4) was 1.4×10^{-5} gmoles/l and from Equation (5) 1.55×10^{-5} g moles/l. The mismatch of the constant volume model fit for A_2 0.3 and A^* prediction are due to the inadequacy of the constant volume model. These facts support the argument for the development of a more comprehensive model, hence the use of variable volumes in the following section.

THE VARIABLE VOLUME MODEL

In this model, an imperfectly mixed batch reactor is thought of as two perfectly mixed batch reactors which exchange fluid at rates F_1 and F_2 as shown in Figure 1. The basic assumptions in the development of the model are that the flow rates F_1 and F_2 can be described by:

$$F_1 = D_1 V_1 \qquad (6)$$

$$F_2 = D_2 V_2, \text{ where } D_1 \text{ and } D_2 \text{ are constants} \qquad (7)$$

and that the reaction is first-order and irreversible. A material balance on zones 1 and 2 yields the equations under isothermal conditions:

$$\frac{dV_1}{dt} = F_2 - F_1 = D_2V_2 - D_1V_1 \qquad (V_1)_0=V_{10} \qquad (8)$$

$$\frac{dV_2}{dt} = F_1 - F_2 = D_1V_1 - D_2V_2 \qquad (V_2)_0=V_{20} \qquad (9)$$

where:

V_1 = the volume of zone 1 at anytime

V_2 = the volume of zone 2 at anytime

then, V_1 and V_2 can be found as functions of time, given the initial conditions: V_{10} and V_{20}.

$$V_1 = \frac{D_2V_T}{D_1+D_2}\left[1-e^{-(D_1+D_2)t}\right]+V_{10}e^{-(D_1+D_2)t} \qquad (10)$$

$$V_2 = \frac{D_1V_T}{D_1+D_2}\left[1-e^{-(D_1+D_2)t}\right]+V_{20}e^{-(D_1+D_2)t} \qquad (11)$$

where $V_T = V_1 + V_2$

A balance on component A in Zone 1 yields the equation:

$$\frac{d(V_1A_1)}{dt} = -kA_1V_1 - F_1A_1 + F_2A_2 \tag{12}$$

Expanding $d(V_1A_1)/dt$ and combining the resulting expression with Equations (6)-(8) gives another relationship for $d(A_1V_1)/dt$

$$\frac{d(V_1A_1)}{dt} = (F_2 - F_1)A_1 + \frac{dA_1}{dt} V_1 \tag{13}$$

Setting Equations (12) and (13) equal gives an expression in which V_1 is not differentiated.

$$(F_2-F_1)A_1 + \frac{dA_1}{dt} V_1 = -kA_1V_1 - F_1A_1 + F_2A_2 \tag{14}$$

Using Equations (6) and (7) and the fact that $V_1 + V_2 = V_T$, Equation (14) can be simplified to:

$$\frac{dA_1}{dt} = -kA_1 + (A_2 - A_1) \left[\frac{D_2V_T}{V_1} - D_2\right] \tag{15}$$

Analogously, in zone 2, the differential equation for A_2 becomes:

$$\frac{dA_2}{dt} = -kA_2 + (A_1 - A_2) \left[\frac{D_1V_T}{V_2} - D_1\right] \tag{16}$$

Equations (10),(11),(15), and (16) describe the variable volume two zone mixing model.

The Early Time Approximation

It is difficult to find an analytical solution to Equations (10),(11),(15) and (16) because V_1 and V_2 are such complex functions. However, if Equations (10) and (11) are linearized in time and other approximations are made, one can derive an analytical solution.

First, linearize V_1 and V_2 about t=0.

$$V_1 = V_{10} + \left.\frac{dV_1}{dt}\right|_{t=0} \cdot t \tag{17}$$

$$V_2 = V_{20} + \left.\frac{dV_2}{dt}\right|_{t=0} \cdot t \tag{18}$$

$$\left.\frac{dV_1}{dt}\right|_{t=0} = D_2V_T - V_{10}(D_1+D_2) \quad D_2V_T \tag{19}$$

$$\left.\frac{dV_2}{dt}\right|_{t=0} = D_1V_T - V_{20}(D_1+D_2) \quad -D_2V_T \tag{20}$$

The approximations in Equations (19) and (20) can be viewed as setting $(D_1)_0=(D_2)_0$ and $V_{10} << V_{20}$ in Equations (8) and (9). Therefore,

$$V_1 = V_{10} + [D_2V_T - V_{10}(D_1+D_2)]t \tag{21}$$

$$= V_{10} + [V_{20}D_2 - V_{10}D_1]t$$

and

$$V_2 = V_{20} + [D_1V_T - V_{20}(D_1+D_2)]t \tag{22}$$

$$= V_{20} + [V_{10}D_1 - V_{20}D_2]t$$

Substitution of Equations (21) and (22) into Equations (15) and (16) yields

$$\frac{dA_1}{dt} = -kA_1 + (A_1-A_2) \left[\frac{D_2V_T}{V_{10} + [V_{20}D_2 - V_{10}D_1]t} - D_2\right] \tag{23}$$

and

$$\frac{dA_2}{dt} = -kA_2 + (A_1-A_2) \left[\frac{D_1V_T}{V_{20} + [V_{10}D_1 - V_{20}D_2]t} - D_1\right] \tag{24}$$

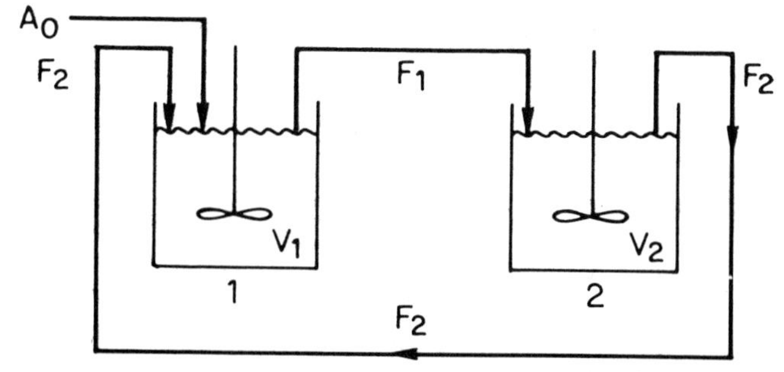

FIGURE 1. Idealized model of a two-zone batch reactor.

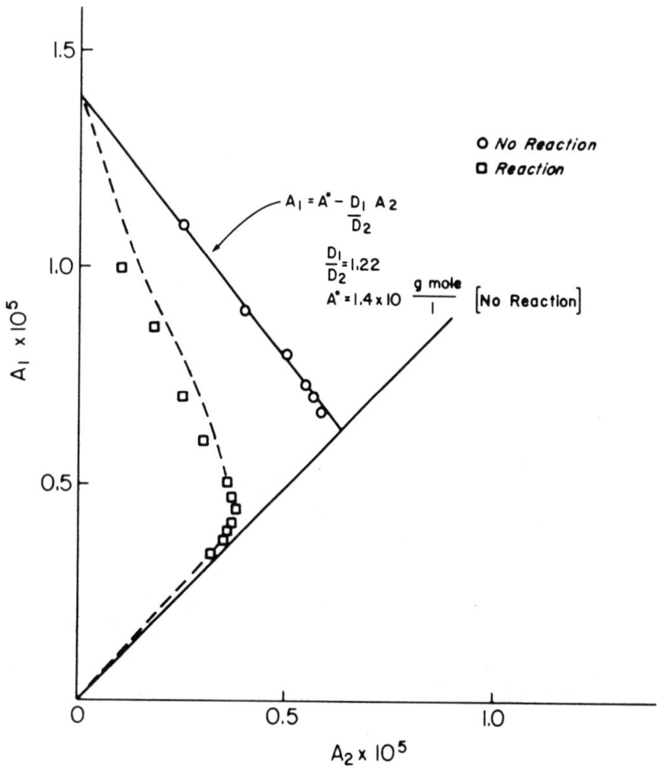

FIGURE 2. Crossplot of concentration of crystal violet dye in zone 1 versus concentration of crystal violet dye in zone 2 for the reaction and no reaction cases.

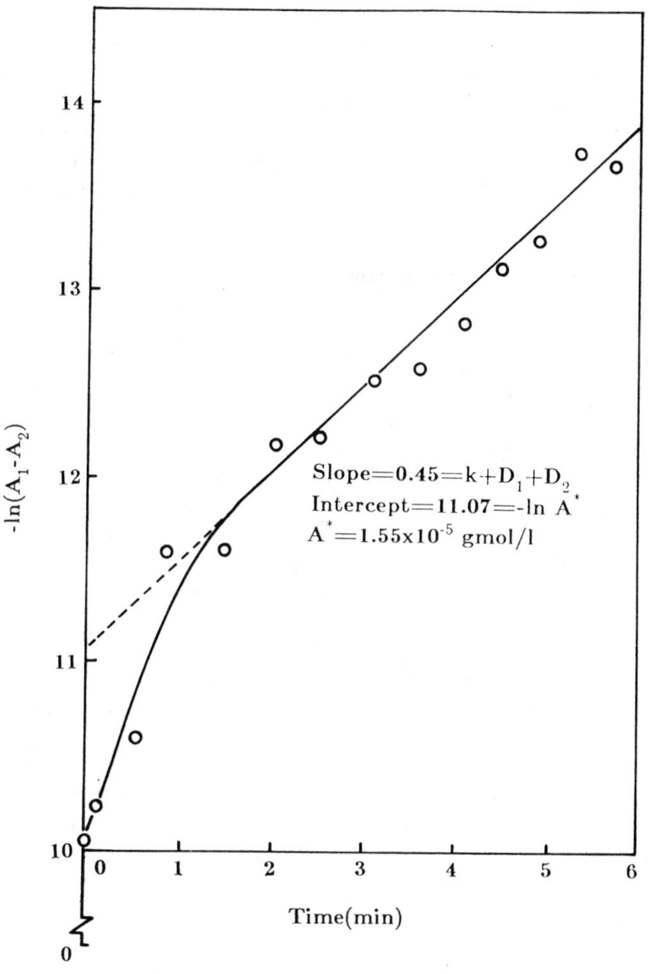

FIGURE 3. Reaction data replotted according to equation (5) to determine $(k + D_1 + D_2)$ which equals the slope in the constant volume model.

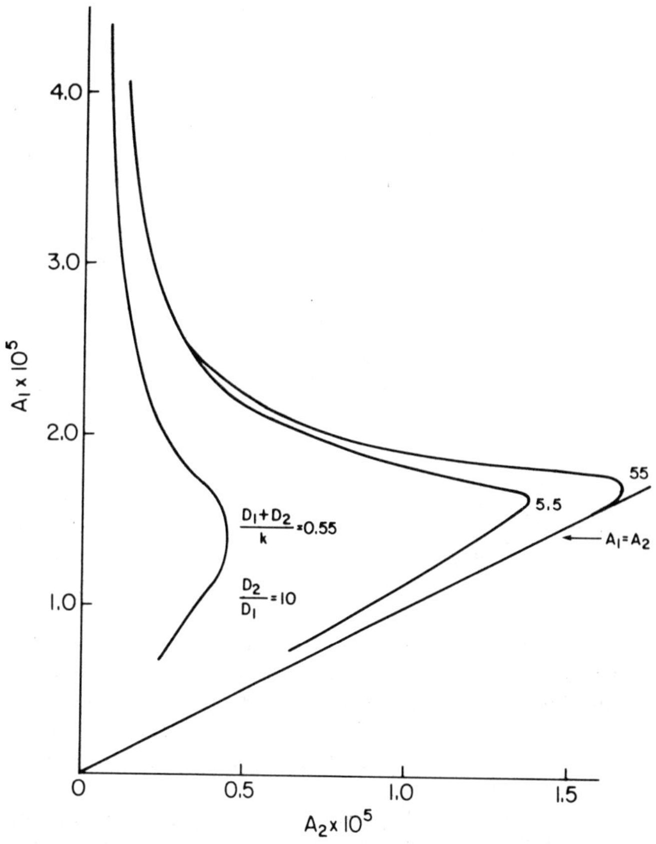

FIGURE 4. Effect of changes in $(D_1 + D_2)/k$ on the crossplot of concentration of A in zone 1 versus concentration of A in zone 2. Simulated variable volume model.

Then, application of Picard's iteration, where

$$A_{10} = A^* \text{ and } A_{20} = 0 \quad \text{gives}$$

$$A_1 = A^* - kA^*t + A^*D_2t - \frac{A_1D_2V_T}{[V_{20}D_2 - V_{10}D_1]} \cdot \ln \frac{V_{10} + [V_{20}D_2 - V_{10}D_1]t}{V_{10}} \tag{25}$$

$$A_2 = -A^*D_1t + \frac{A^*D_1V_T}{[V_{10}D_1 - V_{20}D_2]} \cdot \ln \frac{V_{20} + [V_{10}D_1 - V_{20}D_1]t}{V_{20}} \tag{26}$$

Since the argument of the ln function approaches 1 as time goes to zero, the ln[x] can be approximated by x-1. Therefore, Equations (25) and (26) can be approximated by

$$A_1 = A^* - kA^*t + A^*D_2t - \frac{A_1D_2V_T}{V_{10}}t \tag{27}$$

and

$$A_2 = -A^*D_1t + \frac{A^*D_1V_T}{V_{20}}t \tag{28}$$

Simplification of Equations (27) and (28) yields

$$A_1 = A^* - [k + \frac{V_{20}D_2}{V_{10}}]A^*t \tag{29}$$

$$A_2 = A^*D_1 \frac{V_{10}}{V_{20}} t \tag{30}$$

Thus, if early time data of A_1 and A_2 are available, then the parameters D_1 and D_2 can be estimated from the initial slopes of A_1 and A_2 versus time curves. However, it is generally difficult to obtain enough data at early times to clearly determine values of the initial slopes. Subtracting Equation (30) from Equation (29) gives the analogous relationship to Equation (5) for the variable volume case at early times:

$$\ln\left(\frac{1}{A_1 - A_2}\right) = [k + \frac{V_{20}D_2}{V_{10}} + \frac{V_{10}D_1}{V_{20}}]t + \ln\frac{1}{A^*} \tag{31}$$

Qualitative Features of Crossplots of A_1 vs. A_2

Values for A_1 and A_2 were determined by numerically solving Equations (10),(11),(15), and (16), using the IMSL subroutine DVERK. The numerically simulated equations were used to generate values which could be qualitatively compared to experimental data and used to determine the effect of changing the parameters D_1 and D_2 on the shape of the A_1 vs A_2 crossplot. Typical distinguishing shapes of the A_1 vs A_2 crossplot for the variable volume model are shown in Figures 4 and 5.

Several combinations of values of D_1, D_2, V_{10}, and k were tried in the simulation program and used to develop these crossplots of A_1 vs. A_2. In comparing these curves with the constant volume case in Figure 2, it is evident that the variable volume model simulations lead to a more distinct peak (representing a rapid rise in concentration in the second zone) than does the constant volume model.

Parameter Estimation

If plots of A_1 and A_2 versus time are made for experimental data, the values of A_1, A_2, dA_1/dt and dA_2/dt can be determined for various values of time. The values of the groups $[(D_2V_T/V_1) - D_2]$ and $[(D_1V_T/V_2) - D_1]$ can be determined at various time values. The values of D_1 and D_2 can also be estimated using a computer assisted trial and error solution of Equations (10) and (11) given the values of the above two groups at a specific time.

The experimental set-up used to obtained data to test the variable volume model differed from previous work in that a single stirrer was used as opposed to two stirrers. Two mixing speeds: 106 rpm and 75 rpm were studied, and the reaction vessel volume was larger: 8 liters as opposed to 1 liter. Otherwise, the procedure was similar to that previously used (2). Plots of experimental data and simulations based on values of D_1 and D_2, determined from the trial and error method, are shown in Figures 6 to 8. The simulation corresponds well with one set of

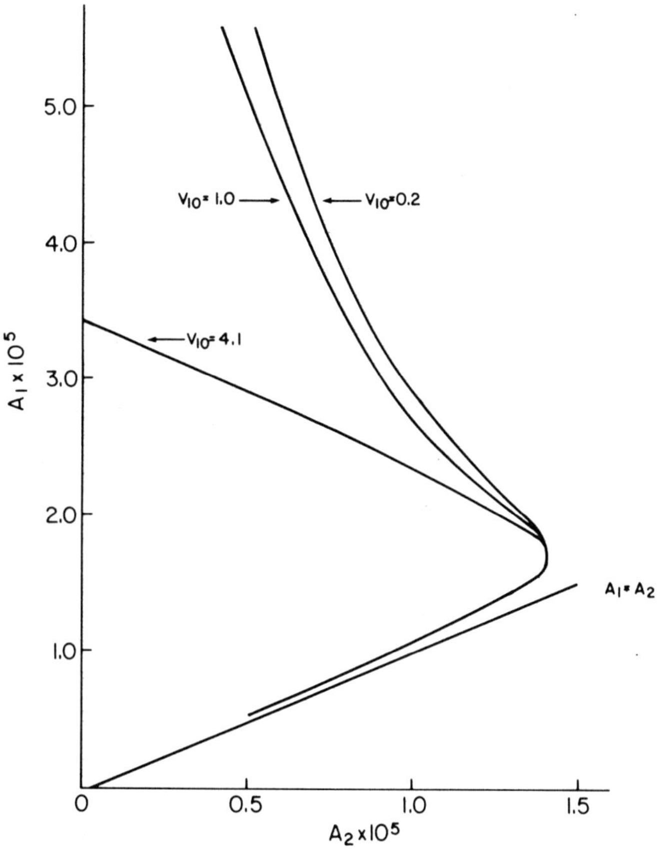

FIGURE 5. Effect of the initial volume of zone 1, V_{10}, on the crossplot of the concentration of A in zone 1 versus concentration of A in zone 2. Simulated constant volume model.

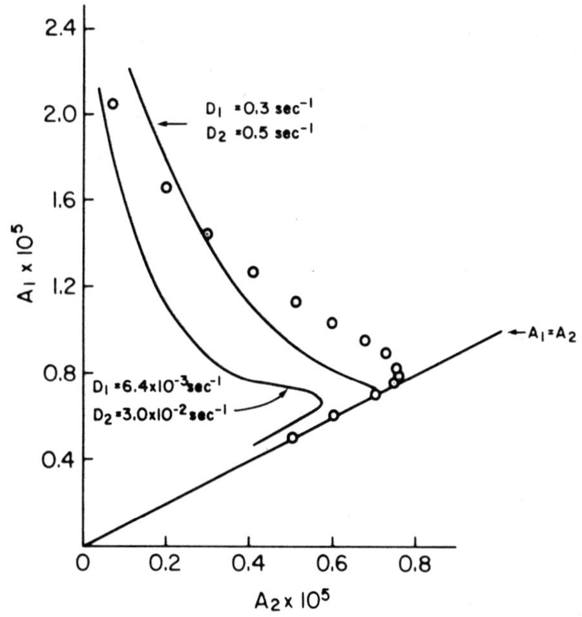

FIGURE 7. Comparison of experimental data for run 2 with simulation values from the variable volume model. A single impeller turning at 106 rpm. $k = 0.005$ sec^{-1}.

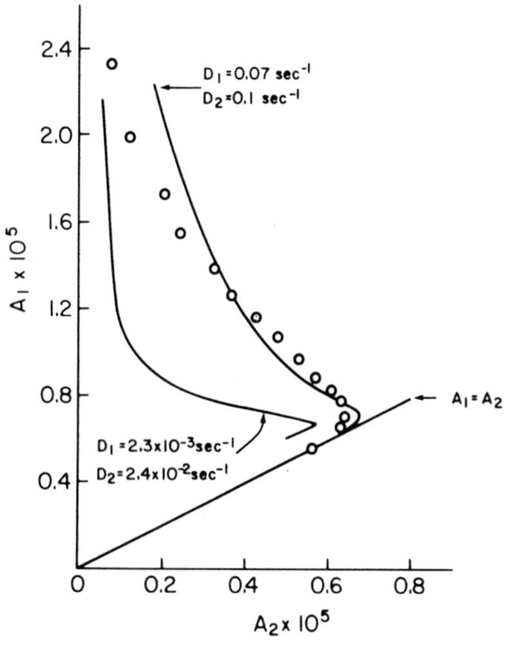

FIGURE 6. Comparison of experimental data for run 1 with simulation values from the variable volume model. A single impeller turning at 106 rpm. $k = 0.005$ sec^{-1}.

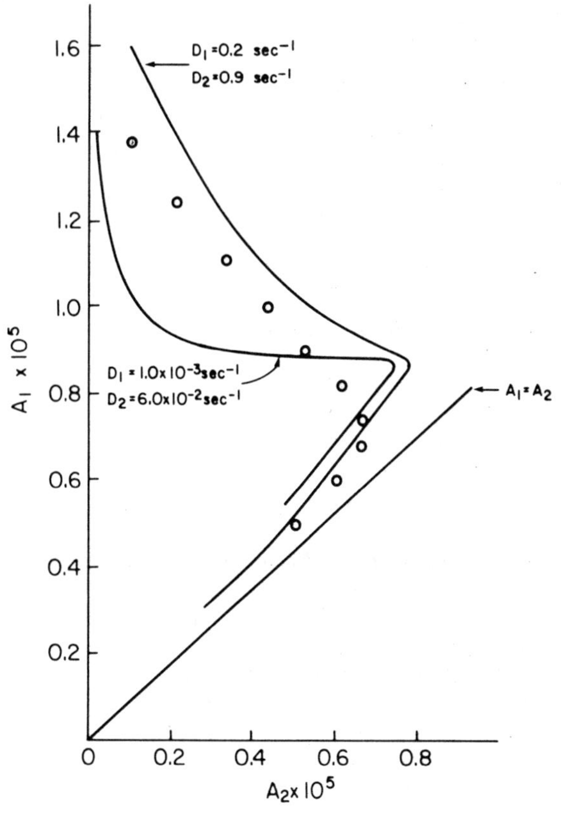

FIGURE 8. Comparison of experimental data for run 3 with simulation values from the variable volume model. A single impeller turning at 75 rpm. $k = 0.01$ sec^{-1}.

data (Figure 6), and qualitatively for the other two sets of data. The shapes of the latter two sets of data seem to conform to the general shape for the variable volume model. The values for D_1 and D_2, obtained by trial and error can be used to bound the estimated values for the experiments. Shown in Figure 9 is a plot of $\ln(1/(A_1-A_2))$ versus time obtained from simulations of the variable volume model showing the qualitative effect of changes in reaction rate constant. As the reaction rate increases the slope of the line for later times increases. During the initial stages of the reaction, the trajectories come together. The shape of the curve is qualitatively similar to the data plotted in Figure 3 indicating that the volume was varying in that experiment.

CONCLUSIONS

The variable volume two zone model provides a better qualitative fit to experimental data than the constant volume model. For the constant volume case, a method has been presented to estimate the dilution rates D_1 and D_2. The variable volume mixing model has been shown to be applicable to a case for a single impeller agitated system. Parameter estimation could be improved by obtaining reliable early time data for the concentrations in each mixing zone.

NOMENCLATURE

A_1	Concentration of reactant A in Zone 1,g mol/l
A_2	Concentration of reactant A in Zone 2,g mol/l
A^*	Initial concentration of reactant A in Zone 1,g moles/l
D_1	Interzone fluid flow rate per volume for Zone 1, min^{-1}
D_2	Interzone fluid flow rate per volume for Zone 2, min^{-1}
F_1	Fluid flow rate from Zone 1 to Zone 2,l/min
F_2	Fluid flow rate from Zone 2 to Zone 1,l/min
K	Reaction rate constant, l^{-1}
V_1	Volume of Zone 1, l
V_2	Volume of Zone 2, l
V_{10}	Initial volume of Zone 1, l
V_{20}	Initial volume of Zone 2, l
V_T	Total volume of reactor = $V_1 + V_2$, l

REFERENCES

1. Tanner, R.D., I.J. Dunn, J.R. Bourne and M.K. Klu, Chem. Eng. Sci. 40, 1213 (1985).

2. Tanner, R.D., K.A. Debelak, S. Rohani, I.J. Dunn and J.R. Bourne, "A Mixing Model for Diagnosing Reacting Tracers and Tracing Reactants in Chemical and Biochemical Processes," Accepted for publication in Can. J. Chem. Eng.

3. Robinson, J.E., K.A. Debelak and R.D. Tanner, "A Two-Zone Model to Describe Mixing Effects in Equilibrium Chemical and Biochemical Processes," Accepted for publication in Chem. Eng. Comm.

4. Lipinsky, E.S., Genetic Engineering News, 6,14 (1986).

5. Cosaro, G., J. Chem. Educ. 41,48 (1964).

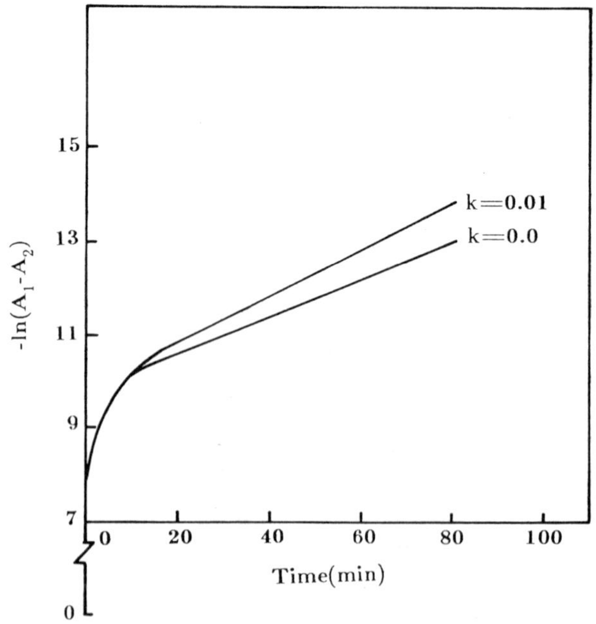

FIGURE 9. Reaction data simulated by variable volume model and plotted according to equation (31).

Stage Models for Mixing in Stirred Bioreactors

RAKESH BAJPAI
PETER U. SOHN
Chemical Engineering Department
University of Missouri–Columbia
Columbia, Missouri 65211

INTRODUCTION:

Circulations of fluid due to its displacement by impellers are a characteristic feature of stirred reactors and have often been used as a basis for quantifying mixing in batch systems. Models of Bryant (1), Khang and Levenspiel (2), Bajpai and Reuss (3), Mukataka et al. (4) belong to this category. These are generally represented as a two parameter log-normal distribution. In terms of Khang and Levenspiel's description, the number of stirred-tanks-in-series, N, is determined by the variance of the circulation-time distribution. These models have successfully been used to represent the dynamics of distribution of tracer in single impeller systems. Reuss and Bajpai (5) used this concept to couple the kinetics of oxygen uptake by microorganisms with mixing behavior in reactors having single and two impellers. The predicted overall behavior was found to be in excellent agreement with the observations (6). However, attempts to use sophisticated multi-reaction kinetics with these models result into serious computational problems, mainly due to large differences between time constants of different reaction steps and the process of mixing (3). This is particularly true when the number of stirred-tanks-in-series larger than five must be considered. Simpler models of mixing, based primarily upon the measurements of residence time distributions in continuous stirred reactors have also been used. Tanner and coworkers (7, 8) have used these models for simulation and characterization of mixing in batch reactors. These models typically have only two stirred-tanks-in-series per impeller through which fluid recirculates. These may be considered to be special cases of circulation-time distribution models where N=2. Naturally, these would fail to reconstruct the measured tracer responses in all but the very special cases; these, on the other hand, offer tremendous computational advantages and deserve attention. The objective of the present communication is to discuss the circumstances under which the simpler two compartment models may be used to represent mixing phenomena in reactors. This has been achieved by simulating Khang and Levenspiel's model (2) and its extensions for multiple impellers, and comparing the results with the simpler case of N=2. Additionally, some thoughts are presented as to how the parameters for the simpler case may be estimated from those already worked out for the circulation-time distribution models.

Theoretical development and results of simulations:

As a representation of recirculating flow in a stirred reactor, Khang and Levenspiel (2) suggested a tanks-in-series model shown in Figure 1. If volumes of all the tanks are considered equal, impulse response for such a system in the Laplace domain is of the form:

$$\frac{\overline{C_N(s)}}{C_\infty} = \frac{\theta}{(1 + \frac{S\theta}{N})^N - 1} \tag{1}$$

which upon inversion becomes

$$\frac{C_N(t)}{C_\infty} = 1 - B\exp(\frac{2Nt}{\theta})$$
$$+2\sum_{k=1}^{M}\exp\{-N(1-\cos\frac{2\pi k}{N})\frac{t}{\theta}\}\cos(\frac{tN}{\theta}\sin\frac{2\pi k}{N} + \frac{2\pi k}{N}) \tag{2}$$

where B = 0 ; N = odd
1 ; N = even

M = (N-1)/2 ; N = odd
N/2 - 1 ; N = even

Thus, measured signal in the vicinity of impeller in a stirred reactor wherein a pulse injection of inert tracer is made near the stirrer-tip, is of oscillatory nature with decaying amplitude provided N>2. For the case of N=2, Equation (2) reduces to

$$C_2(t)/C_\infty = 1 - \exp(-4t/\theta) \tag{3}$$

If the volume of the two vessels are not assumed to be equal, the resulting expression will be

$$C_2(t)/C = 1\text{-}\exp\{-\frac{t}{\alpha(1-\alpha)\theta}\}; \tag{4}$$

$$\alpha = \frac{V_1}{V};\ \theta = \frac{V}{q}$$

Clearly, no oscillations will be observed if N=2. The special case of N=2 is shown in Figure 2 where one compartment may be considered to be the impeller region while the other one consists of the rest of liquid volume. Naturally, the assumption of well mixed second compartment is a gross simplification of the actual picture.

The schematics shown in Figure 1 is applicable only to a symmetrically placed single impeller. When an unsymmetrically placed radial flow impeller is used, any such representation must consider two interconnected configurations, each corresponding to the upper and lower circulation loops (9). In this communication, we will restrict ourselves to symmetric configurations only.

The concept of circulation time distributions can be easily extended to multiple impellers too wherein the recirculating flows of each individual stirrer interact with each other. Using flow-follower techniques, the exchange between impellers has been quantitatively measured by Mukataka et al. (10) and Bajpai and Reuss (5). In the frame work of tanks-in-series model, two problems are encountered. The first deals with the fact that the adjoining circulation loops can interact with each other. Secondly, one must address the question of which tanks-in-series would be involved in the

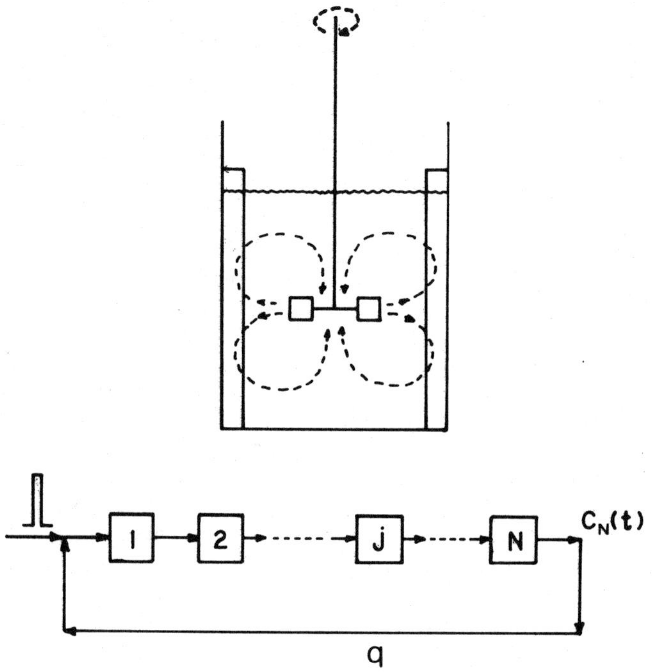

FIGURE 1. Khang and Levenspiel's model [2] for symmetrically placed agitator in a stirred vessel.

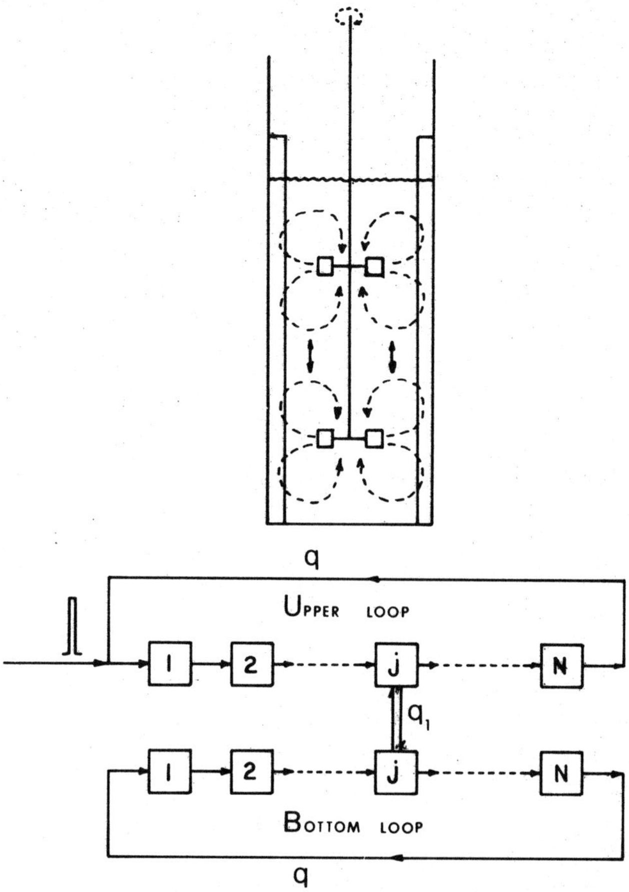

FIGURE 3. Extension of Khang and Levenspiel's model for the case of two symmetrically placed agitators in a stirred vessel.

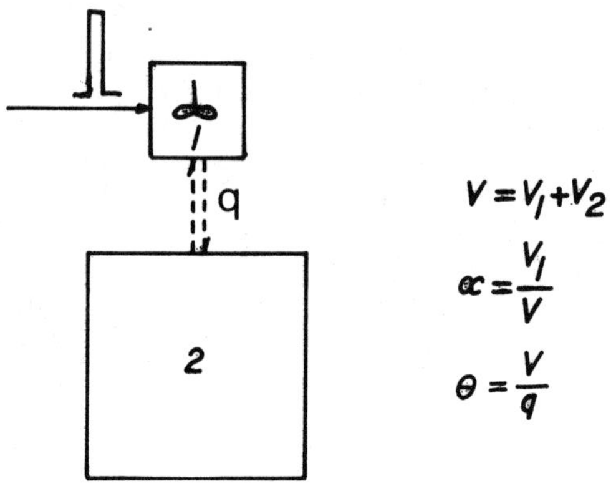

FIGURE 2. A two-compartment model for a single impeller system.

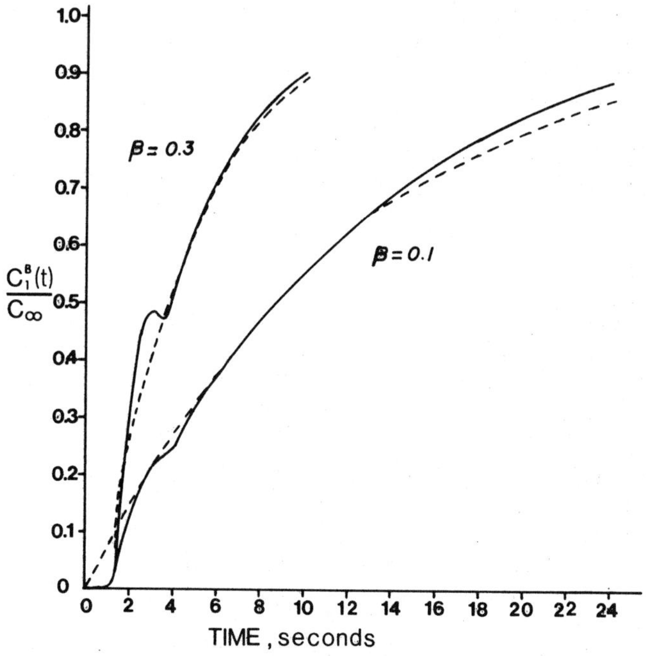

FIGURE 4. Simulations of the Khang and Levenspiel model for two different values of interimpeller exchange parameters (solid lines). $N = 10$. The dotted lines are the predictions of a two-compartment model for the single-parameter optimization values shown in Table 1.

exchange of flow between impellers. As a first approximation, we have considered only a single circulation loop per impeller, as shown in Figure 3 for a two impeller system. Additionally, only the middle tank is assumed to be involved in interimpeller exchange.

For the case of pulse injection of tracer in the vicinity of upper impeller, governing equations for the configuration shown in Figure 3 are

$$\frac{d}{dt} C_1^u = \frac{1}{\alpha\theta} (C_N^u - C_1^u) + 2N\, C_\infty\, \delta(t)$$

$$\frac{d}{dt} C_1^u = \frac{1}{\alpha\theta} (C_{i-1}^u - C_i^u) \quad i=2,--,N;\ i \neq j$$

$$\frac{d}{dt} C_j^u = \frac{1}{\alpha\theta} (C_{j-1}^u - C_j^u) + \frac{\beta}{\alpha\theta} (C_j^B - C_j^u) \tag{5}$$

$$\frac{d}{dt} C_1^B = \frac{1}{\alpha\theta} (C_N^B - C_1^B)$$

$$\frac{d}{dt} C_i^B = \frac{1}{\alpha\theta} (C_{i-1}^B - C_i^B) \quad i=2,--,N;\ i \neq j$$

$$\frac{d}{dt} C_j^B = \frac{1}{\alpha\theta} (C_{j-1}^B - C_j^B) - \frac{\beta}{\alpha\theta} (C_j^B - C_j^u)$$

where $\theta = \frac{V}{2q}$; $\alpha = \frac{V_1}{V/2} = \frac{1}{N}$; $\beta = \frac{q_1}{q}$

For N even and $j=N/2$, its solution for concentration of an inert tracer in the first reactor of the bottom loop, in Laplacian domain is

$$\frac{\overline{C_1^B(s)}}{C_\infty} = \left[\frac{\theta}{(1+\frac{S\theta}{N})^N - 1}\right] \cdot \left[\frac{1}{(1+\frac{S\theta}{N})}\right] \cdot \left[1 - \frac{1}{1 + \dfrac{2N\beta/\theta\ (S+N/\theta)^{N-1}}{(S+N/\theta)^N - (N/\theta)^N}}\right] \tag{6}$$

Although this can be subjected only to numerical inversion, similarities and differences between Equations (1) and (6) can throw light upon the nature of impulse tracer responses in case of one and two impeller systems. Equation (1) represents the oscillatory nature of tracer response in a single impeller system. The first term in Equation (6) is exactly the same as the one in Equation (1) thus suggesting oscillations in case of the two impeller system. Yet, the second and the third terms in Equation (6) are of exponential nature and cause the oscillations to die down fast. As a result, if inert tracer is injected in the vicinity of one impeller and the tracer concentrations are measured in the vicinity of the other impeller of a two impeller system, the oscillations may not be observed under experimental conditions. The results of simulation of tracer responses using Equation (6) are shown in Figure 4 for two different set of parameters. Clearly, very slight oscillations are predicted. When more than two impellers are in use, the oscillations can be expected to be even less observable, except when the measurements are made at the impeller at which tracer is injected. In physical terms, the exchange between impellers plays an important role in case of multiple impeller systems and it reduces the significance of parameter N, the number of tanks-in-series.

For the case of N=2, the configuration reduces to the one shown in Figure 5. If the two impellers are identical and are properly placed, the total liquid volume may be considered to be equally shared by the two impellers. At the same time, volumes of impeller-regions for the two impellers will also be same. In this case, three parameters (q, α, and β) will need to be known. The governing equations can be rewritten as

$$\frac{d}{dt} C_1^u = \frac{1}{\alpha\theta}(C_2^u - C_1^u) + \frac{2}{\alpha} C_\infty \delta(t)$$

(7)

$$\frac{d}{dt} C_2^u = \frac{1}{(1-\alpha)\theta}(C_1^u - C_2^u) + \frac{\beta}{(1-\alpha)\theta}(C_2^B - C_2^u)$$

$$\frac{d}{dt} C_1^B = \frac{1}{\alpha\theta}(C_2^B - C_1^B) \qquad (7)$$

$$\frac{d}{dt} C_2^B = \frac{1}{(1-\alpha)\theta}(C_1^B - C_2^B) - \frac{\beta}{(1-\alpha)\theta}(C_2^B - C_2^u)$$

where $\theta = \dfrac{V_1 + V_2}{q}$; $\alpha = \dfrac{V_1}{V_1 + V_2}$; $\beta = \dfrac{q_1}{q}$

and its solution results into

$$\frac{C_1^B(t)}{C_\infty} = 1 + \frac{1-\alpha}{\alpha} \exp\left\{\frac{t}{\alpha(1-\alpha)\theta}\right\} + \frac{2\beta}{\alpha^2(1-\alpha)^2\theta^3}\left\{\frac{e^{-(x+y)t}}{(x+y)\, 2y\left(\frac{1}{\alpha\theta} + \frac{1}{(1-\alpha)\theta} - x - y\right)} - \frac{e^{-(x-y)t}}{(x-y)\, 2y\left(\frac{1}{\alpha\theta} + \frac{1}{(1-\alpha)\theta} - x + y\right)}\right\} \qquad (8)$$

where $x = \dfrac{1}{\theta}\left(\dfrac{2}{\alpha} - \dfrac{4\beta}{1-\alpha} + \dfrac{2}{1-\alpha}\right)$;

$$y = \frac{1}{2}\sqrt{(2x)^2 - \frac{8\beta}{\alpha(1-\alpha)\theta^2}}$$

If the curves shown in Figure 4 are considered as experimental measurements and the parameters of two compartment model (Equation 8) are estimated, the values shown in Table I are obtained. The estimations were conducted with the help of Nelder and Mead's simplex algorithm. The exchange parameter was kept constant as it is separately measurable using the techniques of Reuss and coworkers (5) and of Mukataka et al. (10). The predictions using Equation (8) are also shown in Figure 4 as dotted lines. The two compartment model can represent the expected behavior quite well. Although no experimental data are presented, the curves of the type shown in Figure 4 are experimentally observed {Jensen et al., (11)} and lend credence to the simulation results.

Similarly, the two compartment concept can be extended to three or more impellers. If the identical impellers are symmetrically located and are mounted on the same shaft, the liquid is equally shared by the different impellers and the system is still characterized by three parameters (q, α, and β). Numerical simulation of the system can be easily performed by writing the governing equations similar to Equation (7). The interimpeller exchange parameter, β , can be experimentally measured. The circulation parameters, θ and α , can be estimated from the measured tracer responses.

Discussion of parameters:

In the absence of any experimental measurement, some ideas are presented in the following for the estimation of parameters, θ and α from the correlations available in Literature. From the measurements of energy dissipation rates in stirred vessels, Liepe et al. (12)

TABLE 1
Estimated Parameter Values for a Two-Compartment Mixing Model

Parameter	Expected Values	ESTIMATED VALUES One Parameter Optimization	ESTIMATED VALUES Two Parameter Optimization
	0.1	0.1 (fixed)	0.1 (fixed)
(sec.)	2.5	2.5 (fixed)	2.2
	-	0.07	0.18
	0.3	0.3 (fixed)	0.3
(sec.)	2.5	2.5 (fixed)	1.95
	-	0.155	0.40

The expected values were used in Khang & Levenspiel's model [Figure 3] with N = 10 to generate experimental data. Twenty equi-spaced points were picked and used for estimation of parameters for the two-compartment model.

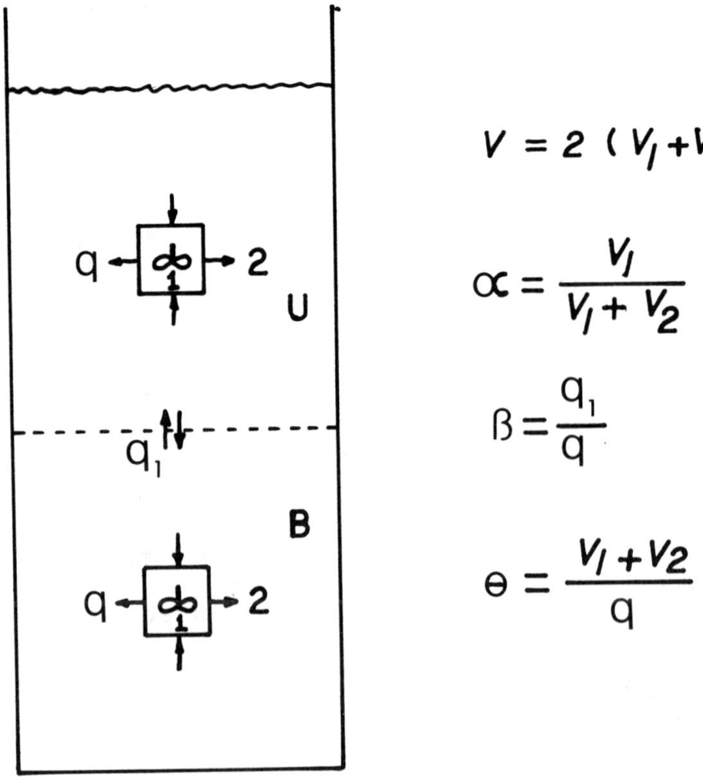

FIGURE 5. Representation of a two-compartment model for a two-impeller system.

have proposed for the maximum energy dissipation rate

$$\varepsilon_{max} = 0.5\,\bar{\varepsilon}\,(d_i/D_T)^{-3} \qquad (9)$$

If it is assumed that all of the energy introduced through an impeller is dissipated uniformly in the impeller region 1 of Figure 2, one may write

$$\alpha = \bar{\varepsilon} \;/\; \varepsilon_{max} \qquad (10)$$

which upon substitution of Equation (9) results in

$$\alpha = 2(d_i/D_T)^3 \qquad (11)$$

Since the nature of energy dissipation function, Equation 9, is expected to be dependent upon the rate of aeration and the rheology (13), Equation (11) should be carefully applied.

An estimation of the parameter, θ, can be obtained as follows:

According to the model of Khang and Levenspiel (2), terminal mixing time to achieve a given degree of homogeneity ($1-A_m$) is given by

$$\theta_{mix} = \frac{1}{K_A}\ln(2/A_m) \qquad (12)$$

where K_A is the amplitude decay parameter and is related to the operating conditions as

$$K_A = 2\,n\,(d_i/D_T)^{2.3} \qquad (13)$$

According to the two compartment model, Figure 2, the same degree of homogeneity is achieved in time

$$t_{mix} = \alpha(1-\alpha)\,\theta\ln(1/A_m\alpha) \qquad (14)$$

Equating t_{mix} to θ_{mix},

$$\theta = \frac{\left(\frac{1}{2n}\right)(d_i/D_T)^{-2.3}\ln(2/A_m)}{\alpha(1-\alpha)\ln(1/A_m\alpha)} \qquad (15)$$

A number of correlations have been proposed for the mean circulation time in an agitated vessel. A correlation proposed by Bajpai and Reuss (5) relating the mean circulation time to the geometry and the operating conditions is

$$\bar{\theta} = \frac{0.3}{n}\,1.64\,(H/D_T)^{0.6}\,(D_T/d_i)^{3.1} \qquad (16)$$

A combination of Equations (15) and (16) results in

$$\frac{\theta}{\bar{\theta}} = (D_T/H)^{0.6}\,(d_i/D_T)^{0.8}\,\frac{1}{\alpha(1-\alpha)}\,\frac{\ln(2/A_m)}{\ln(1/A_m\alpha)} \qquad (17)$$

which in the (d_i/D_T) range of 0.25-0.5 can be reduced to

$$\frac{\theta}{\bar{\theta}} = (D_T/H)^{0.6}\,(d_i/D_T)^{-1.3} \qquad (18)$$

The validity of Equations (11) and (18), however, needs to be checked out experimentally.

CONCLUSIONS:

Mixing models using two stirred tanks in series per impeller are special cases of the more general circulation time distribution models. The simpler models are capable of predicting the observed tracer responses in multiple impeller systems as long as the measurements of concern are those at an impeller other than the one in which tracer is injected. This is also a common mode for measurements. As already stated, multi-impeller systems may be characterized with the help of three parameters which depend a great deal upon the geometric and operating conditions. From a single trajectory of tracer response, two of these parameters, θ and α, can be easily estimated. The exchange parameter, β, should be separately estimated for better resolution.

NOTATION:

Am	minimum fractional homogeneity (e.g. 0.95 or 0.99) specified for a well mixed system.
$\overline{C}(s)$	Laplace transform of concentration
$C_j(t)$	concentration of tracer in the jth tank
C_∞	final concentration of tracer
d_i	impeller diameter
D_T	diameter of vessel
H	height of fluid in the vessel
n	revolutions per second
N	number of tanks in a circulation loop
q	rate of circulatory flow in a loop
V	total volume of liquid in the vessel
t	time
α	volume fraction of impeller region in a circulation loop
β	fraction of circulatory flow exchanged between loops
δ	Dirac-delta
θ	average circulation time in a loop defined by model
$\overline{\theta}$	mean circulation time
$\overline{\varepsilon}$	energy dissipation rate per unit mass of liquid

subscripts:

max	maximum value

superscripts:

u	property in the upper circulation loop
B	property in the bottom circulation loop

LITERATURE CITED:

1. Bryant, J., Adv. Biochem. Eng., 5, 101 (1977)
2. Khang, S.J. and Levenspiel, O., Chem. Eng. Sci., 31, 579 (1976)
3. Bajpai, R.K. and Reuss, M., Can. J. Chem. Eng., 60, 384 (1982)
4. Mukataka, S., Kataoka, H., and Takahashi, J., J.Fermn. Technol., 58, 155 (1980)
5. Reuss, M. and Bajpai, R.K., "Performance of Stirred Bioreactors in the light of Mass and Energy Distributions," manuscript in preparation; 1986

6. Boelcke, C., "Experimentelle Untersuchungen zur Zirkulationsverteilung der Fluessigphase in Ruehrreaktoren," Diplomarbeit, Fachgebeit Biotechnologie, TU Berlin, (1983)

7. Tanner, R.D., Dunn, I.J., Bourne, J.R., and Klu, M.K., Chem. Eng. Sci., 40, 1213 (1985)

8. Debelak, K. and Tanner, R.D., "A Two Zone Mixing Model to Describe Mixing Effects in Bioprocesses," paper 29b presented at the Symposium on Reaction and Deactivation Kinetics in Bioprocesses, held in Seattle AIChE meeting, August 25-28, 1985

9. Berke, J., "Mischzeitcharakteristik in begasten Ruehrkesselfermentern," Diplomarbeit, Fachgebeit Biotechnologie, TU Berlin, 1980

10. Mukataka, S., Kataoka, H., and Takahashi, J., J. Fermn. Technol., 59, 303 (1981)

11. Jansen, P.H. Slott, S.,Guertler, H., "Determination of Mixing Times in Large-Scale Fermenters using Radioactive Isotopes," paper presented at the First European Congress on Biotechnology, Interlaken,Switzerland, Preprints pt. 2, page 80, 1978

12. Liepe, F., Moeckel, H.-D., and Winkler, H., Chem. Tech. (Leipzig), 23, 1971

13. Reuss, M., Bajpai, R.K., and Berke, W. J. Chem. Technol. Biotechnol., 32, 81 (1982)

Power Absorption by New and Hybrid Mixing Systems under Gassed and Ungassed Conditions

EDI D. ELIEZER*
CPC International
Moffett Technical Center
Argo, Illinois 60501

Achievement of high mass transfer at relatively low power input and shear rates in aerobic fermentation processes requires new and efficient reactor-mixing configurations. Two bioreactor-mixing configurations were studied here in a conventional 1000-liter vessel with the water–air system. The first comprised two Maxflo-T, Prochem impellers and the second "Hybrid" system had a flat-blade turbine impeller at the bottom and an A310, Lightnin mixer at the top of the vessel. The classical correlation between the ratio of gassed to ungassed power P_g/P_0 and the aeration number N_a is parametric with respect to agitation or aeration rate, and reactor-mixing configurations. The experimental results obtained here with new and hybrid mixing systems fitted relatively simple correlations expressing P_g/P_0 as a function of the two independent variables, the gas flow rate g and the agitation rate N. The three-dimensional representation of these correlations resulted in the following observations: (*a*) the Maxflo system, under experimental conditions, provided at most a 24% power reduction by gassing ($P_g/P_0 = 0.76$). However, this system can provide stable fluid hydrodynamics and gas dispersion in a wide range of aeration (0–0.52 VVM). (*b*) The Hybrid system exhibited a power reduction due to gassing as high as 79% ($P_g/P_0 = 0.21$). Nevertheless this reactor-mixing configuration can be considered as an unstable system for dispersing high aeration levels, e.g., above 0.46 VVM.

The information on reactors using new commercial mixer designs, developed here, contributed to the design of new bioreactors which proved the potential of improved mass transfer rates and efficiencies in industrial fermentations.

Achievement of high mass transfer at low mixing power input and relatively low shear rates in aerobic fermentation processes requires new and efficient bioreactors. The conventional stirred tank reactor (STR) with multiple flat-blade turbine (FBT) impellers or some non-conventional reactors like the HTPJ fermentor (1) can provide high mass transfer rates but both designs generate high shear rates detrimental to many microbial cultures. The need in the fermentation industry to improve the existing production STR bioreactors leads one to consider retrofitting the conventional reactors with more efficient mixing configurations.

Literature is abundant with studies on multi-stage gas-liquid reactors that have one type of mixer which most of the time is the FBT impeller. Excellent review articles (2,3) present correlations obtained between mixing power under ungassed (Po) or gassed conditions (Pg) and various physico-chemical parameters.

E.D. Eliezer is now with Abbott Laboratories, North Chicago, IL 60064

Rare articles like the one of Kuboi R. and Nienow A.W. (4) provide correlations on power drawn by mixed impeller-type systems under gassed and ungassed conditions. Very few public literature is available on these type correlations for multistage reactors with new or hybrid mixing systems. The present study develops correlations for the power absorption by new and hybrid mixing systems under gassed and ungassed conditions.

CORRELATIONS ON GASSED TO UNGASSED POWER RATIO, (Pg/Po)

Aeration of a mechanically agitated reactor has been shown to reduce the power absorbed by mixers due mainly to reduction of bulk fluid density and formation of characteristic gas cavities behind impeller blades. The decrease in gassed power has been demonstrated to be strongly related to increased cavity size that lowered stirrer friction (5). On the other hand, the power reduction is affected not only by the total amount of gas supplied to the system but also by the degree of dispersion or recirculation of the gas phase in the liquid phase (6). The residence time distribution of the gas phase, measured in Newtonian fluids and non-Newtonian fermentation broths, demonstrated that the gas phase in the bulk phase can be represented by ideally mixed (m), plug-flow (p) or (m/p) combination models (7).

The most widely used parameter to define the complex effect of aeration on mixing power is the dimensionless aeration number Na (8).

$$Na = g/(ND^3)$$

g = gas flow rate (m^3/min)
N = agitation rate (rpm)
D = impeller diameter (m)

The Na number, ratio of [gas flow/ impeller pumping or fluid flow] can inform on the effect on mixing power, of both the amount of gas (g) and the degree of gas dispersion (g/N). Most correlations show a decrease in power or Pg/Po with increasing aeration number Na (8). The ratio (Pg/Po) called also the power reduction or the gassing factor K enables also to compare the degree of gas dispersion between different reactors under identical agitation and aeration conditions, i.e. Na. A low K value, i.e. a significant power reduction by gassing, could be interpreted as a good gas dispersion and recirculation characteristics of the given reactor-mixing configuration.

The mixing power measured here in two bioeactor-mixing configurations, was correlated to the two independant variables, the gas flow rate, g and the agitation rate, N.

BIOREACTOR-MIXING CONFIGURATION

Most of the mixers used in STR reactors have either a radial-flow, axial-flow or a combined flow pattern. Radial-flow impellers like FBT, generate high shear potentially detrimental to some microbial cultures and, can provide efficient but localized high gas-liquid mass transfer. Axial-flow impellers in contrast, generate less shear and, can provide efficient gas-liquid-solid distribution (good bulk mixing) but relatively poor gas-liquid mass transfer.

The bioreactor-mixing configurations studied here were:

1) Maxflo System: two Maxflo-T impellers with curved blades providing both axial and radial-flow patterns.

2) Hybrid System: had one radial-flow FBT impeller at the bottom of the reactor for high gas-liquid mass transfer and one axial-flow impeller at the top to assure an efficient bulk mixing. The axial-flow impeller used here was the A310 impeller, claimed by its manufacturer to save energy and improve mixing performance.

The axial-flow mixers used here, i.e. Maxflo-T and A310, were installed for downward pumping.

EXPERIMENTAL

The experiments were performed in a 1000-liter microprocessor-controlled bioreactor (Abec, Lehigh Valley, PA), fitted with a ring type gas sparger, four baffles and a bottom- driven agitator with counter-clockwise rotation. The mixing devices used in this study are represented in Figure A and their geometry, along with reactor configurations are represented in Table 1 and Figure B.

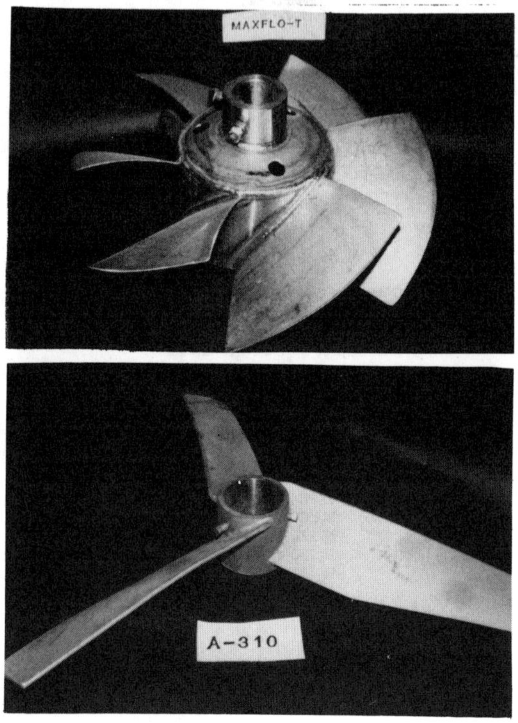

FIGURE A.

Lightnin Company (Rochester, N.Y.) provided the A310 and the flat-blade turbine (FBT) impellers. Prochem Company (Brampton, Ontario) provided the Maxflo-T impellers.

The operating conditions, detailed in Table 1, consisted of a working volume of 690 liters of tap water at 30°C, air flow rates (measured by a mass flow-meter) from 0 to 325 l/min (0 to 0.45 VVM, superficial air velocities of 0 to 0.92 cm/sec), and agitation rates from 100 to 280 rpm's.

TABLE 1
Bioreactor Configurations for Mixing-Power Studies

PARAMETERS	MAXFLO	HYBRID
T(")	34.00	34.00
SPARGER		
Ds(")	15.00	15.00
[Ds/Db]	[0.83]	[0.94]
BOTTOM MIXER		
TYPE	MAXFLO-T	FBT
Db(")	18.00	16.00
[Db/T]	[0.53]	[0.47]
Nb	[6]	[6]
[D/L/W]	VARIABLE	[20/5/4]
PITCH	VARIABLE	90
TOP MIXER		
TYPE	MAXFLO-T	A-310
Du(")	18.00	20.00
[Du/T]	[0.53]	[0.59]
Nb	[6]	[3]
[D/L/W]	VARIABLE	VARIABLE
PITCH	VARIABLE	VARIABLE
MIXER-SPACINGS		
Hb(")	12.50	8.00
[Hb/Db]	[0.69]	[0.50]
Hs(")	18.00	20.00
[Hs/Du]	[1.00]	[1.00]
Hu(")	17.00	19.50
[Hu/Du]	[0.95]	[0.98]
V(l)	690.00	690.00
HL(")	47.50	47.50
[HL/T]	[1.40]	[1.40]

NOTES:
FOR EXAPLANATION OF SYMBOLS OR TERMS REFER TO FIGURE OF REACTOR; OTHERWISE:

1) SPARGER :RING-SPARGER WITH (1/8)" DIAMETER HOLES, SPACED BY 1", FACING DOWN AT 45° ANGLE.
2) MIXER TYPE:FBT=FLAT-BLADE TURBINE IMPELLER.
3) Nb :NUMBER OF BLADES ON MIXER
4) [D/L/W] :RATIO OF [MIXER DIAMETER, D /BLADES LENGTH, L /BLADES WIDTH,W]
5) PITCH :ANGLE OF BLADES ON MIXER, IN (°) OR IN ARBITRARY UNITS (2.1 FOR MAXFLO)
6) V(l) :LIQUID VOLUME (LITERS)

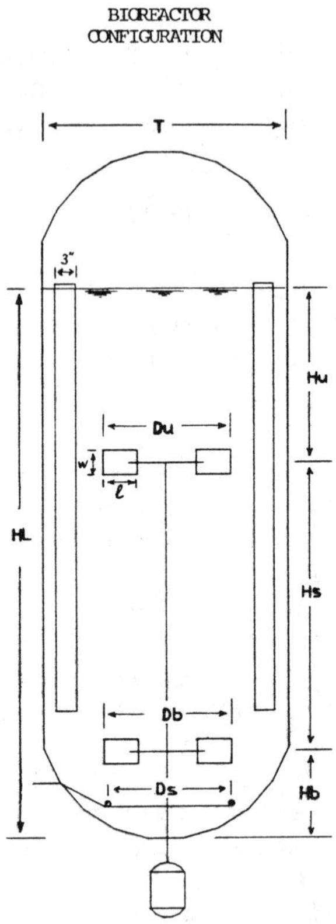

FIGURE B.

The power measurements were possible by voltage (mV) readings on the L400-Seco speed controller which were proportional to mixing shaft torque. The mixing power absorption P was calculated according to the following equation:

$$P = Ko\ N\ (mV - mVe)$$

Ko = Constant of equipment and unit conversions
N = Agitation rate (rpm)
mVe = Voltage reading for empty vessel (no liquid)

The volumetric power input (P/V in Equation (1)) is expressed in HP/1000 gallons which is the most frequently used unit in US fermentation industry.

RESULTS

Maxflo

The power absorption in this system, under various agitation and aeration conditions is represented by Figure 1. The correlations that fitted the experimental values had the following form:

$$P/V = B\,(N)^A \qquad (1)$$

$P/V = \dfrac{\text{Power absorbed}}{\text{Aerated liquid Volume}}$

in (HP/1000 gals)

N = Agitation Rate (rpm)

A,B = Correlation parameters, function of the gas flow-rate (g)

As the gas flow rate was increased, the power absorption with Maxflo, decreased at almost all agitation rates. The exponent A on N, decreased indeed from 3.22 to 3.0 as the air flow rate was increased from 0 to 325 l/min (0.47 VVM), Table 2.

TABLE 2
Results of Mixing-Power Correlations

$$(P/V) = B\,(N)^A \qquad (1)$$

(P/V) : ABSORBED POWER (HP/1000GALS)
N : AGITATION RATE (RPM)
A, B : CORRELATION PARAMETERS

MIXING SYSTEM	VALUES OF A,B	GAS FLOW RATES go	g1	g2	g3	g4
MAXFLO	A	3.222	3.147	3.104	3.001	-
	B	4.98E-07	6.87E-07	8.51E-07	1.32E-06	-
HYBRID	A	2.825	2.747	2.554	2.744	2.836
	B	5.86E-06	7.09E-06	1.64E-05	4.92E-06	2.52E-06

NOTES

1) ALL CORRELATIONS OBTAINED HERE HAVE REGRESSION COEFFICIENTS $R^2 \geq 0.99$
2) VALUES OF GAS FLOW RATES (L/MIN):

MIXING	go	g1	g2	g3	g4
MAXFLO	0	170	227	325	-
HYBRID	0	73	135	170	315

Power correlations obtained for each aeration rate (Table 2) were used to calculate the ratio of gassed (Pg) to ungassed power (Po), i.e. (Pg/ Po). These ratios were plotted as a function of (g/N), similar to the aeration number $Na(=g/ND^3)$, and shown for constant air flow rates in Figure 2.

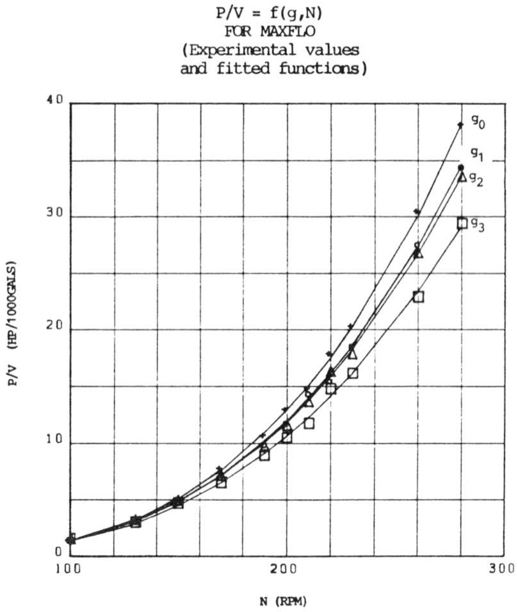

FIGURE 1.

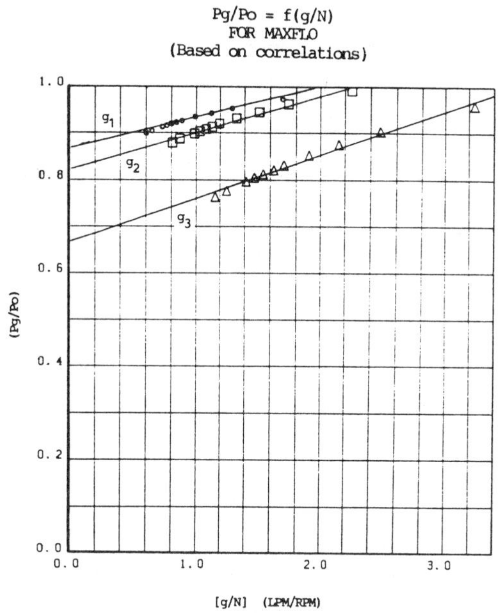

FIGURE 2.

Unlike most of classical correlations where Pg/Po decreases with increasing Na, the Maxflo system shows that, in the present experimental conditions, Pg/Po is increasing with increasing (g/N). This relative increase of gassed power with the ratio of [gas flow/mixing or liquid flow] is similar to results obtained by Dickey (9) with conventional FBT impellers having a large impeller to vessel diameter ratio, i.e. a high bulk mixing system like the Maxflo. Nevertheless the results obtained here show that Pg/Po, at any (g/N), is decreasing with increasing gas flow rate g. The correlation obtained here between Pg/Po and (g/N) can be represented by Equation (2) and detailed in Table 3.

$$Pg/Po = a\ (g/N) + b \qquad (2)$$

TABLE 3
Results of Gassed Power Correlations

(Pg/Po) = a (g/N) + b (2)

Po = UNGASSED POWER (HP/1000GALS)
Pg = GASSED POWER (HP/1000GALS)
g = GAS FLOW RATE (L/MIN) See Table 2
N = AGITATION RATE (RPM)

MIXING	a,b	g1	g2	g3	g4
MAXFLO	a	0.066	0.078	0.093	-
	b	0.869	0.823	0.666	-
HYBRID	a	0.152	0.244	0.046	-0.003
	b	0.744	0.499	0.507	0.460

The correlation parameters (a) and (b) were however function of the gas flow rate (g). They are represented by Equation (3) and detailed in Table 4.

$$(a)\ \text{or}\ (b) = K1\ (g) + K2 \qquad (3)$$

TABLE 4
Correlations Found between Parameters *a*, *b* and Airflow Rate *g*

MAXFLO
a = 1.73E-04 (g) + 0.037
b = -0.001 (g) + 1.109

HYBRID
a = -7.95E-04 (g) + 0.247
b = -9.52E-04 (g) + 0.717

Finally combining Equations (2) and (3), the simultaneous variation of Pg/Po with both the agitation rate (N) and the gas flow rate (g) is represented by Equation (4) and the three-dimensional Figure 3.

$$Pg/Po = 1.73\ 10^{-4}(g^2/N) + 3.74\ 10^{-2}(g/N) - 1.34\ 10^{-3}(g) + 1.109 \qquad (4)$$

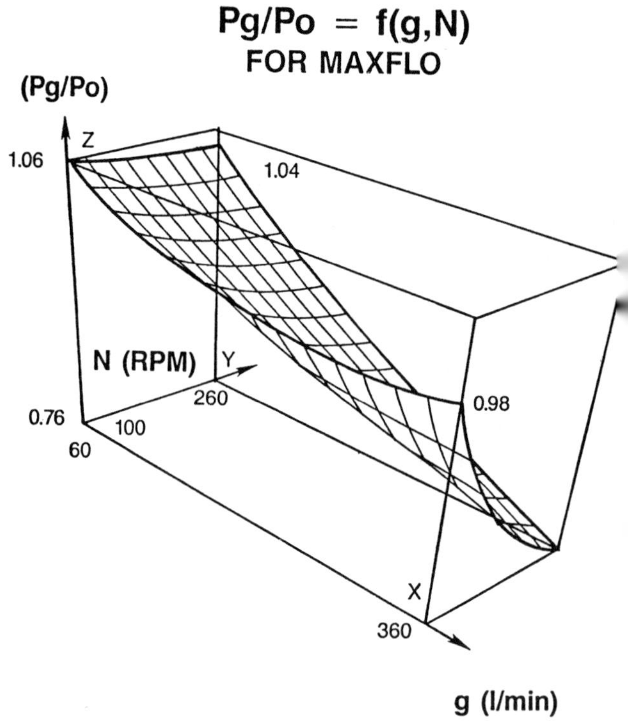

FIGURE 3.

TABLE 5
Some Values of Gassed to Ungassed Power Ratios (*Pg/Po*) for Different Gas Flows (*g*), Agitation Rates (*N*) and Mixing Systems (Based on correlations developed here)

VVM (g/690)	g (l/min)	N (rpm)	(Pg/Po) MAXFLO	(Pg/Po) HYBRID
0.09	60	100	1.06	0.78
0.09	60	260	1.04	0.70
0.52	360	100	0.98	0.21
0.52	360	260	0.76	0.30

Analysis of Figure 3 and some data in Table 5 show that:

a) At Constant Agitation, N (OXZ plane): Pg/Po decreased with increasing aeration g, i.e. larger "amount" of gas flow. This effect is even more drastic at higher levels of agitation which causes a higher degree of gas "dispersion". An interesting observation is that, at low agitation, if the gas flow is increased above a certain value, the power curve tends to be horizontal. This would mean that hydrodynamics of Maxflo became stable: the gas cavities behind or around Maxflo's curved blades reached a stable configuration and the power absorption does not drop any further.

b) At Constant Aeration, g (OYZ plane): Pg/Po decreases with increasing agitation N due to more uniform gas distribution throughout the whole reactor. This effect is more pronounced at higher levels of aeration that lowers bulk fluid density by a higher "amount" of gas phase or larger gas cavities around mixers.

The correlation represented by Equation (4) and Figure 3 is able to express in a simple way the reduction in mixing power as a function of both the total gas flow to the system, g and the degree of gas dispersion or recirculation due to mixing intensity, N. The Maxflo system, under present experimental conditions and in the aeration range of 0 to 0.52 VVM (as shown in Table 5), provided at most a 24% power reduction by gassing (Pg/Po=0.76).

Hybrid

The experimental values of power absorption in the Hybrid system, under various agitation and aeration conditions (Figure 4) were also fitted to Equation (1). The values of the exponent A on the agitation rate N, (Table 2) were in this case lower than the ones for Maxflo system. As the air flow rate was increased from 0 to 315 l/min (0.46 VVM) the mixing power decreased at all levels of agitation.

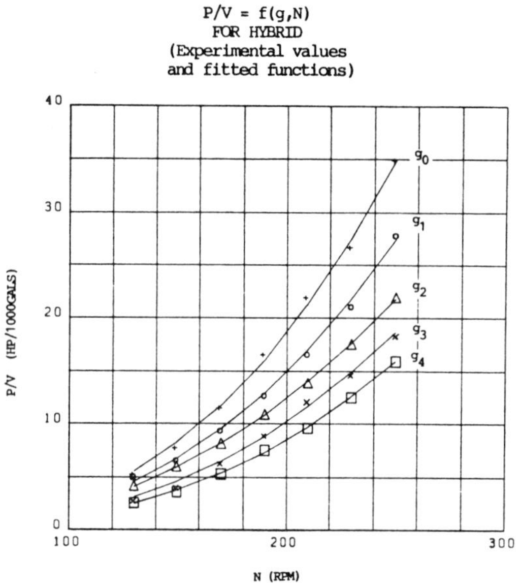

FIGURE 4.

The power correlations obtained above (Table 2), were used as previously, to calculate (Table 3) and plot (Figure 5) the ratio Pg/Po as a function of (g/N). Figure 5 demonstrates that, at all levels of (g/N), the effect of aeration on power reduction in Hybrid is more drastic and different than in the Maxflo system. Comparison of the power reduction factor (Pg/Po), in Hybrid and Maxflo, at similar aeration rates (315-325 l/min or 0.46 VVM), shows that:

a) Pg/Po was 45% in Hybrid as opposed to 65 to 95% in Maxflo, i.e. a more significant power reduction in Hybrid (55% vs. 35% at the most).

b) With increasing (g/N), the Hybrid, in contrast to Maxflo, did not exhibit any change in Pg/Po. A change in agitation rate N, at 0.46 VVM aeration, did not change nor improve gas dispersion characteristics of Hybrid.

The parameters a and b of Equation 2, were also dependant on gas flow rate (Table 3) and fitted Equation 3 which is detailed in Table 4.

Combining Equations 2 and 3, the ratio Pg/Po can be expressed as a function of the two independent variables g and N (Equation 5 and Figure 6):

$$\begin{aligned} Pg/Po = & -7.95\ 10^{-4}\ (g^2/N) \\ & + 0.247\ (g/N) \\ & - 9.52\ 10^{-4}\ (g) \\ & + 0.717 \end{aligned} \qquad (5)$$

The last term of Equation 5, i.e. 0.717, corresponds to Pg/Po for g=0 , which in theory should be equal to 1. This discrepancy is due to poor accuracy and correlations at boundaries (low g,N) with equations 2 and 3 in the Hybrid case. However, the Equation 5 and Figure 6, as will be discussed in the following, are able to demonstrate general trends quite different than Maxflo:

a) At constant agitation N (OXZ plane): as expected, Pg/Po decreases as "amount" of gas flow g increases. As the gas flow rate is increased the power reduction seems however more significant at low speeds than at high speeds : Pg/Po decreases from 0.78 to 0.21 at 100rpm and from 0.70 to 0.30 at 260rpm. It seems that at high agitation and aeration rates the Hybrid system is not very effective to disperse or recirculate the gas phase thus reduce mixing power.

b) At Constant Aeration, g (OYZ plane): Unlike the analysis at constant agitation - where Pg/Po always decreases with g - the variation of Pg/Po with N, at constant g, is function of the aeration level.

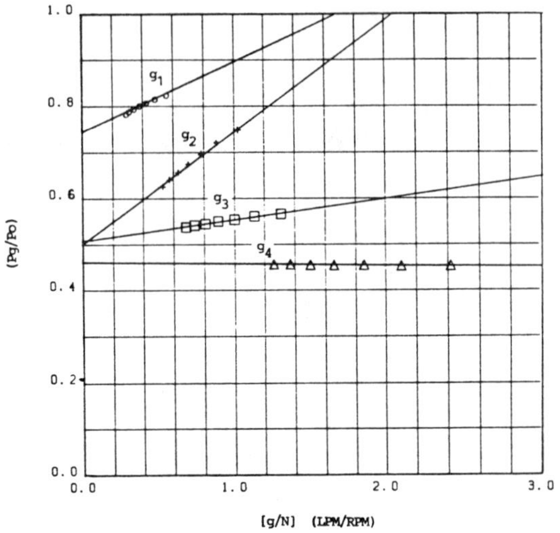

FIGURE 5.

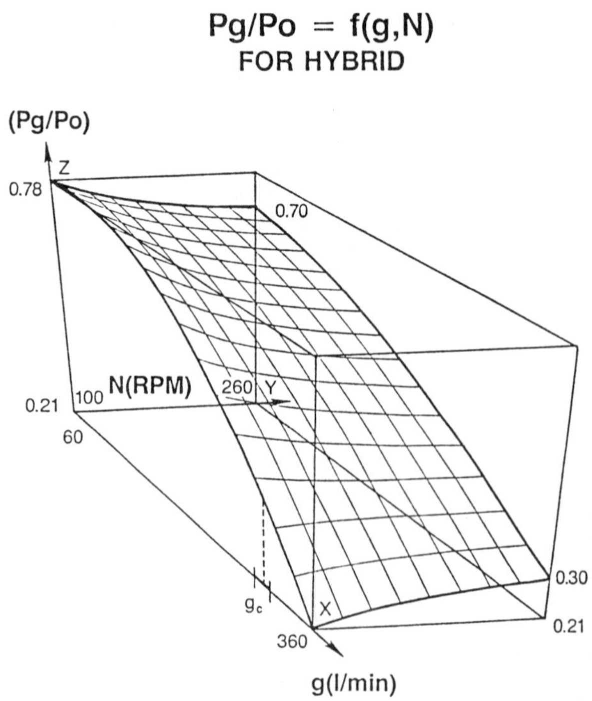

FIGURE 6.

At relatively low aeration levels, below a critical value gc where the Pg/Po=f(N) surface has an inflection point, Pg/Po decreases with increasing N which causes a higher gas "dispersion". At 60 l/min air, by increasing N from 100 to 260 rpm, Pg/Po decreases only from 0.78 to 0.70 (Table 5).

At intermediary aeration levels, around the critical value gc, the ratio Pg/Po seems to be unaffected by changes in agitation N. This situation is similar to results obtained at 0.46 VVM (g=315 l/min in Figure 5) and already discussed above. These observations suggest that the critical aeration level gc, in the present reactor-mixing system, is very close to 0.46 VVM.

At aeration levels higher than gc, Pg/Po increases slightly with increasing N, i.e. the relative mixing power increases faster than the increase of gas dispersion with N. At 360 l/min air (0.52 VVM), by increasing N from 100 to 260 rpm, Pg/Po increases from a low 0.21 to 0.30 (Table 5). The Hybrid reactor, in this case cannot reduce gassed power probably due to inability to disperse large amounts of gas.

It can be assumed that, at higher aeration levels, the downward pumping A310 axial flow impeller cannot overcome the upward movement of the gas and its buoyancy forces. The latter phenomena have been also observed with similar reactor configuration where the upper axial-flow impeller was a pitched-blade impeller (4). The Hybrid reactor-mixing configuration here can be considered as an unstable system for dispersing high aeration levels, e.g. above 0.46 VVM.

CONCLUSIONS

The experimental results obtained here with new and hybrid mixing systems fitted relatively simple correlations expressing the ratio of gassed to ungassed power Pg/Po as a function of the two independent variables, the gas flow rate g and the agitation rate N. The three dimensional representation of these correlations resulted in following observations:

a) The Maxflo system, under experimental conditions, can provide at most a 24% power reduction by gassing (Pg/Po = 0.76). However this system can provide stable fluid hydrodynamics and gas dispersion in a wide range of aeration (0-0.52 VVM).

b) The Hybrid system exhibited a power reduction due to gassing as high as 79% (Pg/Po = 0.21). Nevertheless this reactor-mixing configuration can be considered as an unstable system for dispersing high aerataion levels, e.g. above 0.46 VVM.

The classical correlation between Pg/Po and the aeration number $Na=g/ND^3$ demonstrates a parametric behaviour: Pg/Po exhibits separate curves for different agitation or aeration rates and reactor-mixing configurations, i.e. mixer-type, D/T ratio, etc. (this study, 9,5,11,10). The dimensionless aeration number Na can be considered inadequate to predict the mixing power reduction under gassed conditions. The shortcomings may be explained by the lack of parameter(s) describing the gas recirculation or residence time distribution throughout different regions of a reactor of given configuration and scale.

The present study developed empirical correlations expressing the gassed to ungassed power ratio Pg/Po as a function of the gas flow rate, g and the agitation rate, N for new and hybrid reactor-mixing configurations. The information on commercially available new mixing systems developed here, contributed to the design of new bioreactor configurations which proved the potential of improved mass transfer rates and power efficiencies in industrial-scale fermentations (unpublished). The power correlations on dual mixing systems obtained here are also expected to complement the in-depth work of Nienow and co-workers.

NOTATION

A,B,a,b, Correlation parameters in Equation 1 and 2.

D Mixer diameter (m, or in.)

g Gas flow rate (l/min)

Ki Constants in Equation 3

N Agitation rate (rpm)

Na Aeration Number (dimensionless)

Pg Gassed Power absorbed per unit liquid volume (HP/1000 gals)

Po Ungassed Power absorbed per unit liquid volume (HP/1000 gals)

P Total mixing power (HP)

V Liquid volume (l)

VVM Volume air/Volume liquid/Minute = g/V, (min^{-1})

LITERATURE CITED

1. Eliezer E.D. and Jones R.T., Proceedings of "Biotech'84-USA" Conference, p. 303, Online Conferences Inc., New York (1984).

2. Midoux N. and Charpentier J.C., Intern. Chem. Eng., 24, 249 (1984).

3. Joshi J.B. et al., Chem. Eng. Sci., 37, 813 (1982).

4. Kuboi R. and Nienow A.W., Proc. "4th. Europ. Conf. on Mixing", p. 247, BHRA Fluid Eng., Cranfield, Bedford (1982).

5. Warmoeskerken M.C.G. and Smith J.M., Proc. "4th. Europ. Conf. on Mixing", p. 237, BHRA Fluid Eng., Cranfield, Bedford (1982).

6. Nienow A.W. and Chapman C.M., The Chem. Eng. J., 17, 111 (1979).

7. Popovic et al., Chem. Eng. Sci., 88, 2015 (1983).

8. Oyama Y. and Endo K., Chem. Eng. (Japan), 19, 2 (1955).

9. Dickey D.S., in Advances in Biotechnology, Ed. by Moo-Young M., Pergamon Press, Vol. 1, p. 483, (1980).

10. Nienow et al., Proc. "2nd. Europ. Conf. on Mixing", p.F1-1, BHRA Fluid Eng., Cranfield, Bedford (1978).

11. Ismail A.F. et al., AIChE J., 30, 487 (1984).

Measurement of Shear Rate on an Agitator in a Fermentation Broth

BALDWIN ROBERTSON
JAROMIR J. ULBRECHT
Chemical Process Metrology Division
Center for Chemical Engineering
National Bureau of Standards
Gaithersburg, Maryland 20899

The shear rate was measured on the front face of a Rushton turbine blade rotating in an aqueous polyox solution that models a fermentation broth. The measurements agree with the theory of stagnation flow on the blade. A formula is given for use in scaling the results up to larger-diameter fermentation vats.

An important variable in the design and operation of a fermentation system is the maximum shear rate of the fluid in the vat. If bacteria are being grown, a large shear rate will disrupt the colony and stop the growth. If fungi are being grown, a large shear rate will prevent the cells from multiplying and forming optimal pellets. Most serious, however, is the problem with mammalian cells, whether they are immobilized on a fixed substrate or on a freely suspended microcarrier. This has led to the design of low-shear reactors for growing mammalian cultures. Yet shear, especially in very large scale fermentation systems, is an unavoidable consequence of the mixing of the broth and the dispersion of air bubbles necessary to encourage growth.

The shear rate at any point in a stirred tank of course may be expected to be dependent on the rheological properties of the fluid in the tank, in particular on the magnitude of the consistency and on the degree of shear thinning. Since a fermentation broth is rheologically complex, this requires that shear rate measurements be made either in the broth itself or in a fluid that simulates the rheological properties of the broth.

Since it is the stirring blade that causes shear in the first place, one may expect that the shear rate will be maximum at or near the blade. If a laminar flow does not separate from the boundary, the maximum shear rate will occur at a solid surface. Now the flow over the front surface of a Rushton turbine blade does not separate until the flow reaches the edge of the blade. As a result, one may expect that, in a fermentation vat stirred with a Rushton turbine, the maximum shear rate on the front face of the blade occurs at the outermost leading edge. Even if another method is used to mix the fluid, such as a jet, the maximum shear rate still may be expected to occur on or near a surface of the draft tube or nozzle.

The shear rate has been measured electrochemically on the front surface of a Rushton turbine in rheologically complex fluids by Wichterle, Kadlec, Zak, and Mitschka (1). They concluded that the maximum shear rate $\dot{\gamma}$ on a turbine blade satisfies the same relation

$$\dot{\gamma}/N = (1 + 5.3\,n)^{1/n}(N^{2-n}D^2\rho/K)^{1/(1+n)} \qquad (1)$$

as the maximum shear rate at the surface of a rotating disk. This formula was derived from the solution for the flow around a rotating disk by Mitschka and Ulbrecht (2). Here n is the flow index and K is the consistency coefficient so that $K\dot{\gamma}^n$ is the shear stress. Also, N is the rotation rate, D is the diameter of the agitator, and ρ is the density of the fluid. Notice that the expression in the second parentheses is a generalized Reynolds number.

Wichterle, Zak, and Mitschka (3) have also measured the shear stress on the walls of a vessel agitated with a Rushton turbine. They find that the maximum value of the shear rate on the vessel wall is nearly 30% of the maximum shear rate on the impeller blade.

In Equation (1) the shear rate is proportional to N to the power

$$p = 3 / (1+n), \qquad (2)$$

where p ranges from 1.5 to 2.5 as the flow index n varies from 1 to 0.2. This dependence on the rotation rate N is in contrast to the estimate for the effective average shear rate

$$\dot{\gamma}_{eff} / N = 11 \qquad (3)$$

given by Metzner and Otto (4) and Metzner and Taylor (5). In Equation (3) the effective average shear rate is proportional to N to the first power. Also, the measured shear rates described by Equation (1) are up to 100 times larger than given by Equation (3).

The difference between these results arose because Metzner and Otto used the power consumed by the agitator to determine the shear rate as a function of the rotation rate, and Metzner and Taylor used tracer particles to obtain the average flow pattern. Their methods yield only an effective average shear rate and not the shear rate at the blade.

Even though Equation (1) correlates the measurements of Reference (1) reasonably well, these measurements are scattered plus or minus approximately 20% around the equation. Thus the complicated dependence on the flow index n in Equations (1) or (2) has not been tested precisely. One important test that can be done with greater precision is to determine the dependence on rotation rate while the other variables are held constant. In this paper we report the measurement of the shear rate at the tip of a Rushton turbine blade and compare our measurements with an approximate theory.

EXPERIMENTAL METHOD

We use the electrochemical technique of Mitchell and Hanratty (6). A platinum wire is imbedded in one blade of the turbine with only the end of the wire exposed at the forward facing surface of the blade. The exposed end of the wire is located near the outermost edge of the blade and has a rectangular shape oriented vertically (Figure 1). With this configuration, the wire tip serves as a working electrode that measures the shear rate for velocities in the radial direction. The working electrode is connected to a constant voltage source and microammeter through a mercury contact bearing (7) that is insulated from its support. Electric current returns through a platinum counter electrode placed near the walls of the vat some distance from the working electrode. The counter electrode has an area much larger than that of the working electrode in order to make the effect of flow near it negligible.

The shear rate is determined by measuring the transport of a molecule or ion to the working electrode. Although the oxygen molecules already in the broth could in principle be used for this, the oxygen reaction is so slow that even tiny traces of impurities can interfere with the measurement of shear rate. We used a model broth containing ferricyanide ions since they react very quickly and since their diffusivity is known more accurately (8).

The electrolyte used was 0.005 M potassium ferricyanide and 0.005 M potassium ferrocyanide in 1 M potassium chloride aqueous solution with 0%, 1/2%, or 1% polyox added.

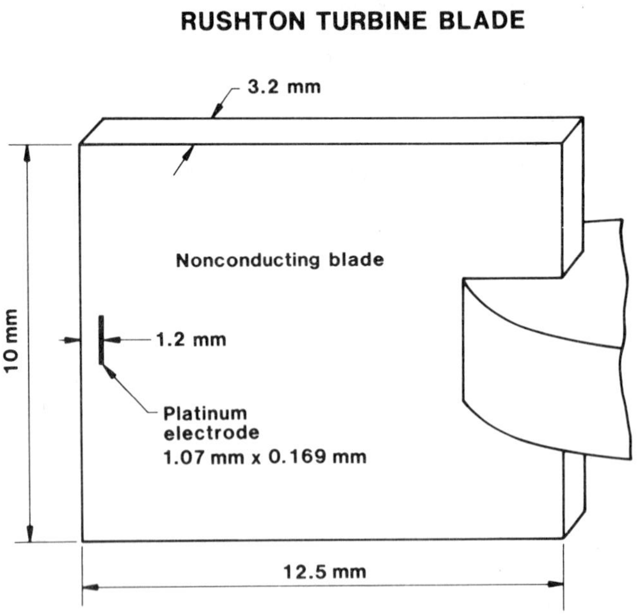

FIGURE 1. Location of working electrode on Rushton turbine blade. The whole turbine is 5 cm diameter and is located at the center of and 5 cm from the bottom of a 10 cm diameter vessel filled to a height of 10 cm.

The ion concentration was set by weighing the salt, which dissociates completely. Argon was bubbled gently through the electrolyte in order to remove the oxygen and to blanket the free surface to keep air away. The temperature of the electrolyte was held at 25 degrees Celsius. The turbine rotation rate was determined by measuring the pulse rate from a 1500 line/revolution encoder.

When a negative voltage is applied to the working electrode, the ferricyanide ions are reduced there. These ions are replaced by identical ions diffusing from the bulk of the fluid. The electric current that flows through the electrode is hence proportional to this diffusion current. As the voltage is made more negative, the current increases, and the concentration of ferricyanide ions at the working electrode surface decreases. When the voltage is sufficiently negative, the concentration becomes zero, and the current is limited by the rate of transport of the ions from the bulk of the fluid.

The effect of flow on the limiting electric current can be understood as follows. When the fluid flow tangential to the working electrode is increased, the transport of unreacted ions to the electrode will be increased, so the electric current will increase. Thus the limiting electric current is a measure of the flow by the electrode and hence of the shear rate at the electrode.

For a flow that has a shear rate $\dot{\gamma}$ and that flows tangentially to an electrode of area A and length L (=0.169 mm) in the direction of flow, the limiting current is (7)

$$i = 0.807\ F\ A\ C^{*}\ (D_i/L)\ (\dot{\gamma}\ L^2/D_i)^{1/3}, \qquad (4)$$

where F = 96,485 coulombs/mol, C^* is the bulk concentration of the reacting ion, and D_i is the ion diffusivity, which we assume is 7.63×10^{-6} cm^2/sec independent of polyox concentration (8). This equation is easily solved for the shear rate.

The quantities with the largest uncertainties in the resulting expression for the shear rate are the ion concentration and diffusivity. These uncertainties are of particular concern because the ion concentration and diffusivity may vary from one solution of a given polymer concentration to another. For this reason, the most accurate measurement that we can make is to determine the dependence of the shear rate on the turbine rotation rate for a fixed polymer concentration.

THEORY

In the following, we perform an order-of-magnitude calculation of the maximum shear rate in the flow of a Newtonian fluid over the surface of a Rushton turbine blade. The calculation will be correct dimensionally and may be expected to give the correct dependence on the rotation rate N, the agitator diameter D, and the kinematic viscosity ν. However, the result will have a numerical coefficient that will be correct only within an order of magnitude.

On the front surface of the blade, the flow can be very roughly approximated by a two-dimensional stagnation flow. If the stagnation point is at x = y = 0, the normal velocity far from the surface at y = 0 is (9)

$$v = -\,a\,y, \qquad (5)$$

and the tangential velocity near the surface is

$$u = 1.2326\ (a/\nu)^{1/2}\ y\ a\ x, \qquad (6)$$

where a is a scale factor. The shear rate is found from the last equation to be

$$\dot{\gamma} = 1.2326\ a\ (a/\nu)^{1/2}\ x. \qquad (7)$$

This shows that the shear rate is proportional to the distance x from the stagnation point.

The maximum shear rate will occur at the edge of the blade, where the electrode is located. Although the stagnation point will probably not be located exactly at the center of the blade, we can obtain an order-of-magnitude estimate assuming that it is. Then the distance of the electrode from the stagnation point is approximately

$$x = D / 8. \qquad (8)$$

Notice that it is only the factor 1/8 in this expression that is approximate. It is clear that x must be proportional to the agitator diameter.

To obtain an order-of-magnitude value for a, use y = D / 4 in Equation (5) to get

$$a = -\,4\,v / D. \qquad (9)$$

Now the velocity of the center of the blade (with respect to the fixed frame) is

$$-\,v = \pi\ (3D/4)\ N. \qquad (10)$$

Insert this expression for v into Equation (9) to get

$$a = 3 \pi N, \qquad (11)$$

and insert the last result and Equation (8) into Equation (7) to get

$$\dot{\gamma} / N = 4.5 (N D^2 / \nu)^{1/2}. \qquad (12)$$

Note that the numerical coefficient is correct only within an order of magnitude. This expression and its experimental confirmation are the principal results of the present paper.

EXPERIMENTAL RESULTS

Equation (12) shows that, for constant kinematic viscosity ν, the maximum shear rate $\dot{\gamma}$ on the surface of the blade is proportional to the agitator diameter D and to the rotation rate N to the 3/2 power. This dependence on N is confirmed with a very small uncertainty by the measurements reported in Figure 2.

Our experimental results for the 0% polyox solution also confirm Equations (1) or (2) with n=1. However, these equations are not confirmed for n different from 1 since the experimentally observed slope of $\dot{\gamma}/N$ versus N is 0.5 for the 1% polyox solution. The data imply that n = 1 must be used in these equations for the 1% solution as well as for the 0% solution.

Although Equation (12) was derived for a Newtonian fluid, one might try to apply it to a shear thinning fluid as follows. For a shear thinning fluid, the kinematic viscosity decreases with increasing rotation rate. When the kinematic viscosity decreases, the shear rate given by Equation (12) increases. Hence for a shear thinning fluid, one might expect the slope to be somewhat larger than the 0.5 slope for the 0% solution This is not confirmed by the 1% polyox measurements in Figure 2, where N ranges from 20 to 600 rpm, and $\dot{\gamma}$ ranges from 17 to 2700 per sec. In order to give agreement with experiment, the kinematic viscosity used in Equation (12) must be independent of shear rate and rotation rate.

The dependence in Equation (12) on diameter and kinematic viscosity can be derived for a Newtonian fluid using dimensional analysis. Note that the quantity in parentheses in Equation (12) is just the Reynolds number Re. Since $\dot{\gamma}$/N and Re are the only dimensionless groups in the problem, $\dot{\gamma}$/N is a function of just Re. This and our experimental result that $\dot{\gamma}$ is proportional to N to the 3/2 power give Equation (12) except for the value of the numerical coefficient.

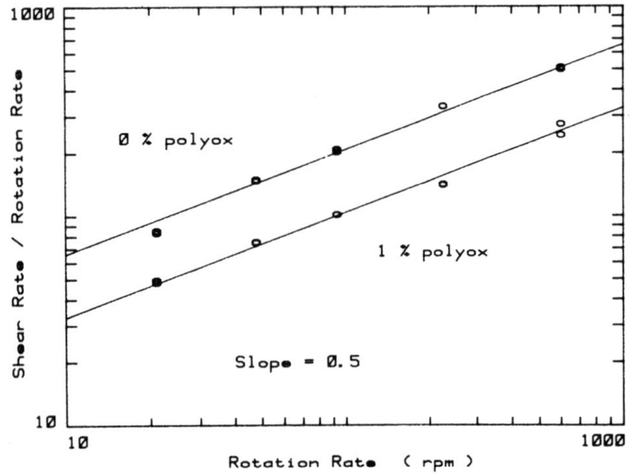

FIGURE 2. Dependence of the shear rate on the rotation rate for a Rushton turbine in an aqueous electrolyte solution and in a 1% polyox solution. For both solutions the shear rate is proportional to the rotation rate to the 1.5 power.

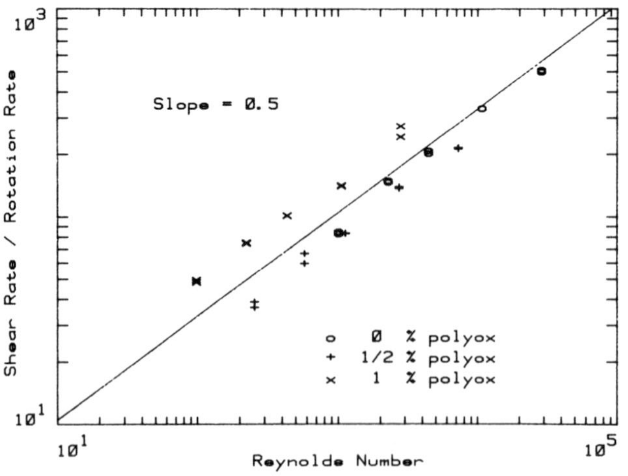

FIGURE 3. Dependence of the shear rate/rotation rate on Reynolds number for a Rushton turbine in an aqueous electrolyte solution and in two polyox solutions.

This relation is also approximately confirmed by the data presented in Figure 3. The viscosities used to compute the Reynolds numbers for this graph are zero-shear viscosities measured with Cannon-Fenske viscometers. The collection times and kinematic viscosities obtained were 926 sec and 0.0344 stokes for the 1/2% polyox solution and 524 sec and 0.0891 stokes for the 1% polyox solution.

The systematic deviation of the data from the line in Figure 3 is due to inaccuracy in determining the ion concentration and zero-shear viscosity and because the diffusivity may depend on polyox concentration. The inaccuracy in the concentration probably is due mostly to evaporation. If the ion concentration times the diffusivity for the 1/2% polyox solution were changed by -5.2%, the 1/2% data would fit quite closely on the line. Similarly, if the ion concentration times the diffusivity for the 1% polyox solution were changed by +17%, these data would fit the line closely.

CONCLUSION

The maximum shear rate $\dot{\gamma}$ on a Rushton turbine blade is given by

$$\dot{\gamma} / N = 3.3 \, (N D^2 / \nu)^{1/2} , \qquad (13)$$

where N is the rotation rate, D is the diameter, ν is the zero-shear kinematic viscosity, and the numerical coefficient is obtained experimentally. This correlation has been confirmed from Re = 100 to 29,000 using 0%, 1/2%, and 1% polyox solutions. The uncertainty in the numerical coefficient is plus or minus 20% due to uncertainties in the ion concentration and diffusivity used to determine the shear rate and due to uncertainty in the kinematic viscosity.

This correlation will be useful for designing large fermentation systems and for scaling shear rate effects from the laboratory scale up to the production scale.

LITERATURE CITED

1. K. Wichterle, M. Kadlec, L. Zak, and P. Mitschka, Chem. Eng. Commun. 26, 25-32 (1984).
2. P. Mitschka and J. J. Ulbrecht, Coll. Czech. Chem. Commun. 30, 2511-2526 (1965).
3. K. Wichterle, L. Zak, and P. Mitschka, Chem. Eng. Commun. 32, 289-305 (1985).
4. A. B. Metzner and R. E. Otto, A.I.Ch.E. Journal 3, 3-10 (1957).
5. A. B. Metzner and J. S. Taylor, A.I.Ch.E. Journal 6, 109-114 (1960).
6. J. E. Mitchell and T. J. Hanratty, J. Fluid Mechanics 26, 199-221 (1966). A review of this work is given by T. J. Hanratty and J. A. Campbell in Fluid Mechanics Measurements, edited by R. J. Goldstein, Hemisphere Publishing Co. (1983), Chapter 11, pp 559-615.
7. C. Deslouis, I. Epelboin, C. Gabrielli, and B. Tribollet, J. Electroanal. Chem. 82, 251-269 (1977).
8. C. Deslouis and B. Tribollet, Electrochemica Acta 23, 935-944 (1978). A Caprani, C. Deslouis, S. Robin, and B. Tribollet, "AC Electrochemical Measurements in Rheological Systems", paper presented at the "Second Conference of European Rheologists" Prague (June, 1986).
9. H. Schlichting, Boundary-Layer Theory, 6th edition, McGraw-Hill (1968), Chap. 5, Sec. 9, pp 88-91, Figures. 5.9 and 5.10, and Table 5.1. The original work was by K. Hiemenz (1911).

A Model for the Dynamic Rheological Behavior of Pelleted Microbial Suspensions

TIMOTHY OOLMAN
DUNCAN YU
Department of Chemical Engineering
University of Utah
Salt Lake City, Utah 84112

The dynamic rheological properties of broths of *Streptomyces tendea* have been measured and modeled. A modified Casson model is used to characterize the apparent viscosity under steady shear conditions. A modified Maxwell model, which incorporates the proposed steady shear model, is shown to fit the experimental results for various dynamic shear experiments.

SUMMARY
The dynamic rheological properties of broths of *streptomyses tendea* have been measured and modeled. This work is unique from previous investigations in that the dynamic, rather than the steady-state, rheological behavior of the microbial broths was studied. Since mass tranfer in a biochemical reactor occurs predominantly in dynamic flow conditions, the dynamic rheological behavior of the broth will be an important factor in determining reactor productivity.

BACKGROUND

The biochemical reactor is, in general, a multiphase chemical reactor containing liquid media, cells, and air bubbles. In most commercial operations, this entire multiphase suspension is agitated to suspend the solid phase and disperse the gas phase, thereby enhancing the transfer of reactants and products. The resulting complex hydrodynamic conditions determine the rates of mass transfer and mixing in the reactor.

The transfer of mass and heat are important factors in all microbial systems. Microorganisms must extract all nutrients required for their growth and metabolism from the surrounding medium. At the same time, potentially toxic metabolic products and heats of reaction must be excreted and dissipated.

The productivity of a biochemical reactor is often limited by the rate with which metabolic reactants are transferred to the microorganisms. This limitation is particularly relevant for aerobic bioreactors. Oxygen is typically supplied to such systems by sparging air through the reactor. Because of the low solubility of oxygen in aqueous solution, the rate of oxygen transport from the gas phase often controls the rate of microbial growth and metabolism in the reactor.

The interrelationships between broth rheology, reactor design and operation, mass transfer and mixing, and reactor productivity are illustrated in Figure 1. The broth rheology,

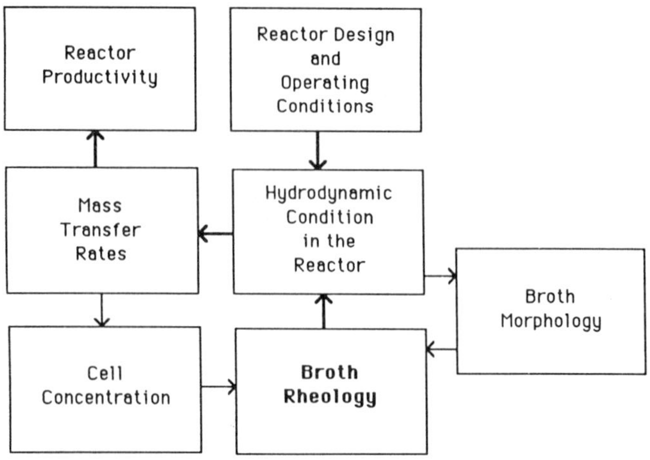

FIGURE 1. Schematic of factors controlling the productivity of a biochemical reactor.

along with the design and operating conditions of the reactor, directly influences mass-transfer rates and, thus, reactor productivity. The reactor operating conditions also influence the concentration and morphology of the cells in the reactor, thereby indirectly influencing the rheological behavior of the broth. The general objective of the present work is to illucidate the individual effects illustrated in Figure 1. This knowledge is necessary for the development of a systematic approach to reactor design and operation.

Primary resistances to oxygen transfer in an air-sparged biochemical reactor are encountered in the liquid film at the surface of the air bubble and within the bulk liquid phase. (Diffusion through the microbial mass can also be significant and is given consideration elsewhere(1).) Several previous investigators have proposed models for predicting the magnitude of these resistances(2-9). The more generally applicable models require a detailed description of the rheological properties of the liquid phase. In particular, the empirical models of Yagi and Yoshida(10) for agitated vessels and of Nakanoh and Yoshida for bubble columns(11) show the shear thinning behavior and elasiticity of the reactor fluid to have a very significant influence on mass transfer rates in air-sparged reactors. Thus, the determination of optimum reactor design and operating conditions requires an understanding of the dynamic rheological behavior of the microbial suspension.

A second aspect of bioreactor performance where the broth rheology is significant is in the determination of the power input required to mix the microbial suspension. The optimum operating condition is to induce the minimum energy input required to maintain turbulent mixing throughout the reactor. However, if a shear-thinning broth is being cultivated in a conventional stirred-tank reactor, an excessive power input must be introduced in the vicinity of the impeller in order to maintain turbulence through the broth. Alternate bioreactor designs which are optimized to the rheological behavior of the broth could greatly reduce the energy costs of the biochemical process.

The significance of broth rheology is not limited to the operation of the biochemical reactor. The flow behavior of the broth is also a very important consideration in the subsequent "downstream processing" of a mature microbial culture. In particular, the most appropriate method by which a desired product can be recovered is dependent on the rheological behavior of the broth. One example where the non-Newtonian behavior of the broth can be put to advantage is in the unit operation of ultra-filtration(12). Whereas filtration is typically not advisable with a highly viscous medium, the shear thinning nature of the microbial broths combined with the very high shear rates induced in an ultrafiltration device makes ultrafiltration an attractive alternative in the purification of biochemical products.

CHARACTERIZATION OF MICROBIAL BROTH RHEOLOGY

The rheological behavior of a microbial broth depends on the concentration and morphology of the microorganism as well as the composition of the suspending medium. The rheology of the broth is a strong function of cell concentration; increases in cell concentration result in higher viscosities and larger deviations from Newtonian behavior. When cell concentrations are high, the morphology of the cells also has a large effect on the rheological properties of the broth. Suspensions of spherical cells will be less viscous and approach Newtonian behavior much more closely than suspensions of mycelial organisms. Organisms that produce extracellular polysaccharides will significantly increase the viscosity and non-Newtonian behavior of the suspending medium and, thus, of the microbial suspension.

Steady State Models:

Various rheological models have been proposed to describe the steady shear-flow behavior of various microbial broths. The simplest model is Newton's law of viscosity:

$$\tau_{xy} = -\eta \gamma_{xy} \, . \tag{1}$$

This equation defines the viscosity, η, as the ratio of shear stress, τ_{xy}, to shear rate, $\gamma_{xy} = dv_x/dy$, for a simple shear flow. For a Newtonian fluid, η is independent of shear rate and duration of shear.

Suspensions of spherical particles, such as yeast cells or spores, exhibit Newtonian behavior for concentrations up to 14 volume percent, with a concentration-dependent viscosity correlated by the Vand equation[13]:

$$\eta = \eta_L(1 + 2.5\varphi + 7.25\varphi^2) . \qquad (2)$$

Here η_L is the viscosity of the suspending medium and φ is the volume fraction of the suspended particles.

Suspensions with a high cell concentration or complex cell morphology exhibit rheological behaviors that deviate significantly from Newtonian. An appropriate model for non-Newtonian rheological behavior is Equation 1 with η being redefined as a function of shear rate. This shear-dependent viscosity is usually referred to as the 'apparent viscosity'. For most non-Newtonian microbial systems the apparent viscosity is a decreasing function of shear rate ($d\eta/d\gamma < 0$); this type of behavior is referred to as pseudoplastic or shear thinning.

Two different models have been commonly used to describe the shear thinning behavior of microbial broths. The most frequently used is the power law model,

$$\tau_{xy} = -K|\gamma|^{n-1}\gamma, \text{ or } \eta = -K|\gamma|^{n-1} \qquad (3)$$

where K and n are empirical constants. For a pseudoplastic suspension the value of K is positive and n lies between 0 and 1. The advantage of this model is its mathematical simplicity. This model fits the measured rheological behavior of extracellular microbial polysaccharide solutions over a wide range of shear rates[14,15,16]. However, for microbial suspensions with a rheological behavior dictated by the morphological structure of the suspended organisms this model is not appropriate[17].

A model commonly applied to filamentous suspensions is the Casson Equation:

$$\tau^{1/2} = \tau_0^{1/2} + k_c\gamma^{1/2}, \text{ or}$$

$$\eta^{1/2} = (\tau_0/\gamma)^{1/2} + k_c \qquad (4)$$

where τ_0 is the yield stress of the suspension and k_c is an empirical constant. Both τ_0 and k_c are functions of the cell concentration. At low values of γ the behavior of this model is dominated by the yield stress term; at high values of γ the apparent viscosity is predicted to approach a constant value. Such a variation in the apparent viscosity of a microbial suspension could be attributed to the variation in interparticle structuring within the broth. The yield stress, which is approached as γ approaches zero, is due to a continuous structuring of the microbial particles throughout the suspension. As γ increases, the interparticle structuring decreases; at very high values of γ the interparticle structuring is totally broken down, and a limiting lower value of η is approached. This model fits the rheological behavior for concentrated suspensions (greater than 10 volume percent) of filamentous organisms quite well[17]. Filamentous morphology is typically observed with mycelial organisms cultivated in stirred-tank cultures.

In the present work we propose a modified version of the Casson model for the characterization of mycelial suspensions which grow in a pelleted morphology. This model presents the empirical constant k_c as a function of shear rate as well as of cell concentration:

$$k_c = [(\eta_0 + \eta_\infty\lambda\gamma)/(1 + \lambda\gamma)]^{1/2} \qquad (5)$$

λ is a time constant and η_0 and η_∞ are limiting viscosity parameters at low and high shear rates, respectively. This model was found to work very well for suspensions of *Streptomyces tendae* which grew with a pelleted morphology. Pelleted morphology is usually observed with mycelial organisms that are grown in low shear environments. It is proposed that the variation in k_c with shear rate is due to a change in the internal structuring (ie. the shape) of the suspended pellets.

A representative comparison between experimental measurements and the predictions of Equations 4 (with substitution of Equation 5) are shown in Figure 2. For each of the data sets four parameters of the rheological model were fit to the experimental data; these were the two limiting viscosities, η_0 and η_∞, the yield stress, τ_0, and the time constant, λ. Figure 3 shows the best fit values of η_0, η_∞, and τ_0 as a function of cell concentration for 10 *Streptomyces* suspensions with pelleted morphology. The best fit values of λ showed only random variation with various samples and was assumed to have a constant value of 0.1 sec. With this assumption, a fairly smooth correlation between each of the other three parameters and the cell concentration is obtained. The τ_0 values were also measured by a more direct method. Data from the direct measurements are also reported in Figure 4a. An average deviation of less than 20 percent between the respective measured and fit values of τ_0 is observed.

The same model has also been fit to limited experimental data obtained for filamentous suspensions of *Streptomyces tendae*. In this case the respective fitted values of η_0 and η_∞ were very nearly equal.

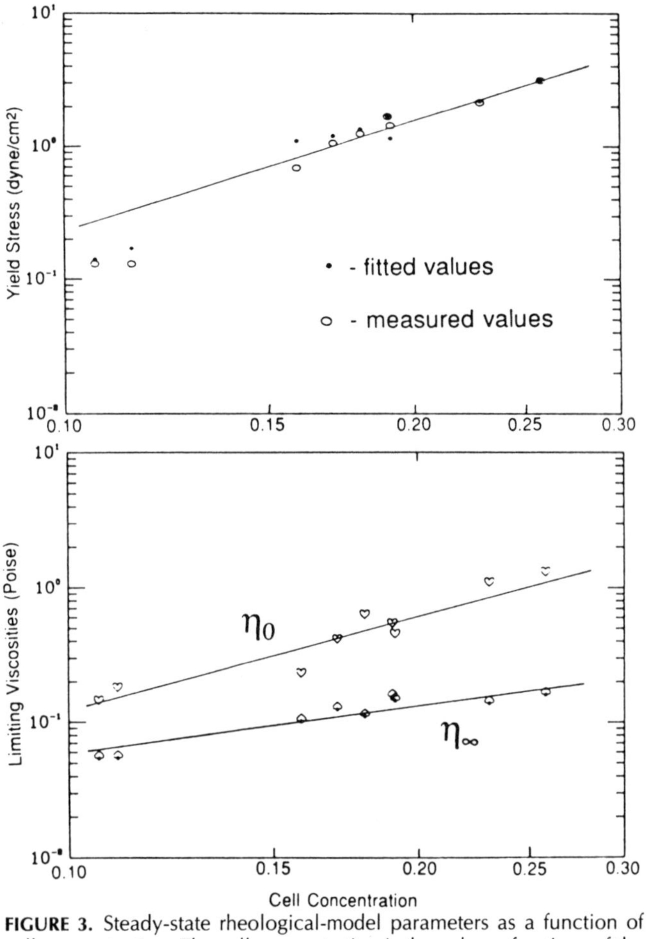

FIGURE 3. Steady-state rheological-model parameters as a function of cell concentration. The cell concentration is the volume fractions of the cells determined by centrifugation.

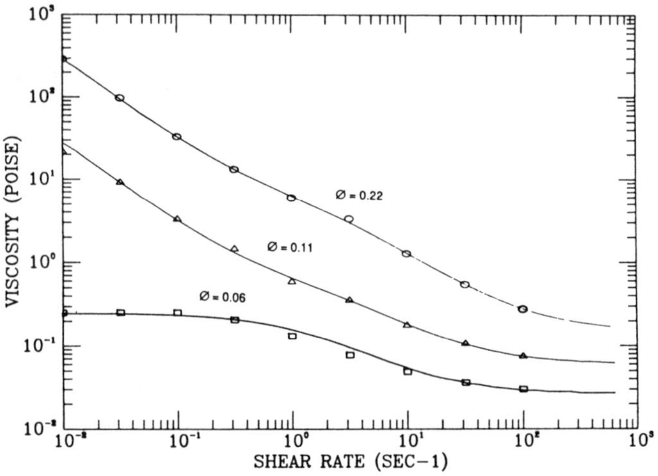

FIGURE 2. Comparison of measured and predicted values of the steady shear viscosity of pelleted suspensions of *Streptomyces tendea*. The ϕ-values represent volume fractions of the cells as measured after centrifugation.

An attractive feature of the data reported in Figures 2 and 3 is that the rheological parameters are presented as a function of cell concentration alone, with no dependence on the size distribution of the pellets in the suspension or the age of the broth. This can be interpreted to mean that the rheological behavior is independent of the conditions under which the cultures were grown. Such behavior is not observed, however, for suspensions of microbes with filamentous morphology (ie., grown in stirred vessels). For these broths, the rheological behavior appears to be strongly dependent on the shear environment in the vessel (i.e., the impeller speed) and the age of the broth. More data is needed before a quantitative relationship between these parameters can be developed for filamentous cultures.

Dynamic Rheological Model:

The rheological behavior of a microbial suspension cannot be totally characterized by steady-state models. Microbial suspensions display various time-dependent effects, including elasticity and restructuring of the suspended particles. The dynamic model presented below considers both of these effects.

The hydrodynamic condition in a gassed, stirred-tank reactor is inherently not at a local steady state with respect to points moving with the fluid. Each individual element of fluid in the reactor experiences a continually changing shear environment. For example, a large jump in the shear rate is experienced as a fluid element passes through the impeller blades. More moderate variations in the shear field are experienced as a gas bubble rises past a fluid element. Thus, dynamic flow behavior will have a significant effect on the rates of mixing and mass transfer in a biochemical reactor.

Previous to the present author's work, no systematic attempt has been made to characterize the dynamic rheological behavior of microbial suspensions. The literature that most closely relates to this field are the publications on blood rheology(18).

In the present work we show that a modified Maxwell model represents the experimental data for dynamic rheological measurements of *Streptomyces tendae* cultures very well. The proposed model is shown in Table 1. The Maxwell Equation with the time parameter λ_0 is shown on the top line. This analytical model is analogous to the mechanical model of a spring and dash pot in series. The analogy with the mechanical model can be extended to the structuring which occurs within a microbial suspension. The individual microbial particles tend to form a continuous structure throughout the broth which exhibits the elastic properties of a solid structure, such as a spring. However, this structuring can also break down, allowing individual particles to slip past each other; this results in a viscous flow which can be characterized by a dashpot. λ_0 is defined on the next line of Table 1 as the ratio of the instantaneous values of the viscosity and elasticity of the suspension. The instantaneous viscosity, η, is defined in an analogous manner to the steady-state viscosity, the model for which is discussed above. The parameter χ is defined as a function of the limiting viscosities, η_0 and η_∞

TABLE 1
Dynamic Rheological Model for Stress Buildup and Stress Relaxation Experiments

$$\tau + \lambda_0 (d\tau/dt) = \eta(t)\dot{\gamma}$$

$$\lambda_0 = \eta(t)/G(t)$$

$$\eta(t)^{1/2} = \eta_\infty^{1/2}(1-(1-\chi)\beta_2)^{1/2} + (\tau_0/\dot{\gamma})^{1/2}$$

$$\chi = (\eta_0/\eta_\infty)$$

$$\tau_0 = \tau_y \beta_1$$

$$G(t) = G_0 \beta_1$$

$$\beta_i(t) = \beta_i^{eq} + (\beta_i^0 - \beta_i^{eq}) \exp(-t/\lambda_i), \quad i = 1,2$$

$$\beta_i^0 \text{ is the value of } \beta_i \text{ at time } = 0$$

$$\beta_1^{eq} = \beta_2^{eq} = 1/(1+\lambda_3\dot{\gamma})$$

$$\lambda_1 << \lambda_2$$

An additional dynamic character is induced in this model by expressing the instantaneous viscosity and elasticity as functions of the structural parameters, β_1 and β_2. These parameters reflect the particle structuring within the suspension: β_1 reflects the interparticle structuring and β_2 reflects intraparticle structuring (i.e., β_2 is a shape factor). The β values in a stagnant fluid are, by definition, equal to 1. If the fluid is suddenly exposed to a shear flow, the β–values will decrease as the structuring changes. The rates with which the structures within the suspension change are characterized by the time constants, λ_1 and λ_2. After a sufficiently long time at a constant shear rate, the structural parameters approach constant values, β_i^{eq}, which are functions of the shear rate. Substitution of the definition for β_1^{eq} into the definition for $\eta(t)$ will regenerate the steady state rheological equation given by Equations 4 and 5. The structural changes hypothesized in this model are assumed to be totally reversible.

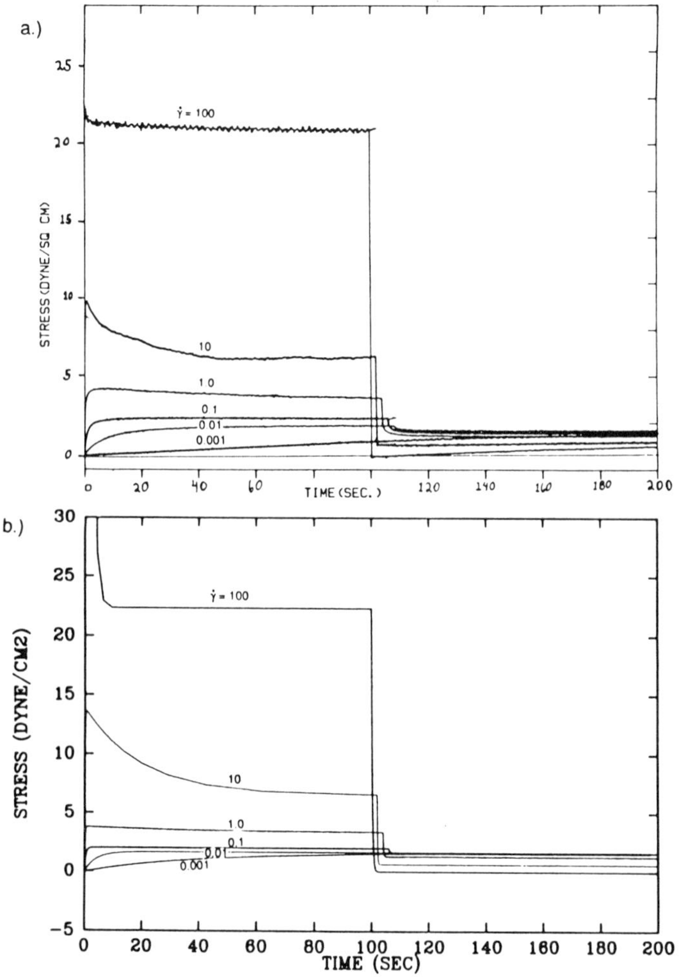

FIGURE 4. Comparison of measured and predicted values of the dynamic shear-stress response to step changes in the shear rate. The step changes occurred at 0 and 100 s. Figure a shows experimental data and Figure b shows the model predictions. The cell concentration for this sample was 20 vol. % as measured by centrifugation.

The stress responses predicted by the model given in Table 1 are compared with experimental data in Figure 4 (for a sample with pelleted morphology). The model predictions are shown above the experimental results. Each of the lines represents an experiment where the shear rate was raised from zero to some constant value at time equal to zero, held at that value for 100 seconds, and then dropped back to zero for the remainder of the experiment.

The model shows good agreement with experimental results. The largest deviation occurs in the short time period following the initiation of high shear rates. In this range the theoretical stress 'overshoot' is significantly larger than the experimentally measured value. This deviation is most likely a consequence of the microbial suspension slipping along the wall of the experimental apparatus. The rheological model assumes there to be no net velocity between the wall of the rheometer cup and the adjacent fluid. The experimentally observed wall slip is a result of the large inertial force which must be overcome during rapid acceleration of the fluid. This behavior results in a diminishing of the elastic buildup of stress within the suspension.

When a constant shear rate is induced for a sufficiently long period of time, both the theoretical and experimental curves must approach the steady state behavior. It can be clearly seen that the Maxwell model simplifies to the steady state model when the shear stress is constant with time. For this reason the steady state rheological model is an important component of the dynamic model. Once an appropriate steady state model has been developed, only a few additional parameters (namely, an elasticity and two time constants) need be determined for the dynamic model.

LITERATURE CITED

1. Moo-Young, M and H. Blanch, "Kinetics and transport phenomena in biological reactor design" in Foundations in Biochemical Engineering, ed. Blanch, Papoutsakis, and Stephanopoulos, p. 335, ACS Symposium Series 207, Washington D.C. (1983).

2. Bhavaraju, S. M., R. A. Mashelkar, and H.W. Blanch, AIChE J., 24, 1063 (1978).

3. Moo-Young, M., and T. Hirose, Can. J. Chem. Eng., 50, 128 (1972).

4. Schaftlein, R.W. and T. W. F. Russell, Ind. Eng. Chem., 60(5) 12 (1968).

5. Cichy, P.T., and T. W. F. Russell, Ind. Eng. Chem., 61(8) 15 (1969).

6. Joshi, J. B., A. B. Pandit and M. M. Sharma, Chem. Eng. Sci., 37, 813 (1982).

7. Kawase, Y., R. A. Mashelkar, and J. J. Ulbrecht, Int. J. Multiphase Flow, 8, 433 (1982).

8. Oolman, T., H. W. Blanch, "Non-newtonian fermentaion systems." in Critical Reviews in Biotechnology, pp. 133-184, CRC Press, Inc., Boca Raton, FL (1986).

9. Höcker, H. and G. Langer, Rheol. Acta, 16, 400 (1977).

10. Yagi, H., and F. Yoshida, Ind. Eng. Chem. Preocess Des. Dev., 14, 488 (1975).

11. Nakanoh, M. and F. Yoshida, Ind. Eng. Chem. Preocess Des. Dev., 19, 190 (1980).

12. Ho, L., R. Taylor, and R. Jasper Eur. Pat. Appl. EP 69,523, Pfizer Inc. (1981).

13. Deindoerfer, F. H. and J. M. West, Adv. Appl. Microbiol., 2, 265 (1960).

14. Thomson, N., and D. Ollis, Biotech. Bioengin., 22, (1980).

15. Chang H.-T., and D. Ollis, Biotech. Bioengin., 24, 2309 (1982).

16. Charles, M., Adv. Biochem. Eng., 8, 1 (1978).

17. Roels, J. A., J. van den Berg, and R. M. Voncken, Biotech. Bioengin., 16, 181 (1974).

18. Quemada, D., Biorheology, 20, 635 (1983).

Bubble-Column and Airlift Bioreactors

Non-Newtonian Broths in Airlift Bioreactors

MARK A. YOUNG
RUBEN G. CARBONELL
DAVID F. OLLIS
Department of Chemical Engineering
North Carolina State University
Raleigh, North Carolina 27695-7905

The airlift reactor depends critically upon buoyancy-driven hydrodynamics for its perfomance. The major analytical model requirements for a hydrodynamic description of airlift operation are delineated. An initial view of our general modeling approach is given. A qualitative study of pilot-scale fermentor operation using increasingly pseudoplastic solutions is discussed. It is concluded that a strongly pseudoplastic solution characterized by a zero-shear viscosity of approximately 0.1 Pa-s constitutes the high viscosity limit for bubble flow operation in fermentors of the type examined.

Airlift fermentors are noted for high mass transfer and good mixing with relatively low shear and low power input. Due to these features, airlift reactors are currently employed in certain large scale industrial applications, most notable for single cell protein fermentation. Recently, researchers have utilized the airlift design for applications which develop non-Newtonian broths, such as fermentation of filamentous fungi (1,2) and plant cell culture (3). Onken and Weiland (4) cited other possible applications, including culture of animal cells and production of citric acid and xanthan gum.

While the construction and operation of airlift reactors are particularly simple, the complexity of recirculating two-phase flow hampers design optimization and scale-up. The complexity is, of course, compounded when non-Newtonian effects are included. Over the last decade a number of hydrodynamic airlift models have been presented in the literature (e.g., 5 to 8). Most currently available models are based upon a macroscopic balance approach and are limited in their practical applicability owing to their rheological simplicity, restriction to zero gas recirculation, and/or dependence upon highly specific correlations. We seek to develop an airlift model of greater general applicability and to investigate the feasibility of the airlift design for various non-Newtonian fermentations. At this early stage, we outline the features that a more general model would necessarily include; we indicate our general modeling approach and report qualitative results from our initial experiments involving pseudoplastic solution in a pilot-scale airlift fermentor.

FEATURES OF A COMPLETE MODEL

A complete hydrodynamic airlift model would incorporate at least five features not found in currently available models.

Separated flow treatment. Many investigators have assumed plug flow of a strictly homogeneous two-phase mixture in the riser. Clearly, there exists a relative velocity between gas and liquid, with attendant frictional losses, which homogeneous models ignore. Further, Merchuk and Stein (6) reported significant variations in gas holdup with axial position in the riser. Visual observations of our vessel seem to indicate preferential flow of gas bubbles near the riser center, with this phenomenon becoming more pronounced with high gas flows and/or pseudoplastic liquids. Hence, the presence of axial and perhaps radial variations dictates the use of separate transport equations for each phase, the so-called separated flow approach (9).

Effects of surfactants. Surfactants will be present to some degree in most fermentation broths, and many important fermentation products are highly surface active. Even small quantities of surfactants may influence gas/liquid behavior substantially, including changes in equilibrium bubble size, tendency to coalesce and terminal rise velocity of a bubble swarm. All the early airlift models have assumed pure water for the liquid phase.

Recirculation of gas in the downcomer. Gas recirculation occurs when the downcomer liquid velocity exceeds the bubble terminal rise velocity. Mixing and mass transfer requirements will frequently dictate the use of sparging rates which give rise to partial recirculation of the gas. Additionally, in an industrial scale fermentor, the fluid residence time in the return leg will likely require significant oxygen transfer to occur in that section in order to avoid anoxia of the biocatalyst. All current models are limited to the regime of zero downcomer gas holdup and, by extension, regimes of low gas sparging rates.

Range of broth rheologies. Many industrially important fermentation broths are pseudoplastic. The range of pseudoplasticity over which the airlift design will function effectively has not been well documented in the literature. We seek to establish these limits and incorporate the shear thinning effects into the liquid phase momentum equation.

Judicious use of correlations. While the use of empiricism in a model of this nature is unavoidable, correlations should be, to the greatest extent possible, scale and geometry independent. Several simple models previously offered for airlift reactors depend upon highly specific correlations developed by the investigators. A model of general utility would incorporate only correlations of broad applicability, and, preferably, ones requiring only relatively convenient laboratory measurements.

GENERAL MODELING APPROACH

A separated flow modeling approach for bubble flow in riser and downcomer sections entails writing averaged continuity and momentum equations for each phase:
Continuity:

$$\nabla \cdot \varepsilon<\rho><v> = 0$$

Momentum:

$$\nabla \cdot \varepsilon<\rho><v><v> = -\varepsilon\nabla<P> + \nabla \cdot \varepsilon<\tau> + \varepsilon<\rho>g + F\left[\frac{A_{ab}}{V}\right]$$

where < > denotes a volume-averaged quantity, F is the drag force vector per unit interfacial area and all other symbols have their usual meaning and are listed in the notation section. In writing these equations, we have assumed steady, isothermal bubble flow, no mass transfer, nearly spherical bubbles and small deviations in ρ and v relative to their mean values. (For a discussion of the volume-averaged, two-phase transport equations, see Sáez and Carbonell (10).) Although it is of principle interest in analyzing fermentor performance, mass transfer is neglected here since the associated momentum flux is insignificant and since this approximation has negligible effect upon the continuity equation. The degassing section at the top of the column will be modeled as a well-mixed region.

Model development will proceed in stages of increasing flexibility and generality. Initially, the flow will be considered to be one dimensional and will be cross-sectionally averaged. These restrictions result in momentum equations of the following form:

$$\varepsilon<\rho><v_z>\frac{\partial<v_z>}{\partial z} = -\varepsilon\frac{\partial<P>}{\partial z} + \frac{2}{R^2}\left[\varepsilon\mu r\frac{\partial<v_z>}{\partial r}\right]_{r=R} + F_z\left[\frac{A_{lg}}{V}\right] + \varepsilon<\rho>g_z$$

When evaluated at the outer radius, the second term on the right hand side represents the wall frictional losses, τ_w. In order to solve the continuity and momentum equations, additional relationships for $<P_g>=f(<P_l>)$, $P_g=f(<P_g>)$, $r_b(z)$, F_z and τ_w are needed. Of these, the last three are the most problematic. The functional dependence of bubble radius upon axial position might be adequately described using a momentum balance on a representative bubble flowing along the riser or downcomer. A similar technique of analyzing a representative particle for two-phase flow modeling has been successfully employed by Crowe et al. (11) for solid/gas systems.

The interfacial drag term might be successfully developed using a form analogous to the Ergun equation for pressure drop through packed beds, i.e., an equation of the form:

$$\frac{F}{1-\varepsilon_g}\left[\frac{A_{gl}}{V}\right] = A(\mu,\varepsilon_g,r_b)\left[<v_g> - <v_l>\right] + B(\rho_l,\varepsilon_g,r_b)\left|<v_g> - <v_l>\right|\left[<v_g> - <v_l>\right]$$

Such an approach might be fruitful due to the similarity in the liquid flow pattern around spherical packing particles and around spherical bubbles. New forms for the viscous coefficient A and the inertial coefficient B must be developed. Nevertheless, the Ergun equation form represents an attractive starting point, particularly since the viscous term has been successfully generalized by Sadowski (12) for non-Newtonian solutions in packed beds.

Wall frictional losses have commonly been treated using correlations, either a two-phase friction factor approach or the two-phase multiplier approach of Lockhart and Martinelli. The former, which is more suitable for our separated flow approach, takes the general form (9):

$$\tau_w = \tfrac{1}{2} C_f \rho_m v^2$$

where C_f = two phase friction factor for pipe flow and $v = (Q_g + Q_l)/A$.

EXPERIMENTAL APPROACH

Data for evaluating model predictions will be collected using a 0.2 cubic meter external-loop airlift system. The fermentor is designed as a hydrodynamic research tool employing mock fermentation broths and will not be used for actual fermentations. As shown in Figure 1, the system is comprised of the airlift vessel and instrumentation for measuring liquid pressure, void fraction and liquid velocity. The acrylic vessel is approximately three meters in overall height with a 0.19m ID riser and interchangeable 0.19, 0.14 and 0.09m ID downcomers.

Pressure can be measured at eight points along the riser using an inclined manometer and gas-bubbler arrangement. By using nitrogen gas to balance the liquid pressure at a given tap location and to transmit that pressure to the manometer fluid, very sensitive pressure measurements may be made on a wide range of broth rheologies.

Cross-sectionally averaged gas holdup may be measured at any point along the riser axis using a translating gamma densitometer. Determination of void fraction using the gamma beam attenuation principle allows nonintrusive measurements to be made on any broth rheology. Unlike manometer techniques for void measurement, this arrangement allows complete separation of the void fraction and pressure measurements.

The liquid velocity profile will be measured in the downcomer by traversing the tube with a hot-film anemometer. These measurements will indicate the degree of deviation from plug flow and permit calculation of the liquid recirculation rate. Combining these measurements with data on void fraction along the riser, local average liquid velocity at any point in the riser is known.

FERMENTOR OPERATING LIMITS FOR PSEUDOPLASTIC SOLUTIONS

Initial qualitative studies have been performed to ascertain the gross operating limits of our pilot-scale unit using increasingly viscous pseudoplastic solutions. Xanthan gum, an exobiopolymer produced by *Xanthanomonas campestris*, was chosen to produce test solutions of increasing pseudoplasticity. As shown in Figure 2, xanthan gum produces highly pseudoplastic aqueous solutions at low solute concentrations. Since commercial xanthan fermentations proceed to approximately 3 wt% at completion, xanthan broths represent an extreme case of viscous, non-Newtonian fermentations. In addition to pseudoplasticity, xanthan solutions in the 1 to 3 wt% range may exhibit viscoelastic behavior (14) and liquid crystal formation (15).

Fields *et al*. (16) conducted the only published study of xanthan solutions in airlift vessels. Using a concentric tube configuration having an 0.085 m ID draft tube, the flow regime was observed to shift from bubble to slug flow over a concentration range of zero to 0.5 wt%. It was anticipated that our larger 0.19 m ID riser might operate at higher xanthan concentrations before the transition to slug flow. However, our results essentially parallel those of Fields *et al*..

Increased coalescence near the sparger was observed at xanthan concentrations as low as 0.05 wt%. A small but noticeable fraction of the approximately 0.005 m diameter bubbles produced by the sparger rapidly coalesced, forming bubbles in the 0.01 to 0.02 m diameter range. This churn-turblent behavior occurred at modest superficial gas velocities of approximately 0.04 m/s. All further increases in xanthan concentration led to increased coalescence. At 0.10 wt% large, irregularly shaped bubbles of approximately 0.05 m diameter formed. When operating at 0.2 wt%, the majority of the gas was observed to coalesce in the sparger region to form one large bubble, which passed through the riser without breaking up.

Paralleling these changes in bubble coalescence were marked changes in liquid stagnation and foaming. A stagnant layer developed on the wall of the 0.14 m ID downcomer when the xanthan concentration reached 0.07 wt%. Flow along the riser wall slowed substantially at a concentration of 0.15 wt%; the flow direction would reverse momentarily as liquid was displaced by a large passing bubble. At 0.3 wt% flow reversal was prevalent in the riser, and the liquid velocity in the downcomer was very low (on the order of 0.02 to 0.04 m/s). Over the concentration range 0.1 to 0.4 wt%, an increasingly thick layer of wet, highly stable foam developed on the free surface of the liquid.

From this investigation, we conclude that a pseudoplastic solution characterized by a zero shear viscosity of approximately 0.1 Pa-s constitutes the upper viscosity limit for bubble flow behavior. At higher liquid viscosities, moderate gas flows generate churn-turbulent flow behavior, the modeling of which would necessarily take a different form.

ACKNOWLEDGEMENT

This work was supported, in part, by a grant from Celanese Corporation.

NOTATION

A_{ab}	interfacial area
F	drag force per unit interfacial area
g	gravitational acceleration
P	pressure
r	radial coordinate
r_b	bubble radius
R	radius of tube
V	averaging volume
v	velocity

Greek Letters

ε	volume fraction
μ	viscosity
ρ	density
τ	viscous stress tensor
τ_w	frictional losses at wall

Subscripts

g	gas
l	liquid
z	axial coordinate

LITERATURE CITED

1. Barker, T. W. and J. T. Wogan, Eur. J. Appl. Microbiol. Biotechnol., 13, 77 (1981).

2. Malfait, J. L., D. J. Wilcox, D. G. Mercer and L. D. Barker, Biotechnol. Bioeng., 23, 863 (1981).

3. Townsley, P. M., F. Webster, J. P. Kutney, P. Salisbury, G. Hewitt, N. Kawamura, L. Choi, T. Kurihara, and G. Jacoli, Biotechnol. Lett., 5, 13 (1983).

4. Onken, U. and Peter Weiland, "Airlift Fermenters: Construction, Behavior and Uses," in Dev. in Biotech. Processes, Vol. 1, A. Mizrahi and A. van Wezel (Eds.), Alan R. Liss, New York (1983).

5. Merchuk, J. C., Yehuda Stein and R. I. Mateles, Biotechnol. Bioeng., 22, 1189 (1980).

6. Merchuk, J. C. and Yehuda Stein, Biotechnol. Bioeng., 23, 1309 (1981).

7. Hsu, Y. C. and M. P. Duduković, Chem. Eng. Sci., 35, 135 (1980).

8. Jones, A. G., Chem. Eng. Sci., 40, 449 (1985).

9. Wallis, G. B., One-Dimensional Two-Phase Flow, McGraw-Hill, New York (1969).

10. Sáez, A. E. and R. G. Carbonell, AIChE J., 31, 52 (1985).

11. Crowe, C. T., M. P. Sharma, and D. E. Stock, J. Fluids Eng., June, 325 (1977).

12. Sadowski, T. J., Trans. Soc. Rheology, 9, 251 (1965).

13. Whitcomb, P. J. and C. W. Mascoko, J. Rheology, 22, 493 (1978).

14. Thurston, G. B. and G. A. Pope, J. Non-New. Fluid Mech., 9, 69 (1981).

15. Lim, Timothy, J. T. Uhl and R. K. Prud'homme, J. Rheology, 28, 367 (1984).

16. Fields, P. R., F. R. Mitchell and N. K. Slater, Chem. Eng. Commun., 25, 93 (1984).

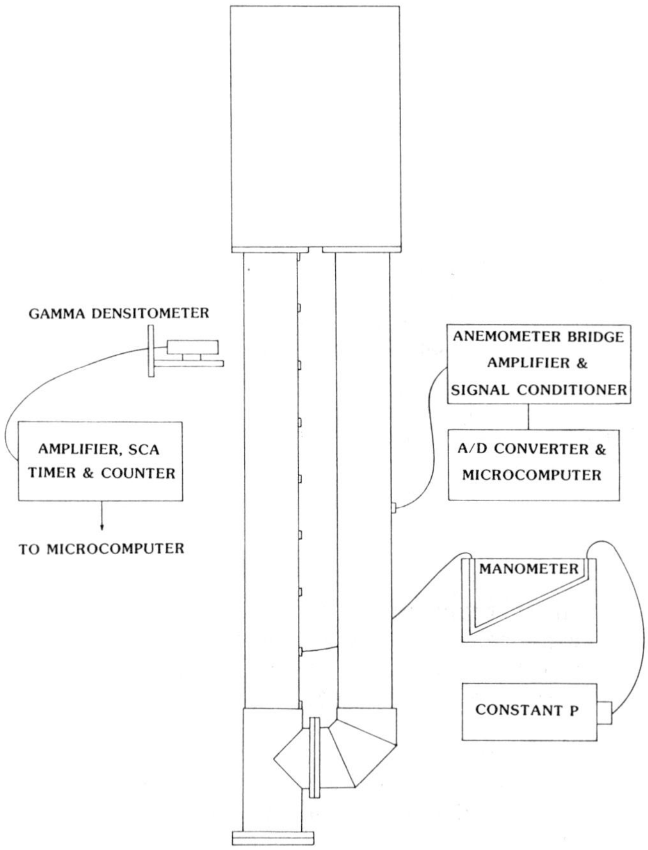

FIGURE 1. Experimental apparatus.

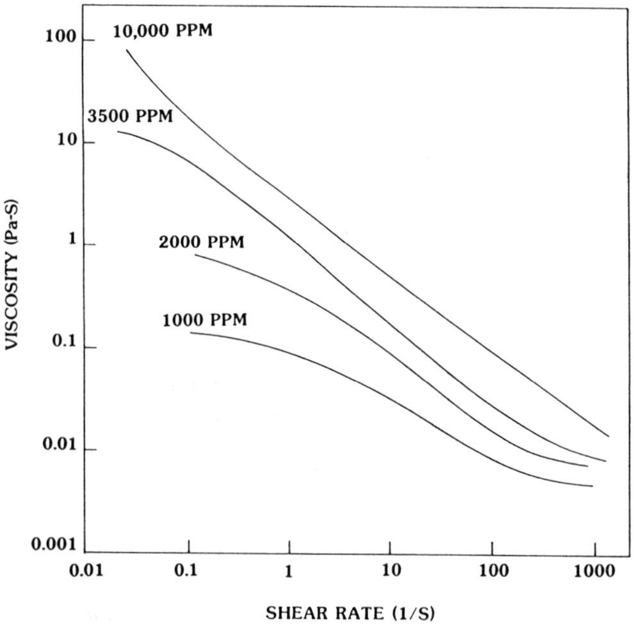

FIGURE 2. Xanthan viscosity. (Data are from reference *13*. Used by permission.)

Liquid Circulation in Airlift Fermentors

C.H. LEE
L.A. GLASGOW
L.E. ERICKSON
S.A. PATEL
Department of Chemical Engineering
Kansas State University
Manhattan, Kansas 66506

A mathematical model based on the macroscopic mechanical energy balance is developed to predict liquid circulation velocity in airlift fermentors. This analysis takes into account effects of energy dissipation due to wakes behind the bubbles in the upflow zone and energy dissipation due to the upflow motion of large bubbles with respect to the liquid in the downflow zone. The results indicate that each of these modes of energy dissipation accounts for 30% to 60% of the total for the draft-tube and split-cylinder airlift fermentors examined. The liquid turnaround in the top and bottom regions of the airlift fermentor and other resistance to liquid recirculation in the system account for the rest of the energy dissipation.

Since its discovery in 1954, airlift fermentation has increasingly been explored as an important and potentially superior alternative to stirred tank fermentation. The special feature that distinguishes an airlift fermentor from the more common stirred tank reactor is that the pressurized gas which is furnished for mass transfer also provides the necessary circulatory drive. The main advantages of this design are the capacity for satisfying the high oxygen demand of hydrocarbon fermentation (1) without mechanical moving parts, low energy input, especially in large configurations (2, 3), and the possibility of easier removal of the heat generated in the fermentation process.

Airlift fermentors can be designed in one of two basic configurations. Either the upflowing gassed liquid and the downflowing liquid are separated by a draft tube or a baffle plate suspended inside the reactor (internal loop), or the two streams flow in separate pipes connected at the top and bottom (external loop). It has been shown that, because of the introduction of gas, a strong liquid circulation pattern is developed. The liquid circulation rate is considered to be one of the key parameters in the design and scale-up of the airlift reactor. It affects mixing characteristics of the reactor, gas hold-up, volumetric mass and heat transfer coefficients and uniformity of reaction temperature and cell concentration; thus, it affects the overall performance of the reactor.

Fundamental studies of airlift reactors have been reported by Hatch (2), Hatch et al. (4), and Orazem and Erickson (5), among others. The problem of modeling an airlift reactor has been considered by Hatch (2), Ho et al. (6) and Merchuk et al. (7). The main difficulty in such modeling is the lack of information on the hydrodynamics of the airlift reactor.

One of the earliest studies of the draft-tube airlift column was by Lamont (8) who predicted the liquid circulation rates in a Pachuca tank using airlift pump theory. In this semiempirical model the energy transferred as the air expands in rising through the vessel is corrected for energy dissipation due to bubble slip and the remainder gives rise to velocity heads associated with liquid flow in the vessel.

Rietema and Ottengraf (9) later presented a theoretical model based on minimum entropy production; a momentum balance was developed to predict the diameter of the "bubble street" and laminar liquid circulation velocities in a gas sparged column filled with glycerol. Whalley and Davidson (10) also modeled the liquid circulation in a shallow bubble column based on an energy balance method and showed that the liquid circulation velocities calculated by the energy balance method are closer to reality than those predicted by pressure balance. Joshi and Sharma (11) modified the energy balance procedure and successfully developed a multiple cell circulation model to predict liquid circulation velocities and fractional gas hold-ups in tall bubble columns. More recently, Jones (12) proposed a simple model based on airlift pump theory to predict liquid circulation velocities in draft-tube bubble columns. Reasonable agreement was found only for draft-tubes $\leq$ 0.121 m in diameter at gas flow rates $\leq$ 0.0004 m^3/s; significant deviations were found with larger draft-tube diameter.

The primary objective of this work was to develop a mathematical model based on the macroscopic energy balance and determine the applicability of this model for the prediction of liquid circulation velocities in airlift fermentors.

THEORY

Energy Balance

The basis of the energy balance method is to equate:

(1) the rate of energy input to the fermentor by the gas to
(2) the rate of energy dissipation by the fluid motion.

For buoyancy-driven two phase flows, the supply of energy results from introduction of the dispersed phase. The energy input rate for gas-liquid dispersion is given by the following equation:

$$E_i = QP\ \ln[1+\rho_L gH_L/P] \quad (1)$$

where
- E_i: energy input rate (W),
- Q: volumetric gas flow rate (m^3/s),
- P: pressure at the top of the column (N/m^2),
- ρ_L: density of the liquid phase (kg/m^3),
- H_L: ungassed liquid height (m),
- g: acceleration due to gravity (m/s^2).

The energy dissipation in an airlift column is due to the following factors:

(1) Energy dissipation in the upflow zone due to wakes behind the bubbles, E_{d1}.
(2) Energy dissipation in the top and bottom regions due to liquid turnaround and other resistance to liquid recirculation in the column, E_{d2}.
(3) Energy dissipation in the downflow zone due to the upflow motion of bubbles with respect to the liquid, E_{d3}.
(4) Energy dissipation due to the viscous drag at the walls of the column, E_{d4}.

An order of magnitude calculation for a 15 cm diameter column indicates that the dissipation due to viscous drag, E_{d4} makes up only 1 to 2% of overall energy input for liquid viscosities up to 20 mPa s. Therefore, the energy dissipation due to this mode is negligible.

The energy dissipation rate, E_{d1} in the bubble wakes can be evaluated by considering a bubble cloud with uniform fractional hold-up E_{G1}, ungassed liquid height H_L, liquid density ρ_L, horizontal cross-sectional area in the upflow zone A_1, and liquid and gas superficial velocities U_{L1} and U_G. E_{d1} is the result of three terms, namely the rate of loss of pressure energy of the liquid = $A_1\rho_L gH_L U_{L1}$, the rate of loss of pressure energy of the gas = $A_1\rho_L gH_L U_G$, and the rate of gain of potential energy of the liquid = $A_1\rho_L gH_L U_{L1}/(1-E_{G1})$, giving

$$E_{d1} = A_1\rho_L gH_L E_{G1}[U_G/E_{G1}-U_{L1}/(1-E_{G1})] \quad (2)$$

The slip velocity, the relative velocity of the bubbles with respect to the liquid, is equal to $U_G/E_{G1}-U_{L1}/(1-E_{G1})$. Using Turner's (13) assumption that slip velocity is equal to the terminal velocity of bubbles rising in a stagnant liquid, $V_{b\infty}$, gives

$$E_{d1} = A_1\rho_L gH_L E_{G1}V_{b\infty} \quad (3)$$

As might be expected, gas and liquid flow rates do not appear in this equation; the energy dissipation due to wakes behind the bubbles is mainly determined by the gas hold-up.

Evaluation of the energy dissipation E_{d2} in the airlift column gives

$$E_{d2} = \sum_1^2 \frac{1}{2}\rho_L V_i^3 e_i A_i(1-E_{Gi}) \quad (4)$$

where e_i: friction loss factor due to liquid turnaround and other resistance to liquid recirculation in the column,

V_i: interstitial liquid velocity = $U_{Li}/(1-E_{Gi})$ (m/s),
A_i: horizontal cross-sectional area (m^2),
1,2: represent the upflow zone and downflow zone, respectively.

The energy dissipation E_{d3} can be evaluated by considering the macroscopic energy changes of the liquid between the top and bottom regions of the downflow section. If we neglect the change of kinetic energy, E_{d3} is the result of two terms, namely the rate of loss of potential energy = $A_2\rho_L gH_L V_2$ and the rate of gain of pressure energy = $A_2\rho_L gH_L(1-E_{G2})V_2$, giving

$$E_{d3} = A_2\rho_L gH_L E_{G2}V_2 \tag{5}$$

The energy balance, therefore, takes the following form:

$$E_i = E_{d1} + E_{d2} + E_{d3} \tag{6}$$

If we substitute the expressions of E_{d1}, E_{d2}, and E_{d3} into Equation (6) we get

$$E_i = A_1\rho_L gH_L E_{G1}V_{b\infty} + \sum_1^2 \frac{1}{2}\rho_L V_i^3 e_i A_i(1-E_{Gi}) + A_2\rho_L gH_L E_{G2}V_2 \tag{7}$$

Since there is no net liquid flow from the column the primary liquid volumetric flow rates in upflow and downflow zones are equal and so from continuity:

$$V_1A_1(1-E_{G1}) = V_2A_2(1-E_{G2}) \tag{8}$$

Hence, the interstitial liquid velocities V_1 and V_2 in a given geometry of airlift column may now be predicted from Equations (7) and (8) for a given volumetric gas flow rate, Q, and ungassed liquid height, H_L, provided the friction loss factors, e_1 and e_2, are known and the characteristic bubble rise velocity, $V_{b\infty}$, and gas hold-ups, E_{G1} and E_{G2}, can be either estimated from appropriate correlations or measured from the experiment.

Bubble Rise Velocity

The bubble rise velocity, $V_{b\infty}$, can be predicted by the method of Mendelson (14), for example, from the expression:

$$V_{b\infty} = [2\sigma/\rho_L d_e + gd_e/2]^{0.5} \tag{9}$$

where σ is the surface tension and d_e is the equivalent bubble diameter in the gas-liquid system.

The influence of equivalent bubble diameter and surface tension on the terminal rise velocity of bubbles in three different air-water systems based on Equation (9) is shown in Table 1. Although the bubble rise velocity increases gradually with surface tension the difference diminishes with increasing bubble diameter. So, the influence of surface tension on bubble rise velocity can be neglected unless the water contains significant amounts of impurities. It can also be seen from this table that the bubble rise velocity does not change too much in the range of equivalent bubble diameter from 0.002 m to 0.01 m for the three systems examined. Since under normal operating conditions the diameters of most of the bubbles generated in an airlift column lie in this range, we assume that the terminal rise velocity of bubbles in the range of interest is independent of bubble size and approximately equal to 0.23 m/s.

TABLE 1
Influence of Equivalent Bubble Diameter and Surface Tension on the Terminal Rise Velocity of Bubbles in Air-Water System Based on Mendelson's Wave Analogy

	$V_{b\infty}$ (m/s)		
d_e(m)	σ=0.055 N/m	σ=0.065 N/m	σ=0.072 N/m
0.001	0.339	0.367	0.380
0.002	0.255	0.274	0.286
0.003	0.227	0.241	0.250
0.004	0.217	0.228	0.236
0.005	0.216	0.225	0.231
0.006	0.218	0.226	0.231
0.007	0.224	0.230	0.234
0.008	0.230	0.235	0.239
0.009	0.237	0.242	0.245
0.010	0.245	0.249	0.252
0.020	0.322	0.323	0.324
0.030	0.388	0.389	0.390
0.040	0.446	0.446	0.447
0.050	0.500	0.500	0.500

Friction Loss Factor

The friction loss factors, e_1 and e_2, can be determined by one of two methods:

(a) simultaneous solution of the macroscopic balances or (b) experimental measurement. Theoretically, these values will vary with flow condition and column configuration. However, due to the lack of information of the experimental data for e_1 and e_2 we will assume the friction loss factors are constants and roughly equal to 2 to 4 depending on the resistance to liquid recirculation in the system.

DISCUSSION

The energy balance model described above was solved numerically using Newton-Raphson iteration and was tested by comparing the liquid circulation velocities predicted by this model with the experimental data reported in the literature from three different types of airlift columns.

Draft-Tube Airlift Column

The first set of data tested was from Hatch's Ph.D. dissertation (2) where the liquid used was fermentation medium. The dimensions of the airlift column and the results are shown in Table 2 and 3, respectively. The experimental data reported were the gas hold-up in the draft tube and annulus regions and liquid circulation velocity as a function of volumetric gas flow rate. It can be seen from Table 3 that the liquid circulation velocities predicted by the energy balance model are

TABLE 2
Draft-Tube Airlift Column Dimensions*

Column Diameter	0.30 m
Column Height	5.00 m
Ungassed Liquid Volume	0.20 m^3
Ungassed Liquid Height	2.80 m
Draft-Tube Height	2.60 m
Draft-Tube Diameter	0.206 m
Height of Draft-Tube above Ferementor Base	0.033 m
Downflow/Upflow Area Ratio	1.12
Liquid Height/Diameter Ratio	9.33

* From Hatch, R. T.: Experimental and Theoretical Studies of Oxygen Transfer in Air-Lift Fermentor, Ph.D. Dissertation, M.I.T. (1973).

TABLE 3
Comparison of Liquid Velocities Estimated from Energy Balance Model with the Experimental Data Obtained by Hatch (1973) in Draft-Tube Airlift Column, Where We Assume $e_1 = e_2 = 2.0$.

U_G (m/s)	E_{Gi}*	U_{L1}** (m/s)	E_i (W)	E_{d1}	E_{d2}	E_{d3}	$V_{b\infty}$ (m/s)
0.0985	0.172 0.154	0.223 (0.215)	79.73	36.18 (45.4%)	0.98 (1.2%)	42.57 (53.4%)	0.23
0.1277	0.196 0.177	0.267 (0.265)	103.31	41.23 (39.9%)	1.79 (1.7%)	60.30 (58.4%)	0.23
0.1610	0.224 0.194	0.312 (0.334)	130.27	47.12 (36.2%)	3.08 (2.4%)	80.06 (61.5%)	0.23
0.1970	0.244 0.211	0.360 (0.386)	159.46	51.33 (32.2%)	5.00 (3.1%)	103.14 (64.7%)	0.23
0.2803	0.282 0.232	0.469 (0.409)	226.84	59.32 (26.1%)	12.32 (5.4%)	155.21 (68.4%)	0.23
0.3358	0.299 0.233	0.551 (0.431)	271.76	62.90 (23.1%)	21.21 (7.8%)	187.66 (69.1%)	0.23

* The values in the first and second rows of E_{Gi} under each operating condition represent the gas hold-up in the upflow and downflow zones, respectively.
** The first value of U_{L1} is the superficial liquid velocity estimated from the energy balance model, and the second value of U_{L1} in parenthesis is the experimental data obtained by Hatch (1973).

very close to the experimental data obtained by Hatch and the difference is within 7% for superficial gas velocities below 0.2 m/s. However, the difference between the predicted values and the experimental data increases with increasing superficial gas velocities above 0.2 m/s. This probably results from the transition in flow regimes from bubbly to slug flow. As the transition is approached, the gas phase tends to migrate to the center of the column. Because of bubble proximity the frequency of collisions increases, raising the rate of coalescence and skewing the bubble size distribution. This results in higher terminal rise velocity of bubbles and greater energy dissipation in the upflow zone where we assume the bubble size is in the range of 0.002 m to 0.01 m and the corresponding terminal rise velocity of bubbles approximately equals to 0.23 m/s. Thus, the liquid circulation velocity predicted is always higher than the actual velocity in slug flow region unless we adjust the bubble rise velocity to that corresponding to the characteristic bubble size in this region. If we arbitrarily assume the equivalent bubble diameters for superficial gas velocities of 0.280 m/s and 0.336 m/s are 0.02 m and 0.04 m, respectively, the predicted values of liquid circulation velocity shown in Table 4 approach the experimental values.

The experimental data obtained by Jones (12) in draft-tube airlift column where the liquid medium was tap water were also tested. The dimensions of the airlift column and the results are shown in Tables 5 and 6, respectively. Since Jones did not measure the gas hold-up in the annulus region, we use the empirical equation proposed by Bello et al. (15) which relates the gas hold-up in the downflow zone of the draft-tube airlift column to that in the upflow zone by the following equation:

$$E_{G2} = 0.89E_{G1} \tag{10}$$

TABLE 5
Draft-Tube Airlift Column Dimensions*

Draft-Tube Airlift Column Dimensions*	
Column Diameter	0.25 m
Ungassed Liquid Volume	0.06 m^3
Ungassed Liquid Height	1.33 m
Draft-Tube Height	1.22 m
Draft-Tube Diameter	0.146 m
Height of Draft-Tube above Fermentor Base	0.10 m
Downflow/Upflow Area Ratio	1.3
Liquid Height/Diameter Ratio	5.32

* From Jones, A. G.: Liquid Circulation in a Draft-Tube Bubble Column, Chem. Eng. Sci., 14, 449 (1985).

TABLE 4
Comparison of Liquid Velocities Estimated from Energy Balance Model with the Experimental Data Obtained by Hatch (1973) in Draft-Tube Airlift Column, Where We Assume $e_1 = e_2 = 2.0$.

U_G (m/s)	E_{Gi}	U_{L1} (m/s)	E_i (W)	E_{d1}	E_{d2}	E_{d3}	$V_{b\infty}$ (m/s)
0.0985	0.172 0.154	0.223 (0.215)	79.73	36.18 (45.4%)	0.98 (1.2%)	42.57 (53.4%)	0.23
0.1277	0.196 0.177	0.267 (0.265)	103.31	41.23 (39.9%)	1.79 (1.7%)	60.30 (58.4%)	0.23
0.1610	0.224 0.194	0.312 (0.334)	130.27	47.12 (36.2%)	3.08 (2.4%)	80.06 (61.5%)	0.23
0.1970	0.244 0.211	0.360 (0.386)	159.46	51.33 (32.2%)	5.00 (3.1%)	103.14 (64.7%)	0.23
0.2803	0.282 0.232	0.404 (0.409)	226.84	85.11 (37.5%)	7.90 (3.5%)	133.84 (59.0%)	0.33*
0.3358	0.299 0.233	0.424 (0.431)	271.76	117.59 (43.3%)	9.68 (3.6%)	144.50 (53.2%)	0.43*

* The terminal rise velocities of bubbles in the slug flow region have been adjusted to those corresponding to the equivalent bubble diameters of 0.02 m and 0.04 m for the superficial gas velocities of 0.2803 m/s and 0.3358 m/s, respectively.

TABLE 6
Comparison of Liquid Velocities Estimated from Energy Balance Model with the Experimental Data Obtained by Jones (1985) in Draft-Tube Airlift Column, Where We Assume $e_1 = e_2 = 2.0$.

U_G (m/s)	E_{Gi}*	U_{L2} (m/s)	E_i (W)	E_{d1}	E_{d2}	E_{d3}	$V_{b\infty}$ (m/s)
0.0134	0.024 0.021	0.159 (0.152)	2.76	1.20 (43.5%)	0.25 (9.2%)	1.31 (47.3%)	0.23
0.0149	0.026 0.023	0.165 (0.159)	3.07	1.31 (42.7%)	0.28 (9.2%)	1.47 (48.1%)	0.23
0.0167	0.028 0.025	0.174 (0.171)	3.43	1.41 (41.1%)	0.34 (9.8%)	1.69 (49.1%)	0.23
0.0179	0.030 0.027	0.176 (0.183)	3.68	1.51 (40.9%)	0.35 (9.5%)	1.82 (49.6%)	0.23
0.0194	0.032 0.028	0.182 (0.183)	3.98	1.60 (40.2%)	0.38 (9.6%)	2.00 (50.2%)	0.23
0.0209	0.034 0.030	0.186 (0.189)	4.29	1.70 (39.6%)	0.41 (9.6%)	2.18 (50.8%)	0.23
0.0224	0.036 0.032	0.188 (0.196)	4.60	1.81 (39.4%)	0.43 (9.4%)	2.36 (51.2%)	0.23
0.0239	0.039 0.035	0.186 (0.201)	4.90	1.96 (39.9%)	0.42 (8.6%)	2.53 (51.5%)	0.23

* Where we assume $E_{G2} = 0.89E_{G1}$

Table 6 indicates that good agreement between predicted and experimentally measured liquid circulation velocity was obtained with a maximum error of about 8%.

It can also be seen that the percentage of energy dissipated due to liquid turn-around (E_{d2}) for Jones' draft-tube column (8 to 10%) is higher than that for Hatch's draft-tube column (1 to 4%). This is probably due to the higher aspect ratio (height/diameter) for Hatch's draft-tube column. As can be seen from Equation (7) the energy dissipation rates, E_{d1} and E_{d3}, are directly proportional to the ungassed liquid height, H_L, whereas the energy dissipation E_{d2} is not a function of H_L since it depends only on the friction loss factors, liquid velocity, and gas hold-up. Thus, an increase in H_L will cause a proportional increase in energy dissipation E_{d1} and E_{d3}, while E_{d2} will remain unchanged as long as E_{Gi} and V_i do not change significantly. Therefore, for geometrically similar airlift columns the higher the aspect ratio the lower is the percentage of energy dissipation E_{d2}.

TABLE 7
External-Loop Airlift Column Dimensions*

Riser Diameter	0.14 m
Riser Height	4.05 m
Downcomer Diameter	0.14 m
Downcomer Height	4.05 m
Downflow/Upflow Area Ratio	1.0

* From Merchuk, J. C., and Y. Stein: Local Hold-Up and Liquid Velocity in Air-Lift Reactors, AIChE J., 27, 377 (1981)

External-Loop Airlift Column

Merchuk and Stein (16) measured liquid circulation velocities in an external-loop airlift column; the column dimensions are given in Table 7. The comparison of the predicted and experimental liquid circulation velocities is shown in Table 8. Since they used a gas-liquid separator in the top section of the airlift device to prevent gas from recirculating through the downcomer the gas hold-up E_{G2} is negligible and the energy

TABLE 8
Comparison of Liquid Velocities Estimated from Energy Balance Model with the Experimental Data Obtained by Merchuk and Stein (1981) in External-Loop Air-Lift Column, Where We Assume $e_1 = e_2 = 3.4$.

U_G (m/s)	E_{Gi}	U_{L1} (m/s)	E_i (W)	E_{d1}	E_{d2}	E_{d3}	$V_{b\infty}$ (m/s)
0.0469	0.049 0.000	0.663 (0.610)	24.18	6.89 (28.5%)	17.29 (71.5%)	0.00 (0.0%)	0.23
0.0782	0.076 0.000	0.774 (0.750)	40.32	10.74 (26.6%)	29.58 (73.4%)	0.00 (0.0%)	0.23
0.1095	0.100 0.000	0.853 (0.855)	56.45	14.05 (24.9%)	42.40 (75.1%)	0.00 (0.0%)	0.23
0.1407	0.118 0.000	0.920 (0.915)	72.54	16.65 (23.0%)	55.89 (77.0%)	0.00 (0.0%)	0.23
0.1720	0.136 0.000	0.972 (0.980)	88.68	19.13 (21.6%)	69.55 (78.4%)	0.00 (0.0%)	0.23
0.2033	0.154 0.000	1.014 (1.045)	104.81	21.67 (20.7%)	83.14 (79.3%)	0.00 (0.0%)	0.23

dissipation rate E_{d3} is therefore zero. In addition to a liquid flowmeter they also placed many small diameter tubes which acted as straightening vanes in the upper portion of the downcomer for the purpose of changing the resistance to liquid recirculation. They also installed the gas sparger in such a way that it acted as an additional flow obstruction in the riser. Consequently, the friction loss factors are increased from 2.0 for the previous analysis in draft-tube air-lift column to 3.4 for both e_1 and e_2 in order to take the additional flow resistance into account. It can be seen from Table 8 that the predicted values of liquid circulation velocity are in good agreement with the experimental data over the entire range of superficial gas velocities examined. For their system E_{d2} is larger than E_{d1} because the circulation velocities are larger.

TABLE 9
Split-Cylinder Airlift Column Dimensions*

Column Diameter	0.15 m
Column Height	1.36 m
Ungassed Liquid Volume	0.019 m^3
Ungassed Liquid Height	1.10 m
Height of Baffle above Fermentor Base	0.055 m
Downflow/Upflow Area Ratio	1.0

* From Patel, S. A.: Investigation of Two-Phase Flow Structures in an Airlift Fermentor, Ph.D. Dissertation, K.S.U. (1985)

Split-Cylinder Airlift Column

The liquid circulation velocities predicted by the energy balance model for the split-cylinder airlift column were compared with those estimated by Patel (17) for two different systems: tap water and salt water (0.6% NaCl). The dimensions of the airlift column and the results are shown in Table 9 through Table 11. It can be seen from Tables 10 and 11 that the predicted values of liquid circulation velocity are close to those estimated by Patel. However, the average error between the predicted and the estimated values in tap water is about 16.7% which is much larger than those in the previous analysis for draft-tube and external-loop airlift columns. This is probably a consequence of the photographic determination of bubble size and the accompanying experimental error. Furthermore, the measurements of gas hold-up in the split-cylinder airlift column were performed manually by measuring the ungassed liquid height and the average dispersed height between the upflow and downflow zones. The gas hold-up obtained in this way is the average hold-up which can not theoretically represent the actual hold-up in the upflow and downflow zones, respectively. Therefore, the estimation of liquid circulation velocity based on the average gas hold-up is subject to error especially when the interstitial liquid velocity is significantly different from the terminal rise velocity of bubbles.

CONCLUSIONS

A mathematical model based on the macroscopic mechanical energy balance is developed to predict liquid circulation velocity in airlift fermentors. The model is general and applicable to either bubbly or slug flow regimes. Figure 1 indicates that the predicted values of liquid circulation velocity are in good agreement with the experimentally measured values for three different types of airlift columns over a wide range of superficial gas velocities.

This analysis takes into account effects of energy dissipation due to wakes behind the bubbles in the upflow zone and energy dissipation due to the upflow motion of large bubbles with respect to the liquid in the downflow zone. The results indicate that each of these modes of energy dissipation accounts for 30% to 60% of total energy input for draft-tube and split-cylinder airlift columns. The liquid turnaround in the top and bottom regions of the airlift column and other resistance to liquid recirculation in the system account for the rest of the energy dissipation.

In the bubbly flow regime, which is of most practical importance in the operation of airlift fermentors, the influence of bubble size on the terminal rise velocity of bubbles can be neglected. Consequently, the energy dissipation due to the bubble wakes is independent of bubble size and directly proportional to the gas hold-up in the upflow zone. Thus, the liquid circulation velocity can be predicted from the measurement of gas hold-up in the upflow and downflow zones, respectively. In the slug flow regime, where the bubble diameter is much larger than that in bubbly flow, the terminal rise velocity of bubbles is strongly affected by the bubble size. Therefore, the prediction of liquid circulation velocity is only possible when the characteristic bubble size for a given operating condition can be measured or estimated beforehand.

ACKNOWLEDGEMENT

This work was partially supported by National Science Foundation grant CBT-8317967.

TABLE 10
Comparison of Liquid Velocities Estimated from Energy Balance Model with Those Estimated by Patel (1985) in Split-Cylinder Airlift Column for Tap-Water System, Where We Assume $e_1 = e_2 = 2.0$.

U_G (m/s)	E_{Gi}	U_{L1} (m/s)	E_i (W)	E_{d1}	E_{d2}	E_{d3}	$V_{b\infty}$ (m/s)
0.0259	0.065 0.065	0.134 (0.180)	2.42	1.46 (60.4%)	0.05 (2.1%)	0.91 (37.5%)	0.23
0.0517	0.079 0.079	0.296 (0.260)	4.84	1.79 (36.9%)	0.56 (11.5%)	2.49 (51.6%)	0.23
0.0776	0.099 0.099	0.365 (0.330)	7.25	2.23 (30.8%)	1.09 (15.0%)	3.93 (54.2%)	0.23

TABLE 11
Comparison of Liquid Velocities Estimated from Energy Balance Model with Those Estimated by Patel (1985) in Split-Cylinder Airlift Column for Salt-Water System, Where We Assume $e_1 = e_2 = 2.0$.

U_G (m/s)	E_{Gi}	U_{L1} (m/s)	E_i (W)	E_{d1}	E_{d2}	E_{d3}	$V_{b\infty}$ (m/s)
0.0259	0.056 0.056	0.187 (0.187)	2.42	1.20 (49.7%)	0.13 (5.6%)	1.08 (44.8%)	0.22
0.0517	0.089 0.089	0.262 (0.284)	4.84	1.93 (39.8%)	0.39 (8.2%)	2.51 (52.0%)	0.22
0.0776	0.111 0.111	0.328 (0.346)	7.25	2.41 (33.2%)	0.81 (11.2%)	4.03 (55.6%)	0.22

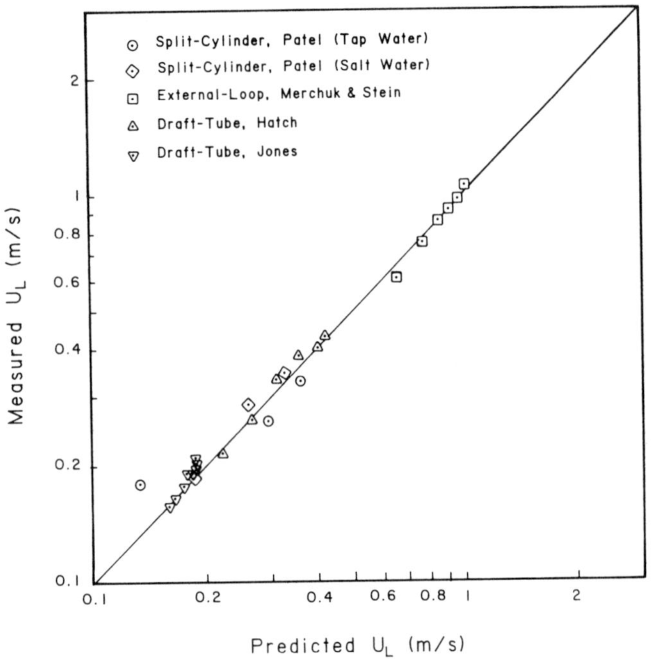

FIGURE 1. Comparison of liquid circulation velocities predicted by the energy balance model with the experimental data obtained in three different types of airlift columns.

NOTATION

A = cross-sectional area, m^2

d_e = equivalent bubble diameter, m

E_{d1} = energy dissipation rate in the bubble wakes in the upflow zone, W

E_{d2} = energy dissipation rate due to liquid turnaround and other resistance to liquid recirculation in the system, W

E_{d3} = energy dissipation rate due to the upflow motion of bubbles with respect to the liquid in the downflow zone, W

E_{d4} = energy dissipation rate due to viscous drag at the walls of the column, W

E_G = fractional gas hold-up

E_i = energy input rate, W

e = friction loss factor

g = acceleration due to gravity, m/s^2

H_L = ungassed liquid height, m

P = pressure at the top of the column, N/m^2

Q = volumetric gas flow rate, m^3/s

U_G = superficial gas velocity, m/s

U_L = superficial liquid velocity, m/s

V = interstitial liquid velocity, m/s

$V_{b\infty}$ = terminal rise velocity of bubbles, m/s

Greek Letters

ρ = phase density, kg/m^3

σ = interfacial tension, N/m

Subscripts

G = gas phase

L = liquid phase

1, 2 = upflow zone and downflow zone, respectively

LITERATURE CITED

1. Wang, D.I.C., Chem. Eng., 75, 99 (1968).
2. Hatch, R. T., "Experimental and Theoretical Studies of Oxygen Transfer in an Air-Lift Fermenter," Ph.D. thesis, Mass. Inst. of Technol., Cambridge (1973).
3. Legrys, G. A., Chem. Eng., Sci., 33, 83 (1977).
4. Hatch, R. T., A. G. Belfield, and G. Goldhahn, "Oxygen Transfer in the Split-Cylinder Air-Lift," Annual Meeting of the American Chemical Society, Chicago (1977).
5. Orazem, M. E., and L. E. Erickson, Biotechnol. Bioeng., 21, 69 (1979).
6. Ho, C. S., L. E. Erickson, and L. T. Fan, Biotechnol. Bioeng., 19, 1503 (1977).
7. Merchuk, J. C., Y. Stein, and R. I. Mateles, Biotechnol. Bioeng., 22, 123 (1980).
8. Lamont, A. G. W., Can. J. Chem. Eng., 8, 153 (1958).
9. Rietema, K., and S. P. P. Ottengraf, Trans. Instn. Chem. Engrs., 48, T54 (1970).
10. Whalley, P. B., and J. F. Davidson, Inst. Chem. Engrs. Symp. Series, No. 38, J5 (1974).
11. Joshi, J. B., and M. M. Sharma, Trans. Instn. Chem. Engrs., 57, 244 (1979).
12. Jones, A. G., Chem. Eng. Sci., 40 (3), 449 (1985).
13. Turner, J. C. R., Chem. Eng. Sci., 21, 971 (1966).
14. Mendelson, H. D., AIChE J., 13 (2), 250 (1967).
15. Bello, R. A., C. W. Robinson, and M. Moo-Young, Biotechnol. Bioeng., 27, 369 (1985).
16. Merchuk, J. C., and Y. Stein, AIChE J., 27 (3), 377 (1981).
17. Patel, S. A., "Investigation of Two-Phase Flow Structures in an Air-Lift Fermentor," Ph.D. thesis, Kansas State University, Manhattan (1985).

Fluid Dynamic Considerations in Airlift Bioreactors

N. H. THOMAS
D. A. JANES
Chemical Engineering Department
Birmingham University
P.O. Box 363, Edgbaston
Birmingham B15 2TT, United Kingdom

Measurements of the dispersion in circuit and zonal transit time statistics of tagging particles in a bench-top airlift bioreactor are attributed to specific phenomenological features of the flow. Most prominent of these in low viscosity fluids are intermittency of close encounters with energetic bubbles in the riser, cycle-to-cycle variations of engagement by eddies and vortices associated with curvature-induced separation of the boundary flows and throttling of the free surface flows, and tortuosity of the transit pathways through the disengager. With highly resistive media representative of cell cultures at biomass loadings of most commercial interest, the fluid motions are largely confined to the riser and display a pattern reminiscent of convectively driven recirculation in a closed cavity.

INTRODUCTION

Practicalities

Our main interest centres on recent biotechnological aspirations for plant cell culture as an industrial operation in the synthesis and recovery of natural products. Although operational feasibilities of the technique have been investigated many times over the past three decades, its economic prospects are presently reckoned to be largely confined to pharmaceuticals and other fine biochemicals of high specific values, e.g. the Mitsui process for the production of Shikonin (Curtin, 1). When it comes to wider exploitation of the techniques, especially to larger-scale production of modestly-valued substances, the major need is now widely recognized to centre on improved strategies for the design and operation of processing equipment (Fowler, 2).

Research efforts to date have mainly been directed at basic studies of the biological and biochemical factors in bench-top assessments, supplemented by engineering appraisals of the practicalities using 'scale-up' evaluations to gauge the performance and control of bioreactors. These traditional approaches have usually been couched in terms of temporal mean drift values of bulk-mean or point-wise monitored quantities adopted as measures of the physico-chemical environment (e.g. nutrients, acidity, temperature, lighting, etc) and the bioreactor performance (viz yield and its dependence on gas fluxes, bubble hold-up, etc); see e.g. Smart (3). This work has drawn attention to the sensitivities and variabilities in sensitivities of different cell species to the conditions and modes of bioprocessing (e.g. stirred tanks versus bubbled tubes).

What these traditional methods are not suited to is diagnosis and analysis of important effects of transient and localized excursions experienced by the cells in the turbulent two-phase flows of conventional bioreactors. There is presently little by way of fundamental information and understanding of how the physics and dynamics of the fluid mechanisms contribute to the accumulation of biomass and recovery of products from cell cultures, including such deleterious aspects as clustering of suspended cells and their attachment to walls and accumulation in foams, or the possibly damaging effects of encounters with bubbles and energetic eddies. It is these facets particularly which define our long-term goals in connection with biotechnological applications of plant cell culture.

The economics of bioprocessing operations are largely determined by the achievable biomass densities and, because of the highly resistive constitutions of concentrated broths, these, in turn, are strongly dependent on the equipment and techniques employed. Although many 'novel' approaches have been advocated, most practical efforts continue to revolve around traditional equipment like stirred tanks of one kind or another and, more recently, bubble tubes and loops; see Martin (4). Some aspects of these developments which bear directly on our work reported herein are briefly outlined below; wider issues, including the need for alternative processing strategies, are reviewed in Thomas and Janes (5).

Performance evaluations and limitations

Wagner & Vogelmann (6) reported a comparative evaluation of five bioreactors, including both a stirred tank and a concentric tube airlift loop. For moderate biomass densities, between ten and fifteen grams per litre (dry weight) of Morinda citrifolia, they obtained highest yields of anthraquinones from the air-lift loop. The shortcoming of stirred tanks, also noted by other workers (e.g. Tanaka, 7), has been attributed to a biologically unfavourable environment within the zones of high shear caused by the rapidly rotating blades of impeller actuators. In the light of these findings, plant biotechnologists have tended increasingly over the past five years or so to favour the stirrer-free, airlift bioreactors (Fowler, 2), although doubts have recently been expressed about deleterious effects of 'over-gassing' the cells (Fowler, 8).

For small-to-medium-scale culturing of plant cells (say up to thirty litres or so), attention has increasingly centred on draught-tube variants of these devices, in which a concentric-cylinder arrangement is employed to achieve internal bulk recirculation. Blenke (9) has delineated some main performance factors of these bioreactors, expressed in terms of mean-value circulation times around the flow loop. He found that circulation times decrease with increasing height of the liquid column and increase with increasing elevation of the bubble sparger above the base of the vessel. Both these results accord with intuitive expectations, and it would be interesting to see how they compare with predictions based on global energy fluxes of bubble buoyancy and frictional dissipation. Blenke found the circulation times to be sensibly independent of geometrical details of the bubble sparger, so this measure of performance might be approximately predicted from such a primitive description.

Configurational factors, in particular, the ratio of diameters of the inner and outer tubes, have been investigated by Weiland (10). In addition to mean-value circulation times, he also measured the performance in terms of one-point time-profiles of the concentration of an injected pulse of salt solution as tracer. Interestingly, he found that the shortest circulation times are realised when the diameter ratio takes a value of about 0.6, where as 0.8 to 0.9 gives rise to the shortest mixing times. This finding nicely distinguishes between the dual practical needs of sufficiently energetic large-scale motions to maintain dispersion of sedimentary materials, such as larger organisms like plant cells, and sufficiently dissipative fine-scale motions to maintain molecular fluxes of dissolved nutrients to the suspended organisms. Although Weiland also assessed some effects of viscosity and rheology of aqueous solutions, using glycerol and carboxy-methyl-cellulose (CMC below) as additives, the effective viscosities he employed appear to be an order of magnitude less than is needed to be representative of plant cell cultures with higher biomass loadings (see below).

Circulation times have also been measured directly using a radio-transmitting particle and found to be consistent with those deduced from the salt-tracer technique (Fields & Slater, 11). Other studies in progress in Birmingham (Kalaher, 12) have demonstrated how sensitive these transit-time measurements are to the size of the tagging particle (as suggested by Fox et al, 13). Our results in these tests (to be reported) indicate that representative estimates are obtained only when the particles are somewhat smaller than the bubbles, a finding which nicely accords with intuitive expectations.

Fields & Slater (11) also paid some attention to the bubble disengagement chamber of their draught-tube flow loop and concluded that this zone is mainly responsible for departures from ideal (plug) flow. Further applications of the RF-tagged particle method were reported by Fields,

Mitchell & Slater (14), this time with dilute aqueous solutions of xanthan gum, to test the effects of pseudoplasticity. They found the measured circulation time increased significantly for concentrations in excess of 0.3% or so, at which values the effective viscosity also increases rapidly with increasing concentration. In contrast to this nicely consistent result, they also found that lower concentrations of 0.1% gave rise to circulation times less than those measured in water. A plausible explanation would be that their particle was more responsive to the bubble-induced motions simply because the bubble sizes were rather larger in the viscous dilute solution as compared with those in gum-free water.

Rheological properties of plant cell suspensions and culture broths have been reviewed by Tanaka (15). Characterised in terms of a power-law constitutitve relation it seems the effective viscosity depends most sensitively on biomass density and increases very dramatically with increasing biomass densities in excess of ten grams per litre (dry weight), or so. Indeed, according to Tanaka the dependence can be described by a power-law with exponent about six and a half! Underlying factors contributing to this rheology in the physical constitution include agglomeration of the cells and their secretion of biopolymers, both of which depend on species and age (and probably history), as well as cell number density of culture suspensions. Wagner & Vogelmann (6) reported that a mature (sixteen-day) broth of M. citrifolia supporting only 5 grams per litre exhibited non-Newtonian pseudoplastic and thixotropic properties, and yield stress behaviour. On the other hand, Kato et al. (16) observed that the cell filtrates of young (four-day) cultures possessed linear constitutive properties and they accordingly argued that non-linearities derive from the presence of suspended cells. However, it is likely that increasing quantities of polysaccharides are secreted into the supporting medium as the culture ages, just as has been found with microbial fermentations (Charles, 17).

Clearly the results which are presently available offer only the haziest of guidelines and certainly fall a long way short of practictioners' needs for databanks on material properties suited to implementation in predictive schemes for bioreactor performance and control. For our part, we anticipate the most fruitful advances will ultimately derive from first principles measurements, physical understanding and modelling of events at the level of the finest-scale (i.e. cellular) hydrodynamics, as has proved to be the case for lasting developments in characterising the rheology of particulate suspensions (e.g. Batchelor & Green, 18).

SCOPE AND METHODS

Our work reported in this paper was undertaken as a familiarization exercise to explore some physical aspects of performance limiting factors in draught-tube airlift bioreactors, particularly departures from ideality in the flows and their basic phenomenologies. Our findings are relevant to the conditions of real fermentation broths in so far as we employed liquids encompassing effective viscosities markedly higher than in the tests reviewed above, such as would be more appropriate for cultures supporting the higher biomass loadings of wider commercial interest. At twenty grams dry weight per litre of biomass, the effective viscosity of a plant cell culture is in the order of 400 cP (according to Tanaka, 15), so this value we adopted as benchmark, using CMC to furnish a representation of the pseudoplastic behaviour manifested at high cell concentrations. Values of the viscosities and densities of this fluid and others we tested (i.e. water, aqueous glycerols) appear in Appendix 1.

A schematic of the 2 litre airlift vessel used in our study appears as Figure 1. Mainly of interest to us here is the bubble disengagement chamber, layout of which is shown enlarged in the sketch of Figure 2. Other information, including dimensions, can be found in Appendix 2. Maximum attainable Reynolds numbers were in the order of 10^4 for water and 10^0 for the viscous liquids. These values are based on a circulation velocity given by the minimum circulation path (equal to twice the height of the draught-tube) and the measured circulation times, the draught-tube diameter and rheometric values of the kinematic viscosity.

Elapsed-time measurements of visually-tagged particles were undertaken in both vessels. Two kinds of particles were employed - 5mm hollow glass beads of approximately neutral buoyancy in each of the test fluids and 2mm portions of dyed grains of boiled rice. The smaller particles had settling speeds about 0.3 mm/s

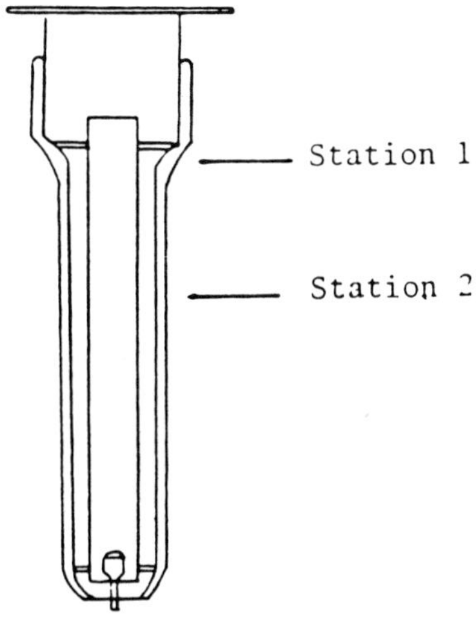

FIGURE 1. Airlift vessel (see Appendix 2 for details).

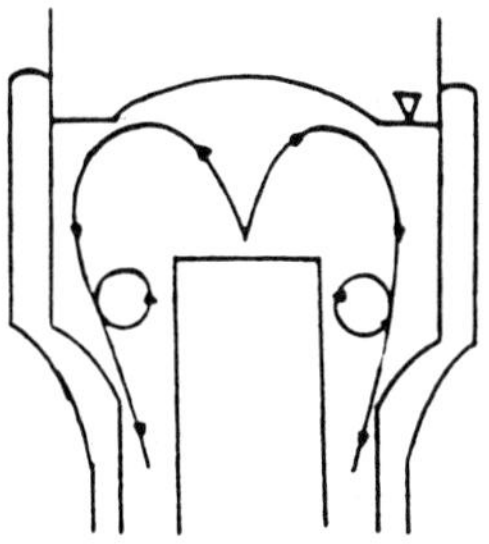

FIGURE 2. Bubble disengagement chamber and impression of a particle trajectory.

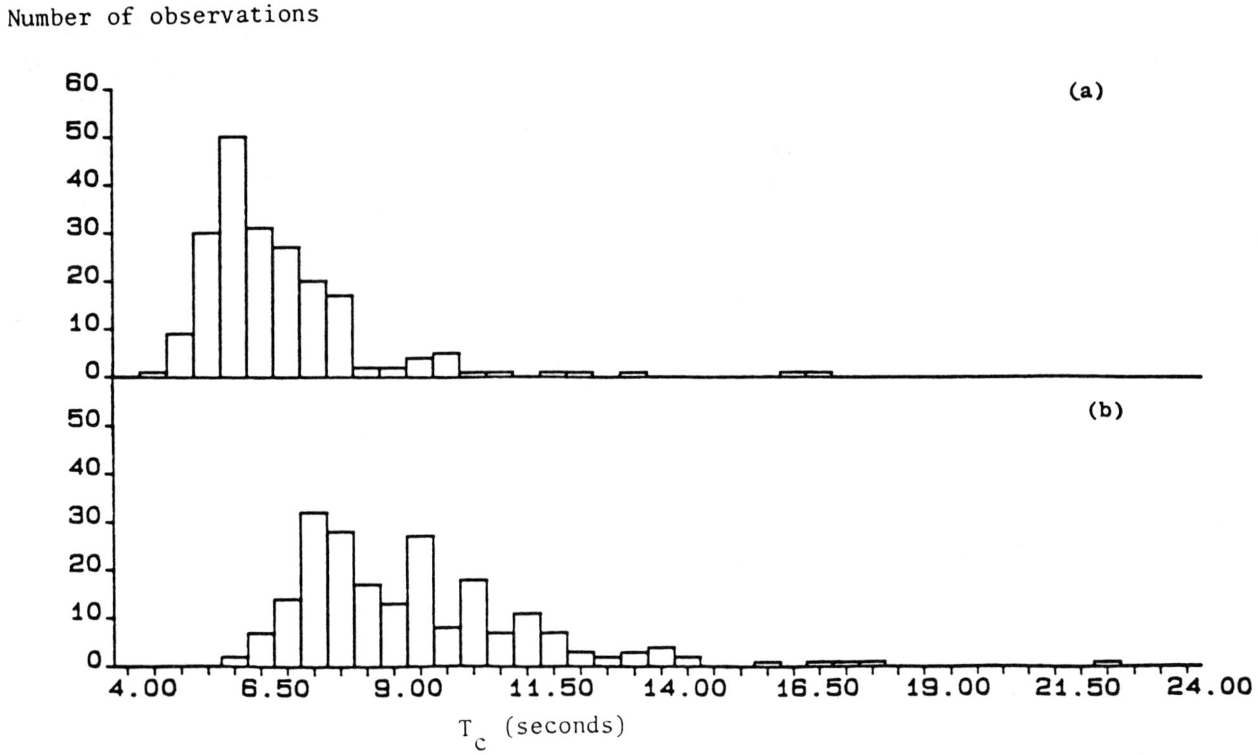

FIGURE 3. Histograms of circulation times T_C measured in the airlift vessel filled with 2100 ml of water at air-flow rates: (a) 1000 ml/min, (b) 400 ml/min.

in CMC solution. They were used because of our suspicions that the energetic scales of bubble-driven eddying motions would not be adequately followed by the larger beads. Circulation times were measured using a digital stop-watch for up to 300 particle passages through station 1 (see Figure 1). Residence times in the disengagement chamber were similarly recorded for transits of the particles between stations 1 in the riser and 2 in the downcomer.

OBSERVATIONS AND INTERPRETATIONS

Circulation times in water

Histograms. Figure 3 shows histograms of the circulation times measured for air fluxes of 400 and 1000 ml/min, both with 2100 ml of water in the vessel. In each case conditions are well-removed from plug-flow, with the larger departures occuring at the lower gas flux. This trend accords with a simple physical picture that the model dispersion derives in large part from cycle-to-cycle variations in the number of close encounters between the tagging particle and the rising bubbles which carry the particle upwards in the draught-tube. Clearly, the mean interval between encounters increases when the bubble-number density declines with decreasing gas flux. On the other hand, the mean duration of each event is determined essentially by the sizes of the bubbles and their slip speeds and, for low voidages, these are both approximately independent of the gas flux. With increasing stochastic intervals between encounters of fixed duration, we should expect an increase in the fractional variability of transit times through the riser, and hence also the circulation times around the loop, as is revealed by the present results.

The observed departures from plug flow also receive contributions from bulk mean shear in the downcomer and, to some extent, the riser as well. When it comes to extended tailing of the histograms shown in Figure 3, of much greater significance is the dispersion deriving from the complex patterns of flow within the disengagment chamber; see below and later in this article. The re-entrant flows at the bottom of the bioreactor are also reckoned to make an important contribution (Fields and Slater, 11). Qualitative observations (Kalaher, 12) suggest this dispersion arises mainly from recirculation of the entrainment flows into the developing bubble plume above the gas sparger, particularly when the narrow plumming flows cannot readily accommodate the tagging particle.

Dependence on liquid volume. Figure 4 shows our measurements of the mean value circulation times obtained with volumes of water ranging from 2000 to 2250 ml, at the two air fluxes employed earlier. Notice that both curves exhibit well-defined minima when the volume of water is about 2150 ml, irrespective of the gas flux. These results, which follow those reported by Blenke (9), can be explained physically in terms of the effects of the disengagement chamber on the flows passing over the draught-tube from the riser into the downcomer, as follows.

Consider first what happens with a reduction in the volume of liquid in the vessel, or equivalently, a lowering of the free surface within the disengagement chamber - corresponding approximately to 5 mm per 50 ml. Eventually this gives rise to weir flow plunging over the top of the draught-tube, the perimeter of which then imposes a hydraulically-critical control section; ultimately, of course, the flows entering the downcomer are entirely suppressed. According to Figure 4, this condition is approached if the volume of liquid falls below about 2000 ml when the air flux is 400 ml/min. Increased hold-up of bubbles and hence increased elevation of the free surface accounts for the rather smaller critical volume realised at an air flux of 1000 ml/min.

Conversely, increasing the volume of liquid and hence elevation of the free surface eventually serves only to enlarge the volume of overlying 'dead-water' in the disengagement chamber. Passages of the rising bubbles through this zone then contribute only localized, recirculatory patterns of pluming and eddying motions, without appreciably affecting conditions in the riser and downcomer. The measured values shown in Figure 4 increase simply because the tagging particle is engaged by these flows, such that its mean path-length is extended without any significant changes in the bulk mean velocities around the circulation loop.

From the practictioner's point of view, these findings are important because of sensitivity in the response of circulation times to modest excursions in the volume of the liquid within this particular vessel. The range displayed on Figure 4 represents excursions of only ± 5% about an 'optimal'

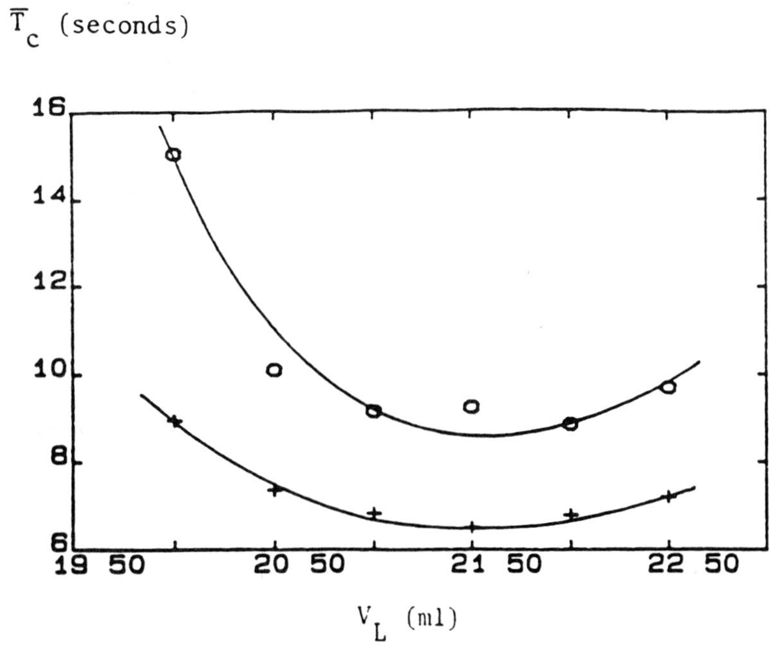

FIGURE 4. Mean-value circulation times $\overline{T}_C$ versus liquid volume V_L at air-flow rates (ml/min) of 400, ○; 1000, +.

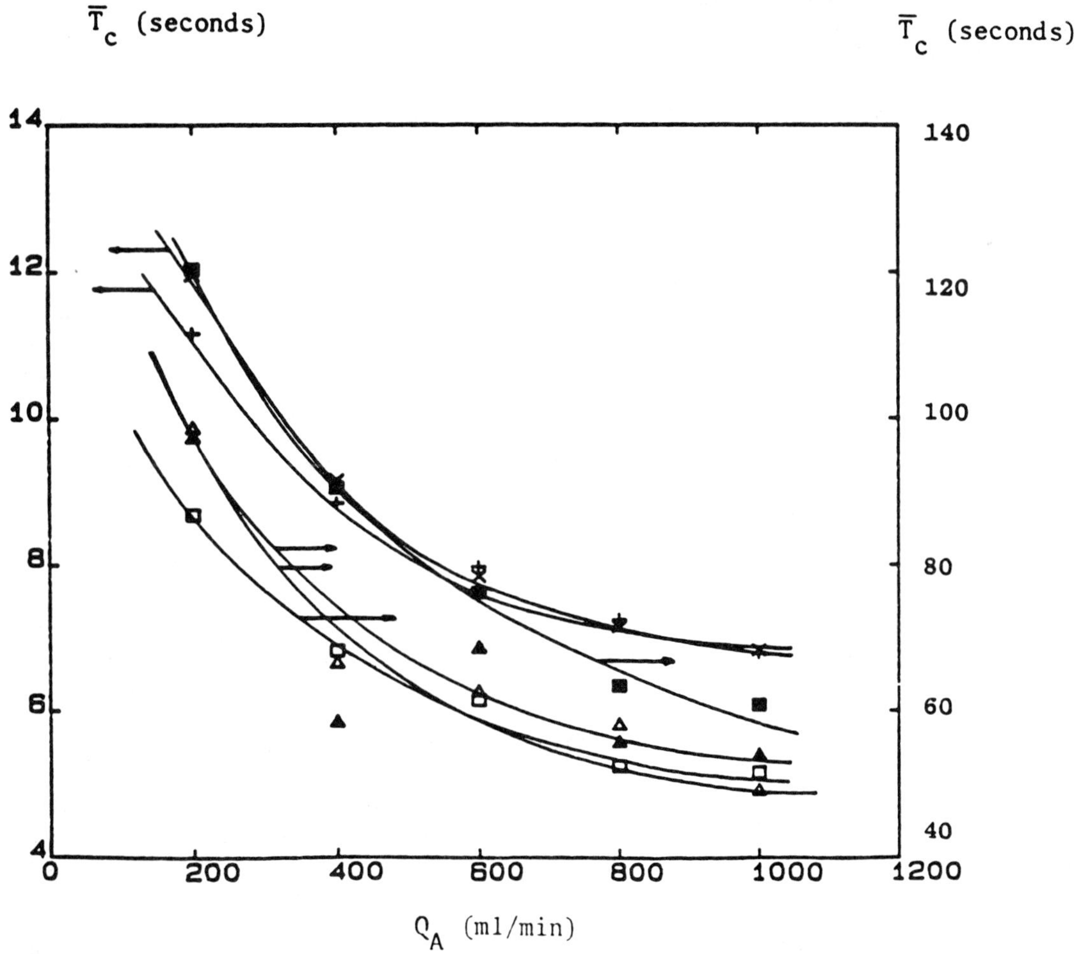

FIGURE 5. Mean-value circulation times T_C versus air-flow rate Q_A with V_L = 2100, 2200 ml respectively of water ×, +; 96% glycerol △, △; 2.5% CMC.

volume of 2150 ml, corresponding to a free surface elevation of about 20 mm above the top of the draught tube. However, the associated increases in circulation times vary from 20% or so, upwards to as much as 60% or more when critical conditions of flow suppression are approached. For this reason, and to more precisely identify these contributions to the extended tailings of the histograms, we undertook some additional tests (to be described later).

Dependence on air flow rate. Figure 5 shows how the mean-value circulation times depend on the air flow rate within a range from 200 ml/min to 1000 ml/min, for two volumes of water (2100 and 2200 ml) in the vessel. The behaviour displayed here generally accords with findings by previous workers (e.g. Fields & Slater and Weiland) - namely, that following an initially substantial reduction as the bubbling rate is increased, the circulation times then level off and appear to approach an approximately constant value, independent of further increases in the bubbling rate. At these higher air-flow rates the circulation times are not so strongly dependent on the volume of water, a finding which is consistent with the results presented earlier in Figure 4. We return to this point again later.

Circulation times in viscous liquids

Figure 5 also demonstrates our results obtained with aqueous solutions of 96% glycerol and 2.5% CMC, both possessing effective viscosities about 400 times that of water. Notice that although the magnitude order of the mean-value circulation times T_C is now ten times that found with water, the trends in T_C with increasing air flow rate are still apparently comparable. However, this superficial similarity masks profound changes in the distribution of flows within the biorector. Thus the riser flows were partitioned into closed loops of upwards and downwards recirculation extending throughout the length of the draught-tube. Only a fraction of the bulk flows passed through the annular downcomer and the vessel performed more as a bubble column enclosed within the loop. Comparable behaviour has also been observed in side-arm loop vessels (Kalaher, 12). Moreover, in contrast to the dispersed flows of 4 mm bubbles obtained at all of the air fluxes employed with water, the bubble sizes were now very substantially larger, for both glycerol and CMC solutions. Accompanying this shift from turbulent to viscous equilibrium between coalescence and fragmentation of the bubbles, and the accumulation of micro-bubbles throughout the vessel, the flows in the riser also exhibited complex unsteady patterns akin to slug or churn regime. Further comments about this behaviour and some striking differences between the glycerol and CMC solutions are offered below.

Observations with glycerol solution. Firstly, a word of caution about interpreting the data on Figure 5. These results represent mean-value elapsed times measured between successive passages past station 1 (see Figure 1) for all circulations of the tagging particle, whether around the bounded loop enclosing the draught-tube or around free loops of recirculating flow within the interior of the draught-tube. Although no substantial differences were recorded between the circulation times following the two routes, in general they can be very different - as Kalaher (12) as shown - because the resistances to motion are different; see also below.

Descending motions of the tagging particle in the bubble-free annular downcomer were smooth and orderly, as with the water flows. In contrast, its rising motions through the draught-tube exhibited vigorous, irregular excursions associated with the passges of disordered slug-churn bubbles. Whilst travelling through the disengagement chamber, the particle was also intermittently entrapped within stagnant surface fluid near to the walls of the vessel. It may be that extended residence times of biological cells within this 'dead' zone are responsible for surface encrustation problems encountered in the processing of concentrated broths (Fowler, 2).

Observations with CMC solution. The recirculating flows again followed both bounded loops and free loops, as described above. The data shown on Figure 5 relate here to circulation times only around bounded loops returning via the annular downcomer. Within the draught-tube, the fluid motions were, by-and-large, confined to a core region, where mobility was sustained by the passages of large slug-churn bubbles and their wakes. Considerable variations in the flow speed with elevation were observed, quite unlike the behaviour found in glycerol solution. A thin sheath surrounding the core region appeared to carry much of the mean shear stress asso-

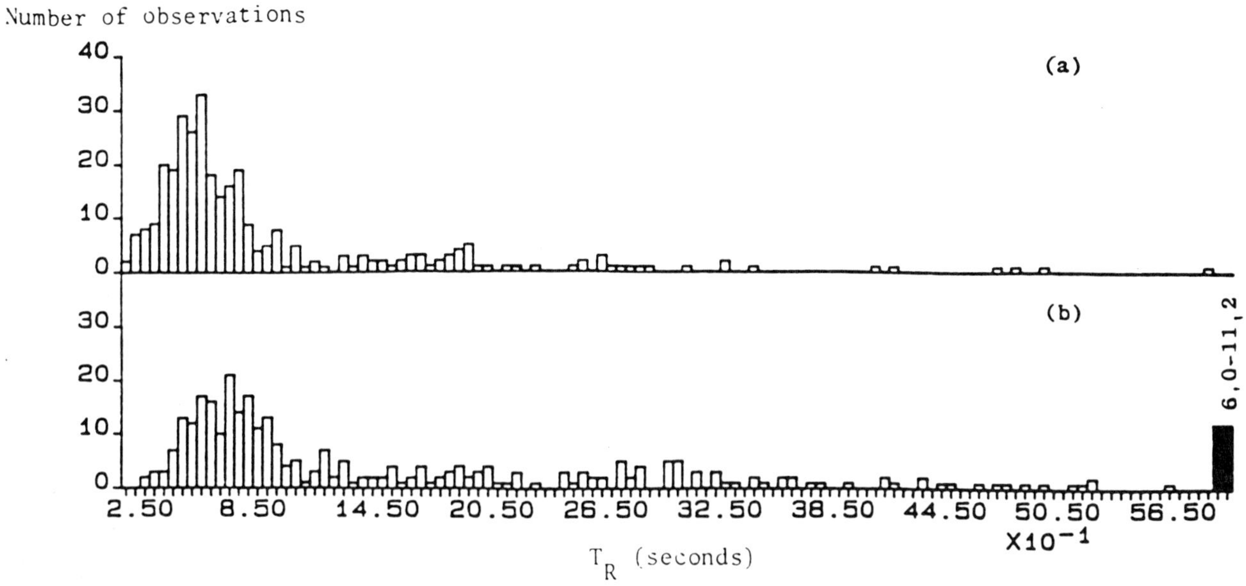

FIGURE 6. Histograms of residence times T_R measured in the disengagement chamber of the airlift vessel; see Figure 3 for conditions.

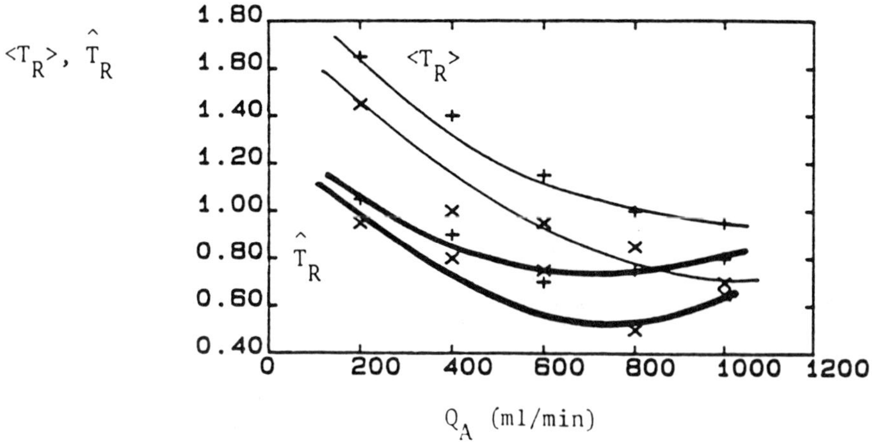

FIGURE 7. Disengager residence times, median and mode values $\langle T_R \rangle$ and $\hat{T}_R$ versus air-flow rate; see Figure 5 for conditions (water only).

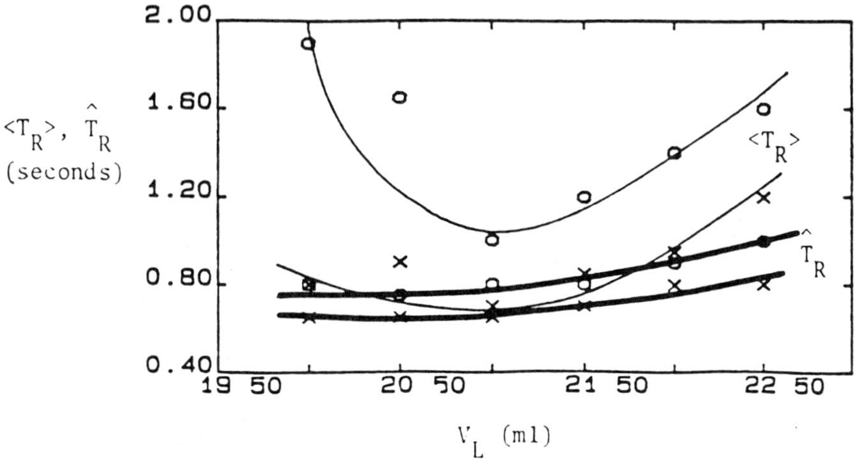

FIGURE 8. Disengager residence times $\langle T_R \rangle$ and $\hat{T}_R$ versus liquid volume; see Figure 4 for conditions.

ciated with the bubble motions. Outside this sheath only languid flows were observed, due to damping by high effective viscosity of the CMC solution. This behaviour is probably akin to 'cavern' phenomena exhibited by mechanically-stirred pseudo-plastic solutions, in which the flows are again sensibly confined to shear zones surrounding the rotating actuator (Solomons et al, 19).

From time to time the tagging particles encountered particularly large bubbles and were displaced into the languid outer region, from which they were not readily re-engaged by the core flows. Indeed, the 5 mm glass bead was never observed to return of its own accord, whilst the 2 mm rice grain was retained for considerably longer periods than a 100 micron speck which chanced to aid our visualisation. Circulation times estimated with the glass bead (between proddings to dislodge it from the outer region) were about twice the values obtained with the rice grain, whilst the speck, although too small to be tracked reliably by eye, travelled significantly faster than the rice particles.

Although the environment seen by our 2 mm particle is very different from that experienced by smaller elements, including material elements of the liquid, it is certainly more representative of that experienced by 2 mm plant-cell clusters, such as are found in aging concentrated broths (Tanaka, 15). Whilst conceding the uncertainties associated with our preliminary studies, they nevertheless serve to emphasize an important general consideration - namely, that the Lagrangian transport statistics of tagging particles of various sizes provide a sensitive indicator of the accommodation scales for dispersed elements within localized and transient heterogeneities of the flows. In the present tests this accommodation scale is set by the thin shear layers between the mobile core flows and the languid wall layers. Very likely, this flow feature will be an important source of distortion between performance evaluations in model and prototype equipment due to Reynolds number dissimilarities between the flows. Other observations of the gas injector zone in a loop bioreactor (Kalaher, 12) have also shown that tagging particles are not readily accommodated by the entrainment flows into the narrow bubble plume emerging from the sparger unless they are significantly smaller than the plume diameter. Such behaviour may well be implicated in the deposition of plant-cell clusters which has been reported as a practical problem in airlift processing of these cultures (Fowler, 2).

Disengager transit times in water

Although all circuits of the 5 mm tagging particles around the water-filled vessel were completed via the annular downcomer, the particle was observed to linger, and sometimes even descend briefly, before passing from the head of riser into the disengagement chamber. These excursions coincided with the emergence of extended clusters of bubbles spiralling out of the turbulent riser flows, so we attribute them to the transient displacement flows associated with the voidgage fluctations. Increasingly frequent and energetic events of this kind were observed with diminishing volumes of liquid in the vessel. We also observed some striking convolutions in the trajectories of particles engaged by the re-entrant flows at the head of the annular downcomer, an impression of which is sketched in Figure 2. These motions are caused by entrainment and hold-up within a torodial eddy which is attached to the outer surface of the draught-tube and is sustained by vorticity shed into the separating riser flows emerging from the lip of the draught-tube. With smaller depths of liquid in the disengagement chamber, we would also expect to find a toroidal standing eddy on the inner surface of the draught-tube, arising from deceleration and separation of the wall-flows in the adverse pressure gradient imposed by the nearby free surface.

Undoubtedly there are other wall-eddies associated with zones of acute boundary curvature imposed on the flows in the disengagement chamber; see Figure 2. In the light of earlier work on the basic dynamics of particle engagement by vortex flows (see Thomas et al., 20) and especially some recent developments in computer simulations by Sene (21; to be reported in Sene et al, 22), we believe that trajectory excursions and intermittent entrapment by these eddies may well be largely responsible for extended tailing of the circulation time histograms presented in Figure 3. That the disengager transit times do make a significant contribution in this respect can be seen from the results of some preliminary tests, as outlined below.

Histograms. Figure 6 shows two

histograms of the transit times recorded for air fluxes of 400 and 1000 ml/min with 2100 ml of water in the vessel, as used in the earlier tests (Figure 3). Notice that the trends are towards diminished tailing at the higher air fluxes, consistent with those graphed on Figure 3. Recall that this behaviour accords with our simple physical picture that fractional variability in the number of bubble-particle encounters should decrease as the bubble-number density increases with air flow rate. Indeed, there is even broadly quantitative consistency, with residence times in the disengagement chamber contributing tails of up to about 5 and 10 seconds respectively at the higher and lower air fluxes.

Dependence on air flux. Figure 7 displays median and mode values of the transit times measured at air fluxes up to 1000 ml/min, with 2100 and 2200 ml of water in the vessel (i.e. the same conditions as for figure 5). Reassuringly, the trends are pairwise consistent, whilst convergence of the median and mode values at higher bubbling rates again reflects a sharpening of the histograms. Notice that the median values depend on air flux in a similar fashion to the mean circulation times shown in Figure 5. Contrast this behaviour with that of the mode values which exhibit well-defined minima. The upturn at higher air fluxes is attributed to hold-up of the particle in the increasingly voided disengagement chamber, as bubbles accumulate prior to breaking at the free surface.

Dependence on liquid volume. Figure 8 shows how the residence times vary with liquid volumes ranging from 2000 ml to 2250 ml for the two air flow rates used in the earlier tests (Figure 4). The median and mode values are by-and-large pairwise consistent, with the median values again following similar trends to the circulation times. Notice also how the departures between median and mode values ultimately increase with depth of water in the disengager. This is in line with our earlier suggestion that close encounters with energetic bubbles occasionally carry the particle into the overlying zone of dead-water where it is then held up by wall eddies and bubble turbulence before escaping into the main body of circulating flow in the lower part of the disengager. Increasing departures between the median and mode values also ultimately occur with diminishing depths of water in the disengager, especially at the lower air flux. According to our earlier interpretation, the flows are now constrained hydraulically by the crest of the draught-tube. Clearly, when the crest depth of flow becomes comparable with particle size, only very close encounters with energetic bubbles provide sufficient impulse to carry the particle into the downcomer. Consistent with this behaviour, the trends of Figure 8 reflect increasing rarity of the increasingly energetic encounters which are required with diminishing crest depths; moreover, reduced bulk voidage, and hence crest depth, accounts for the larger effect obtained at the lower air flux. Indeed, in one realization with 2050 ml of water (at 400 ml/min of air), we recorded an extreme excursions of 29 seconds - that is, 17 times the median value and 39 times the mode value of the transit time!

CONCLUSIONS AND RESOLUTIONS

Our observations and interpretations of the motions and transit time statistics of visually tagged particles in a small draught-tube airlift bioreactor are summarized below.

(a) Circulation times measured in water exhibit substantial departures from plug transport and strong dependencies on gas flow rate and liquid volume, all of which are broadly in line with previously reported findings. The dispersion is attributed here mainly to cycle-to-cycle variations in the number of close encounters with bubbles in the riser tube and to hold-up within vortices and wall-eddies in the disengagement chamber. Dependence on liquid volume derives from sensitivity of the flows to the level of the free surface in the disengagement chamber, associated particularly with hydraulic throttling by weir flows over the draught tube at low levels and meandering trajectories within overlying 'dead-water' at high levels.

(b) Viscous solutions of CMC and glycerol display strikingly different patterns of flow in which much of the energetic motion is partitioned into vertically recirculating closed loops within the riser tube, with only a modest residual flux passing around the annular downcomer. Major sources of dispersion here are attributed to hold-up within languid wall-layers on the riser side of the draught-tube and within stagnant fluid adjoining the perimeter of the free surface in the disengagement chamber. Estimates of the circulation statistics are extremely sensitive to the sizes of tagging particles as compared with the scales of motion in the exchange flows across the shear layers between these inactive zones and the core flows.

In developing the present framework of our findings, special attention is being paid to the following aspects:
(i) Detailed measurements and mathematical modelling of the motions of discrete materials in response to combinations of buoyancy, inertial and hydrodynamic forces, the latter associated particularly with mean flow heterogeneities and transient excursions on energetic scales ranging from steady internal free shear layers (e.g. due to rheologically partitioned flows, sparger pluming flows), through coherent vortices and wall-attached eddies (e.g. as encountered in the disengager flows), to the rising bubbles and their wake flows responsible for dispersive transport in the riser and hold-up in the disengager. Understanding and prediction of the accommodation scales for entrapment or exclusion of bubbles, drops, particles and cells or cell clusters by these flow features is the target here.
(ii) Implementation and evaluation of a reasonant detector technique for enhanced sensitivity in tracking minature tagged particles, ie significantly smaller than the bubble-wake scales of motion for air-in-water dispersed flows. Present RF capsules, about 10 mm or so in diameter (Fields et al, 14), are significantly larger than typical bubble sizes of 4 mm or less.
(iii) Formulation of first principles analyses for the bubble-driven flows in air lift tubes and loops, particularly with regard to dynamical factors (i.e. profiles of buoyancy and mean shear in the riser), material factors (i.e. viscosity and rheology of the liquid phase and effects of dispersed bubbles) and configurational factors (i.e. riser geometry, sparger location, throttling of the return flows by wall friction or free surface controls - e.g. weir flows into the downcomer). The aim is for more reliable and versatile assessments of how they each contribute to determine the global performance (e.g. mode of flow - especially partitioning of the riser flows in viscous liquids, circulation times, etc) than present engineering correlation schemes allow.

ACKNOWLEDGEMENT

We are indebted to Professor Jim Callow of the Plant Biology Department for risking (and suffering) damage to his two-litre airlift bioreactor in making it available for our study. One of us (D.A.J.) was supported by SERC Biotechnology Directorate.

LITERATURE CITED

1. Curtin, M.E., Bio-technology, 1, 659 (1983).

2. Fowler, M.W., Biotech. & Genetic Engrg. Revs ., 2, 41 (1984).

3. Smart, N., Lab. Practice (July, 1984).

4. Martin, S.M., "Mass culture systems for plant cell suspensions" in Plant Tissue Culture as a Source of Biochemicals, Staba, E.J. (Ed.), CRC Press, Florida (1980).

5. Thomas, N.H. & D.A. Janes, BUCHER Memo BEPB-MFTMF/NHT-DAJ (1/1)/8606 (1986).

6. Wagner, F. & H. Vogelmann, "Cultivation of plant tissue cultures in bioreactors and fermentation of secondary metabolities", in Plant Tissue Culture and its Biotechnological Applications, Barz, W., E. Reinhard & M.H. Zenk (Eds.), Springer-Verlag (1977).

7. Tanaka, H., Biotech. Bioeng., 23, 1203 (1981).

8. Fowler, M.W., "Commercial applications and economic aspects of mass plant cell culture", in Plant Biotechnology, Mantell, S.H. & H. Smith (Eds.), Cambridge University Press (1983).

9. Blenke, H., Adv. Biochem. Eng., 13, 121 (1979).

10. Weiland, P., German Chem. Eng., 7, 374 (1984).

11. Fields, P.R. & N.K.H. Slater, Chem. Eng. Sci., 38, 647 (1983).

12. Kalaher, S., "Flow in a Loop Bioreactor", M.Sc. Report, University of Birmingham (1985).

13. Fox, R.I., D.E. Brown & W. Crueger, "Functional characterisation of bioreactors - the investigation of an E.F.B. working party", in Bioprocessing in the Eighties, I.Chem.E. Symposium, Southampton (October, 1982).

14. Fields, P.R., F.R.G. Mitchell, & N.K.H. Slater, Chem. Eng. Comm., 25, 93 (1984).

15. Tanaka, H., Biotech. Bioeng., 24, 425 (1982).

16. Kato, A., S. Kawazoe & Y. Soh, J.Ferment. Tech., 56, 224 (1978).

17. Charles, M., Adv. Biochem. Eng., 8, 1 (1978).

18. Batchelor, G.K. & J.T. Green, J. Fluid Mech., 56, 401 (1972).

19. Solomons, J., T.P. Elson, A.W. Nienow & G.W. Pace, Chem. Eng. Commun., 11, 143 (1981).

20. Thomas, N.H., T.R. Auton, K. Sene & J.C.R. Hunt, "Entrapment and transport of bubbles by transient large eddies in multiphase turbulent shear flows", Paper E1 in Proc. International Conference on Physical Modelling of Multiphase Flow, Coventry (England), BHRA Cranfield (April, 1983).

21. Sene, K.J., "Aspects of bubbly two-phase flow", Ph.D. dissertation, University of Cambridge (1985).

22. Sene, K.J., J.C.R. Hunt & N.H. Thomas, submitted to J. Fluid. Mech. (1986).

APPENDICES

Viscosities and densities of the test fluids

CMC powder and glycerol were supplied by Fisons Scientific. CMC solution (2.5% by weight) was prepared by slowly dissolving the powder in water subjected to high shearing by a polytron (Kinematica, Switzerland). The solutions of CMC and aqueous glycerol werel allowed to stand overnight before use. Effective viscosities, as measured with a Rheomat, and their densities are tabulated below: (a) water, (b) 2.5% CMC, (c) 96% glycerol, (d) 90% glycerol.

	Temperature ^{o}C	Dynamic Viscosity mPas	Density Kg/m^3
(a)	25	0.89	997
(b)	25/24	350/400	1013
(c)	25	350	1248
(d)	19	210	1238

Airlift vessel

The vessel, supplied by LH Fermentatior as a '500 Series Airlift' with packaged air pump and control circuit, was of 2 litres nominal capacity, and main dimensions (mm) as tabulated below: (a) length, (b) internal diameter. The bubble sparger was of fritted glass. Water jacket temperature was maintained at 25°C with a Conair Churchill 'Thermoflow' control unit.

	External Tube	Draught Tube	Disengager
(a)	365	400	130
(b)	80	44	100

Sparger height: 45, Diameter Ratio: 0.59

Reference heights (Figure 1) at stations 1/2: 245/510.

Hydrodynamic and Oxygen Mass Transfer Studies in Bubble Columns and Airlift Bioreactors

M.Y. CHISTI
K. FUJIMOTO
M. MOO-YOUNG
Department of Chemical Engineering
University of Waterloo
Waterloo, Ontario, Canada N2L 3G1

Previous studies of pneumatically agitated bioreactor devices have been confined mainly to two-phase gas–liquid aqueous systems. This work deals with three-phase systems in which fungal fermentation media are simulated using cellulose fiber for microbial biomass. The hydrodynamic and oxygen mass transfer properties of these dispersions are discussed for various reactor configurations: a rectangular bubble column, a similar internal loop airlift, and cylindrical external loop airlift devices.

It is shown that the overall volumetric mass transfer coefficient (k_La_L) and gas holdup (ϵ) correlate satisfactorily according to equations of the form: parameter = $a \cdot Usg^b$. Equations of this type are shown to have a theoretical basis. The coefficients a and b are complicated functions of reactor geometry, fluid properties, and the flow regime. In presence of solids used to simulate basic fungal fermentation media gas holdup and k_La_L were found to decline by up to 80% relative to the values in solid-free aqueous solutions.

Pneumatically-agitated bioreactors (bubble columns and airlifts) are relatively new types of devices (compared to the classical stirred vessels) being applied to aerobic fermentations and waste treatment processes. Despite its limitations the stirred-tank reactor remains the workhorse of the fermentation industry. However, the relative simplicity of the design of pneumatic devices, which leads to mechanical robustness, ease of monoseptic operation over extended periods, low shear fields and low power consumption makes it attractive for new or improved processes in commercial practice. For highly shear sensitive fermentations such as with some mammalian and plant cells where applications are increasing, there may be no choice other than the use of pneumatic reactors.

A large (1000 m^3) airlift aerobic reactor has been operated for a productive (as opposed to waste treatment) fermentation, by Imperial Chemical Industries, U. K., for the manufacture of single cell proteins. However, process engineering information for the design of airlift devices remains limited. Design correlations for the effects of various geometric and operating parameters on reactor performance are not well-established. This uncertainty on performance appears to be the principal factor which restricts wider use of airlift reactor devices in biotechnology industries.

Over the last several years the pneumatic reactors, particularly bubble columns with circular cross sections, have been the subject of much investigation. Nearly all this work has dealt with air-water systems (e.g. 1, 2, 3) which may have little applicability to fungal and viscous non-Newtonian fermentations. In the cases where slurries were examined, the solids used were charcoal or coal particles, calcium carbonate, alumina and other similar material which are quite different from biological solids such as fungal mycelia. In addition, non-cylindrical vessels and external and internal loop airlifts have not found significant industrial applications although the limited bench scale sutdies (4, 5, 6) have indicated good performance - as good or better than the traditional mechanically-stirred reactor types - for a variety of fermentations including bacterial, fungal, tissue culture and plant cell suspension systems. Here we present experimental results on gas holdup and gas-liquid mass transfer carried out in pneumatic devices with simulated fluid systems for yeast, bacterial and fungal cultures.

THEORY

The liquid film control of mass transfer of a sparingly soluble gas from the gas to liquid phase (or in the opposite direction) is well known and the transport rate is given by

$$\frac{d\,C_L}{d\,t} = k_L a_L\,(C^* - C_L) \quad (1)$$

where k_La_L, C* and C_L are the overall volumetric mass transfer coefficient, the saturation concentration of oxygen in the liquid and the liquid phase oxygen concentration at any time, t, respectively. Integration of Equation (1) between the limits $C_L = C_0$ at t = 0, and $C_L = C_L$ at t = t yields

$$\ln \frac{(C^* - C_0)}{(C^* - C_L)} = k_L a_L t \quad (2)$$

which may be utilized for k_La_L determination based on the transient gassing-in method. For point measurements Equation (2) is valid only when the reactor is fully backmixed.

In the transient technique the dynamics of the oxygen electrode itself influence k_La_L results. The electrode delays which are a function of fluid hydrodynamics near its measuring surface may be satisfactorily accounted for by a first-order model (7, 8):

$$\frac{C^* - C_L}{C^* - C_0} = \left[\frac{e^{-t/B}}{S} - \frac{e^{-t/S}}{B}\right] \frac{BS}{(B - S)} \quad (3)$$

$$\frac{C^* - C_L}{C^* - C_0} = \left[\frac{B\,e^{-t/B}}{(B - S)}\right] \quad (4)$$

Thus, a semilog plot of $(C^* - C_L)/(C^* - C_0)$ against t should yield a straight line of slope -1/B (or $-k_La_L$) and intercept ln [B/(B - S)] from which the time delay may be obtained.

Apart from k_La_L another important parameter in bioreactor design is the frational gas holdup, ε. The residence time (t_G) of the gas in liquid, the total dispersion volume, the gas-liquid interfacial area for mass transfer (a_L) and, in airlift reactors, the circulation of liquid, all depend on gas holdup. Thus,

$$a_L = \frac{6\,\varepsilon}{d_B\,(1 - \varepsilon)} \quad (5)$$

$$V_D = \frac{V_L}{1 - \varepsilon} \quad (6)$$

and

$$t_G = \frac{V_L\,\varepsilon}{Q_G\,(1 - \varepsilon)} \quad (7)$$

The possible effects of some fluid properties on gas holdup may be theoretically predicted. Thus, by defination, in a pool of fluid with no net liquid flow the gas holdup is

$$\varepsilon = \frac{A_G}{A} \quad (8)$$

where A_G and A are the actual cross-sectional area for gas flow and the total cross-sectional area of the reactor, respectively. The continuity equation for gas flow is

$$A_G\,U_T = A\,Usg \quad (9)$$

where U_T is the mean terminal rise velocity of gas and Usg is its superficial velocity. Equations (8) and (9) lead to

$$\varepsilon = \frac{Usg}{U_T} \quad (10)$$

The terminal rise velocity of gas bubbles is obtained by equating the bouyancy and drag forces on the bubbles. The bouyancy force, F_B, on a bubble is

$$F_B = (\rho_L - \rho_G)\,\frac{\pi}{6}\,d_B^{\,3}\,g \quad (11)$$

where d_B is the Sauter mean bubble diameter. The drag force, F_D, is (9):

$$F_D = \frac{C_D\,U_T^2\,A_p\,\rho_L}{2} \quad (12)$$

if the bubble is assumed to be a rigid sphere. A_p and C_D are the projected area of the bubble and a dimensionless drag coefficient, respectively. Substitution of

$$A_p = \frac{\pi\,d_B^2}{4} \quad (13)$$

into Equation (12) gives

$$F_D = \frac{\pi C_D\,U_T^2\,d_B^2\,\rho_L}{8} \quad (14)$$

When Equations (11) and (14) are equated the resulting equation can be rearranged to

$$U_T^2 = \frac{4(\rho_L - \rho_G)\,g\,d_B}{3\,C_D\,\rho_L} \quad (15)$$

The general form of the drag coefficient dependence on particle Reynolds number is (9):

$$C_D = \frac{i}{Re_p^{\,j}} \quad (16)$$

where i and j depend on the flow regime. The particle Reynolds number in Equation (16) is

$$Re_p = \frac{(\rho_L - \rho_G)\, U_T\, d_B}{\mu_L} \quad (17)$$

Equation (16) is well known for flow past single spheres. Values of i and j can be analytically calculated for Stokes' law region ($Re_p << 1$) only, and must be empirically determined under other regimes of flow. Substitution of Equation (17) in (16) and of the resulting equation in (15) yields the expression

$$U_T = \left[\frac{4\,(\rho_L - \rho_G)^{1+j}\; g\, d_B^{\;1+j}}{3\,\rho_L\, i\, \mu_L^j}\right]^{1/(2-j)} \quad (18)$$

The bubble size in Equation (18) may be predicted by the well established Kolmogoroff's theory of local isotropic turbulence. Thus, the bubble diameter is given by (10):

$$d_B = \phi \frac{\sigma^{0.6}}{(P_G/V_L)^{0.4}\, \rho_L^{0.2}} \quad (19)$$

where ϕ is a dimensionless constant. P_G/V_L, the power input per unit liquid volume, in pneumatically agitated reactors is predominantly due to isothermal gas expansion and it is analytically related to the superficial gas velocity in the following manner:

$$P_G/V_L = \rho_L\, g\, Usg \quad (20)$$

Substitution of (19) and (20) in (18) leads to

$$U_T = \left[\frac{4\,(\rho_L - \rho_G)^{1+j}\, g}{3\,\rho_L\, i\, \mu_L^j}\right]^{1/(2-j)}$$

$$\times \left[\frac{\phi\, \sigma^{0.6}}{\rho_L^{0.2}}\right]^{\frac{1+j}{2-j}} (\rho_L g\, Usg)^{-0.4(1+j)/(2-j)} \quad (21)$$

In a given fluid under given hydrodynamic conditions Equation (21) reduces to

$$U_T = \psi\, Usg^y \quad (22)$$

Substitution of Equation (22) in (10) leads to equations of the form

$$\varepsilon = a\, Usg^{\,b} \quad (23)$$

where a and b are, respectively,

$$a = \left[\frac{4\, g}{3\,\rho_L\, \mu_L^j\, i} \left(\frac{(\rho_L - \rho_G)\, \sigma^{0.6}\phi}{\rho_L^{0.2}}\right)^{1+j}\right]^{1/(j-2)} \quad (24)$$

and

$$b = 1 + 0.4\,\frac{(1 + j)}{(2 - j)} \quad (25)$$

The theoretical reasoning behind Equation (23) was not established before this although the equation itself has been frequently used to describe the gas holdup behaviour in airlift and bubble column reactors. The above analysis assumed a non-interacting bubbly flow and net liquid flow through any cross-section of the reactor was ignored. In airlift reactors liquid circulation and hence reactor geometric parameters should also influence a in Equation (23). In this work the surface tension and density of slurries used were almost constant and fractional solid contents were used to correlate values of a and b in equations of the same basic form as Equation (23).

EXPERIMENTAL

Studies of gas holdup and overall volumetric mass transfer coefficient were carried out in several different reactors including a rectangular bubble column (R1), an internal loop airlift (R2) of similar dimensions as R1, and external cylinderical loop airlift devices (R3). The details of construction of these reactors, which were made of "plexiglas", are shown in Figure 1 with the main dimensions, including those of the gas spargers, provided in Table 1. Batch liquids or slurry systems were used. Suspension of Solka-Floc (James River Corporation, Grade KS-1016 cellulose fibres) in water or aqueous salt solution (0.15 kmol m^{-3} NaCl) were used to simulate basic fungal fermentation media. The complete range of media used and the surface tension and rheological properties of these mixtures are listed in Table 2. Properties of the Solka-Floc (SF) solids are given in Table 3. Suspensions of Solka-Floc in the aqueous salt solution produce fluids which closely simulate the mycelial fermentation media in which pulp type of morphology occurs and where there is no secretion of polymeric substances.

Volume fraction of gas in the dispersion (gas holdup) was determined by the

TABLE 1
Reactor Dimensions and Sparger Details

Reactor	h_L (m)	L (m)	W (m)	L_r (m)	L_d (m)	dc or equivalent riser dia. (m)	Equivalent downcomer dia. (m)	L_b (m)	L_c (m)	L_t (m)	L_h (m)	A_d/A_r (-)	V_L (m^3) $\times 10^3$	SPARGER n_h (-)	d_h (m) $\times 10^3$	Free area %	Pitch (m)
R1	1.372	0.46	0.155	-	-	0.232	-	-	-	-	-	0	97.8	50	1.0	0.055	0.025 x 0.045 rect-angular
R2	1.372	0.46	0.155	0.17	0.115	0.163 (2 risers)	0.132	1.05	0.095	0.227	-	0.338	97.8	2x20	1.0	0.06	0.025 x 0.045 rect-angular
R3.1	1.75	-	-	-	-	0.152	0.102	1.55	-	-	0.5	0.44	54.2	52	1.0	0.23	Tri-angular
R3.2	1.75	-	-	-	-	0.152	0.076	1.55	-	-	0.5	0.25	44.2	52	1.0	0.23	

TABLE 2
Fluids Used

Fluid	Description	Surface tension mNm^{-1}	K(Pas^n)	n(-)
F1	0.15 kmol m^{-3} NaCl in water*	75	10^{-3}	1.0
F2	1 wt./vol. % SF in 0.15 kmol m^{-3} NaCl in water	75	-	-
F3	2 wt./vol. % SF in 0.15 kmol m^{-3} NaCl in water	75	1.464	0.322
F4	3 wt./vol. % SF in 0.15 kmol m^{-3} NaCl in water	75	6.127	0.237

* Waterloo tap water is drawn from deep wells and has quite consistent properties. Typical concentrations are: total dissolved solids, 325 mg L^{-1} (range: 240-400 mg L^{-1}); total hardness, 320 mg L^{-1}; "M" alkalinity, 260 mg L^{-1}; chlorides, 32 mg L^{-1}; pH 7.4.

TABLE 3
Properties of Solka-Floc Grade KS-1016 (from the Manufacturer)

Average fibre length	= 290 μm
Screen analysis:	
% on 35 mesh	= 10.22
% through 100 mesh	= 38.52
% through 200 mesh	= 17.29
Bulk density	= 18.31 kgm^{-3}
Maximum apparent density	= 190 kgm^{-3}
Minimum apparent density	= 50 kgm^{-3}

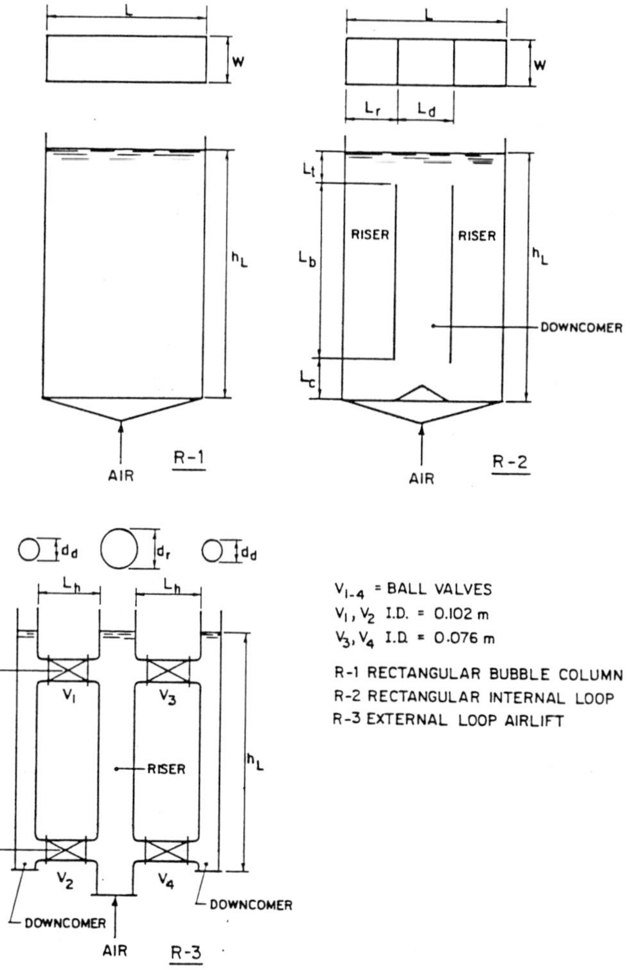

FIGURE 1. Reactor configurations: rectangular bubble column (R1); rectangular internal loop airlift (R2); cylindrical external loop airlift (R3).

volume expansion method or by hydrostatic pressure measurements. Overall volumetric mass transfer coefficient ($k_L a_L$) was determined by the dynamic gassing-in technique (11). A batch of liquid or slurry previously deaerated by bubbling nitrogen was gassed with air and dissolved oxygen concentration in the fluid was followed as a function of time using dissolved oxygen electrodes (Yellow Springs Instruments, YSI 5739 with standard membrane) connected to YSI model 57 dissolved oxygen meter. A constant gas composition and a fully backmixed liquid were assumed. The former assumption is well known for sparingly soluble gases such as oxygen, and the latter was justified because $k_L a_L$ measurements from electrodes located in widely separated areas of the reactors agreed within 6% reproducibility typical of this type of measurement. This was true even for the viscous slurries in which the mixing was visibly poorer because the $k_L a_L$ values were also correspondingly lower. The electrode response time was always less than 10 seconds and $k_L a_L$ data corrected for time delay using a first order model did not differ significantly (<3% difference) from uncorrected values. Thus, electrode lag was ignored.

RESULTS AND DISCUSSION

The particle size (Table 3) of Solka-Floc (SF) cellulose fibres is in a range which is typical for organisms such as Penicillia, Aspergilli, and Streptomyces, i.e., 200 to 300 µm. The particle shape was long thread-like and also resembled that of mycelial and filamentous microorganisms. The SF particles were flexible; absorbed water on wetting and became slightly denser than water (density of wet solids 1.2 gmL^{-1}). The behaviour of SF suspensions was visually the same as those of *Aspergillus niger* or *Chaetomium* broths. The feel and texture of these suspensions was also similar to those of the biological systems mentioned. Further, the SF suspensions displayed power law type of rheology, and as shown in Table 4, the flow behaviour and consistency indicies (n and K, respectively) of these suspensions were very close to the values found in typical mycelia-containing fluids. (Note: homogeneous polymer solutions do not provide such a good simulation of fungal broths).

Gas Holdup

Two methods of gas holdup measurement - the volume expansion and the static pressure technique - were used because either one or the other is more suitable depending on the reactor configuration. For example, separate determination of riser and downcomer gas holdups in internal loop airlifts cannot be done by volume expansion, and manometric measurements are more suitable. In Figure 2 the gas holdup data obtained by the two techniques in various reactors are shown and both holdup measurement methods are seen to provide results which agree within 10%.

The gas holdup data obtained in suspensions in the rectangular bubble column (R1) was correlated by

$$\varepsilon = e^{(0.426\ C_s\ -\ 0.549)} Usg^{(0.29\ C_s\ +\ 0.38)} \tag{26}$$

which applied within ± 10% to the unhindered bubble flow regime ($0.015 \leq Usg\ (m/s) < 0.07$). In the coalesced bubble flow regime ($0.07 \leq Usg\ (m/s) < 0.4$) the following equation was applicable to all fluids in R1:

$$\varepsilon = 1.16\ e^{-(0.273\ C_s\ +\ 0.782)}\ Fr^{0.362} \tag{27}$$

Figure 3 shows excellent agreement (±10%) of Equation (27) with the data. The Froude number in Equation (27) is defined as

$$Fr = \frac{Usg}{(g\ d_c)^{0.5}} \tag{28}$$

where d_c is the equivalent "hydraulic" diameter of the rectangular column. In the rectangular internal loop airlift reactor (R2) the holdup was described by

$$\varepsilon = (1.488 - 0.496\ C_s)\ Usg^{0.892\ \pm\ 0.075} \tag{29}$$

and

$$\varepsilon = (0.371 - 0.089\ C_s)\ Usg^{0.430\ \pm\ 0.015} \tag{30}$$

for the bubble flow and coalesced bubble flow regiems, respectively. Equations (29) and (30) apply to all fluids and the superficial velocity in them is based on the entire reactor cross-section. In all cases it can be seen that mycelia-like solids have a strong negative impact on gas holdup. As much as 80 percent reduction in gas holdup was observed in some cases relative to the solid-free system. In all cases the gas holdup in the airlift reactor (R2) was slightly lower than in the bubble column (R1) mode of operation (Figure 4) under otherwise identical conditions. This is due to the induced liquid circulation in the airlift which enhanced the bubble rise velocities and hence reduced

TABLE 4
Consistency (K) and Flow Behaviour (n) Indicies for *Aspergillus niger* and Solka-Floc Suspensions

Solid content (kg m^{-3})	n(-)	K (Pa s^n)
2	0.55	0.17
5 (20)*	0.33 (0.322)	1.62 (1.464)
10 (30)	0.2 (0.237)	9.06 (6.127)

* The data in parentheses is for the authors' SF (KS-1016) suspensions, the rest is for Aspergillus niger broths of Reuss et al. (14). The impeller method of Reuss et al. (14) was used for the determination of K and n values in all cases.

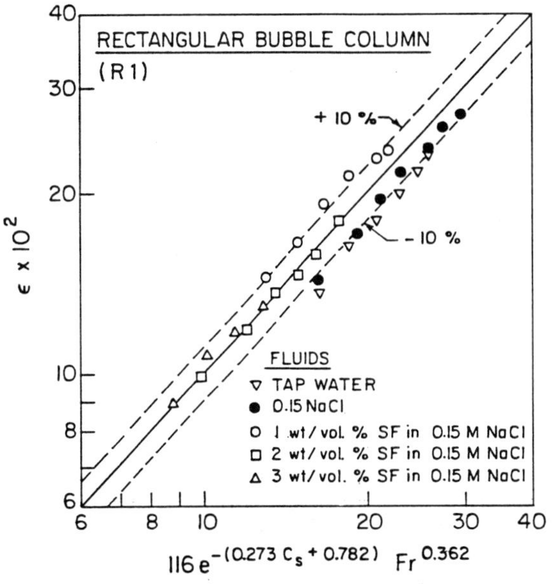

FIGURE 3. Gas holdup in the rectangular bubble column (R1) in the coalesced bubble-flow regime.

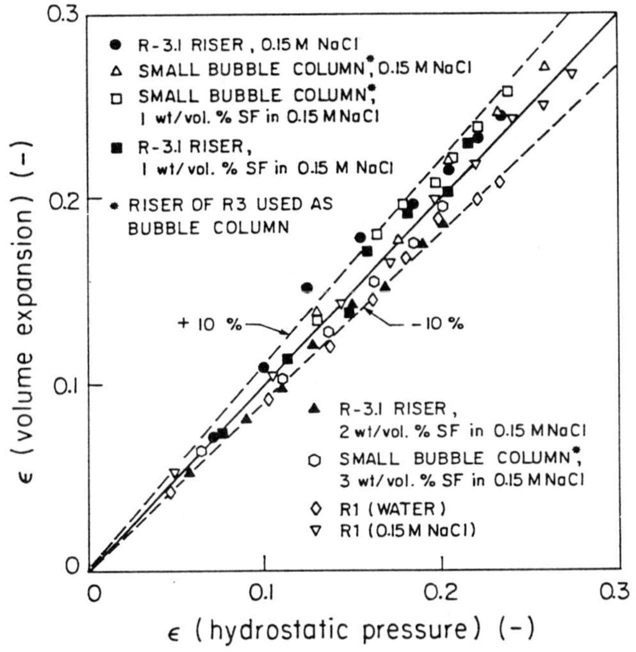

FIGURE 2. Comparison of gas holdup data obtained by volume expansion and hydrostatic pressure techniques in various reactors.

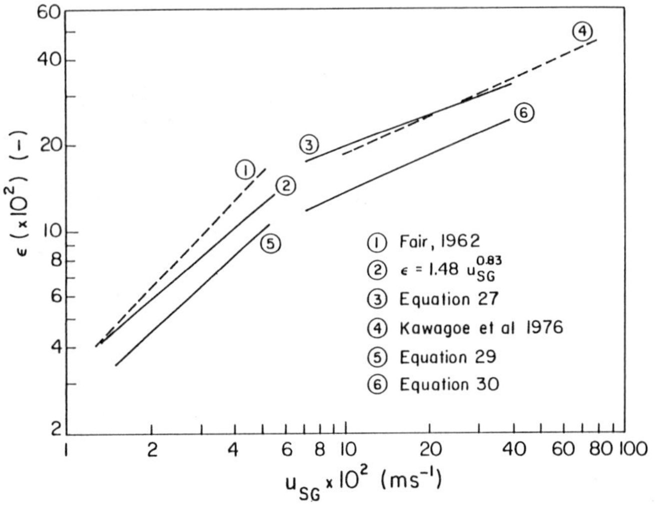

FIGURE 4. Comparison of gas holdup in bubble columns and internal loop airlift reactors. Solid lines are for solid-free salt solution. (1) Air–water correlation of Fair (*12*) ($\epsilon = 3.2\ Usg$) in bubble columns; (2), (5) bubble-flow regime in R1 and R2, respectively; (3), (6) coalesced bubble flow in R1 and R2, respectively; (4) correlation of Kawagoe et al. (*13*) [$\epsilon = Usg/(1.7\ Usg + 0.36)$] in bubble columns.

holdup. The bubble column (air-water) gas holdup equations proposed by Fair (12) and Kawagoe et al. (13) for bubble flow and coalesced bubble flow regiems, respectively, are shown in Figure 4. Satisfactory agreement between the results of these investigators and the data obtained in this study is indicated.

In the external loop airlift devices (R 3.1, 3.2) the overall gas holdup was significantly lower than in the bubble column (R1) or the internal loop (R2), and in this case the riser gas holdup correlated well (Figure 5) with the equation

$$\varepsilon_{riser} = 0.65(1 + \frac{Ad}{Ar})^{-0.258} Usg^{0.603+0.078\ C_s} \quad (31)$$

where the superficial gas velocity ($0.026 \leq Usg\ (m/s) \leq 0.21$) was based on the cross-sectional area of riser. Equation (31) applied to all the fluids used. The increase in the riser gas holdup with declining downcomer-to-riser area ratio resulted from reduced liquid circulation due to the increasing resistance of the liquid circulation path. Under identical gas sparging rates in the risers of the external (R3.1) and internal loop (R2) reactors, the liquid circulation in the external loop was noticeably greater because the geometry of the external loop devices allowed significant gas disengagement in the horizontal connection relative to the internal loops, thereby reducing the gas holdup in the downcomer of the former and leading, consequently, to a greater driving force for liquid circulation.

In the external loop reactors (R3.1, 3.2) the downcomer and riser gas holdups were linearly related:

$$\varepsilon_{riser} = (0.053 \pm 0.008) + (2.172 \pm 0.362)\ \varepsilon_d \quad (32)$$

The satisfactory agreement of the experimental results with Equation (32) is shown in Figure 6, where it is also compared with the equations reported by Bello et al. (2). The available evidence indicates that the relationship between riser and downcomer gas holdups is largely independent of Ad/Ar ratio and of the solid content of the fluid. In the external loop vessels the driving force for fluid circulation, i.e. the gas holdup difference between the riser and downcomer, was found to depend on the absolute value of gas holdup in the riser and increased with increasing riser holdup. Liquid circulation declined with decreasing Ad/Ar ratio and with increasing solid concentration because the frictional losses in the circulation pathway are dependent on these.

Gas-Liquid Mass Transfer

The overall mass transfer coefficient data showed trends very similar to those observed for gas holdup in the pneumatic reactors. For suspensions in the rectangular bubble column (R1) the following correlation was obtained:

$$k_L a_L = (0.317 \pm 0.047)\ C_s^{-(1.781 \pm 0.183)} \times Usg^{(0.637 \pm 0.114)} \quad (33)$$

Equation (33) applied to suspensions only (C_s = 1 to 3 wt./vol. %) for $0.016 \leq Usg(m/s) \leq 0.35$ and it correlated the data within about 20% (Figure 7). The very strong negative effect of solids on mass transfer is once again clear. It is mainly due to gas holdup reduction which occured in presence of solids because of their turbulence dampening and bubble coalescing effects. There was some evidence that k_L is also influenced by the presence of solids. For our external loop reactors (R3.1, 3.2) we obtained:

$$k_L a_L = (1 + \frac{Ad}{Ar})^{-1} (0.349 - 0.102\ C_s) \times Usg^{0.837 \pm 0.062} \quad (34)$$

for all the fluids tested (Figure 8). The influence of solids on overall mass transfer coefficient depends on the type of gas-liquid contacting device and it was found to be stronger in the bubble column (R1) than in the external loop geometry. This is due to the direction of bubble motion being less chaotic in the airlift reactors than in bubble columns and in the former there was less bubble-bubble interaction. Consequently, the effects of such properties as solid content, which enhance coalescence, were less pronounced in the airlift vessels. However, the absolute value of $k_L a_L$ was always substantially lower in the external loops compared to the bubble column and the internal loop which we tested. This may be readily explained in terms of the different gas holdup characteristics of these reactors as discussed earlier.

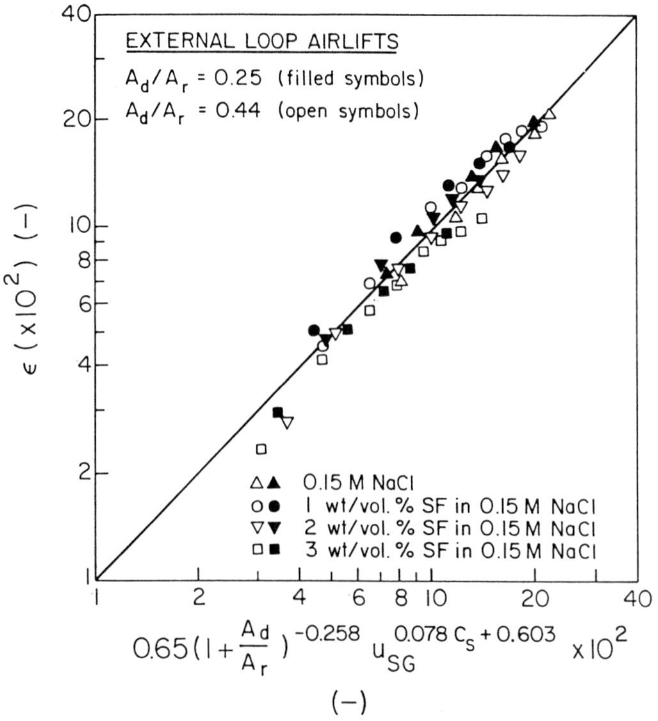

FIGURE 5. Riser gas holdup in the external loop reactors (R3.1, 3.2).

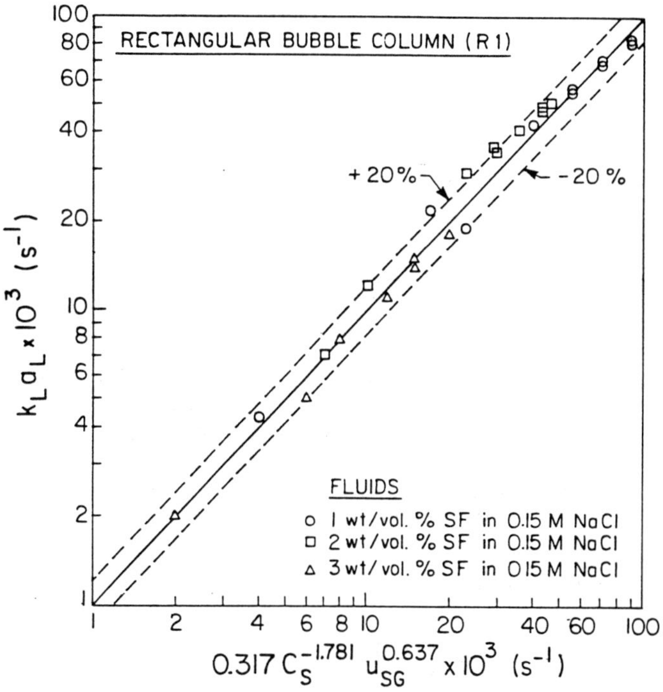

FIGURE 7. Overall mass transfer coefficient for suspensions in the rectangular bubble column (R1).

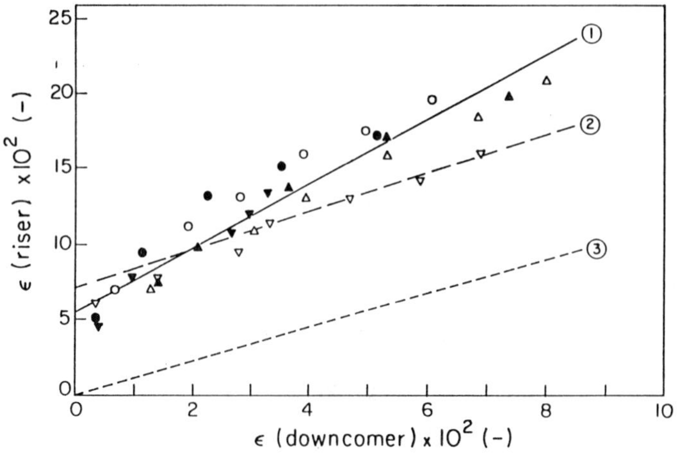

FIGURE 6. Relationship between the riser and downcomer gas holdups in external loop reactors. (1) Equation (32); (2) external loops, air–water: $\epsilon_r = 1.266\, \epsilon_d + 0.072$ (2); (3) internal loops, air–water: $\epsilon_r = 1.124\, \epsilon_d$ (2). See Figure 5 for legend.

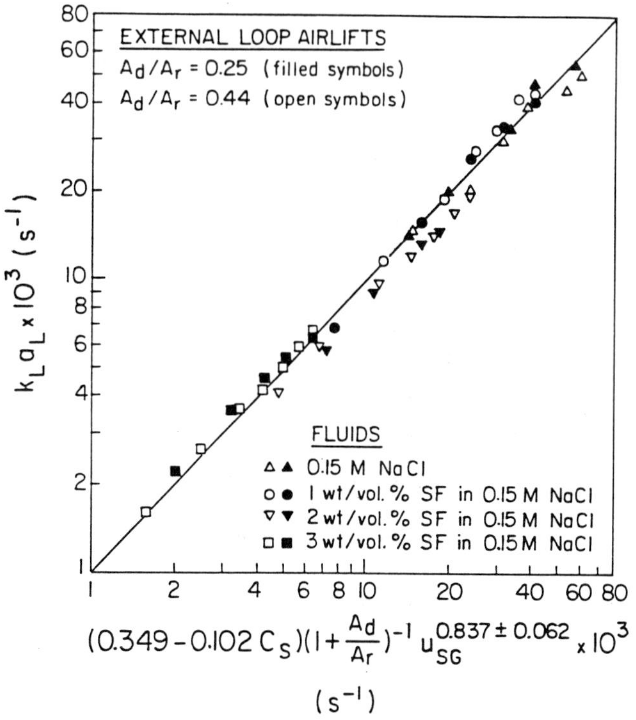

FIGURE 8. Overall mass transfer coefficient in external loop reactors (R3.1, 3.2).

CONCLUSION

In general, the overall mass transfer coefficient and gas holdup in the pneumatic reactors studied are satisfactorily described using equations of the form:

$$k_L a_L = \alpha \, Usg^{\beta} \qquad (35)$$

and

$$\varepsilon = a \, Usg^{b} \qquad (36)$$

where α, β, a and b depend on the contactor geometry and on fluid properties. Equations (35) and (36) have a theoretical basis as explained. Increasing concentrations of solids which simulate fungal mycelia leads to $k_L a_L$ and holdup reduction.

ACKNOWLEDGEMENT

This work was supported by a research grant from the National Science and Engineering Research Council of Canada.

NOTATION

A	Cross-sectional area of reactor, m^2
Ad	Cross-sectional area of downcomer, m^2
A_G	Cross-sectional area occupied by gas, m^2
A_p	Projected area of bubble, m^2
Ar	Cross-sectional area of riser, m^2
a	A parameter $(m/s)^{-b}$
a_L	Gas-liquid interfacial area per unit liquid volume, m^{-1}
B	Reciprocal of $k_L a_L$, s
b	A parameter (-)
C^*	Saturation concentration of oxygen in liquid, kgm^{-3}
C_D	Drag coefficient, (-)
C_L	Concentration of oxygen in liquid, kgm^{-3}
C_0	Initial concentration of oxygen in liquid, kg m^{-3}
C_s	Concentration of solids, dry wt./vol. %
d_B	Sauter mean bubble diameter, m
d_c	Diameter or equivalent diameter of reactor, m
d_d	Diameter or equivalent diameter of downcomer, m
d_h	Sparger hole diameter, m
d_r	Diameter or equivalent diameter of riser, m
F_B	Bouyancy force, kg m s^{-2}
F_D	Drag force, kg m s^{-2}
F_r	Froude number defined by Equation (27), (-)
g	Gravitational acceleration, ms^{-2}
h_L	Unaerated liquid height, m
i	Parameter in Equation (16), (-)
j	Parameter in Equation (16), (-)
K	Consistency index, Pa s^n
k_L	Mass transfer coefficient, ms^{-1}
L	Length of reactor, m
L_b	Height of riser/downcomer, m
L_c	Downcomer clearance from reactor base, m
L_d	Length of downcomer, m
L_h	Length of horizontal connection between riser and downcomer, m
L_r	Length or riser, m
L_t	Distance between top of downcomer and clear liquid height, m
n	Flow behaviour index, (-)
n_h	Number of sparger holes
P_G	Power input due to gassing, W
Q_G	Volume flow rate of gas, $m^3 s^{-1}$
Re_p	Bubble Reynolds number defined by Equation (17), (-)
S	Electrode time lag, s
SF	Solka-Floc
t	Time
t_G	Residence time of gas in liquid, s
Usg	Superficial gas velocity, ms^{-1}
U_T	Ture gas velocity or terminal rise velocity, ms^{-1}
V_{1-4}	Ball valves
V_D	Volume of gas-liquid dispersion, m^3
V_L	Volume of liquid, m^3
W	Width of reactor, m
y	Parameter in Equation (22), (-)

Greek Symbols

α	A parameter
β	A parameter
ε	Overall gas holdup
ε_r	Riser gas holdup
ε_d	Downcomer gas holdup
μ_L	Viscosity of liquid, kg m^{-1} s^{-1}
ρ_G	Density of gas, kg m^{-3}
ρ_L	Density of liquid, kg m^{-3}
σ	Interfacial tension, kg s^{-2}
ϕ	Parameter in Equation (19), (-)
ψ	Parameter in Equation (22), $(m\ s^{-1})^{1-y}$

LITERATURE CITED

1. Bello, R.A., Robinson, C.W., and Moo-Young, M., Can. J. Chem. Eng., 62, 573 (1984).

2. Bello, R.A., Robinson, C.W., and Moo-Young, M., Biotech. Bioeng., 27, 369 (1985).

3. Bello, R.A., Robinson, C.W., and Moo-Young, M., Chem. Eng. Sci., 40, 53 (1985).

4. Malfait, J.L., Wilcox, D.J., Mercer, D.G., and Barker, L.D., Biotech. Bioeng., 23, 863 (1981).

5. Erickson, L.E., Patel, S.A., Glasgow, L.A., and Lee, C.H., Process Biochem., 18(3), 16 (1983).

6. Koenig, B., Seewald, C., and Shugerl, K., in Advances in Biotechnology, Vol. 1, Moo-Young, M. (Ed.), Pergamon Press, Toronto, (1981), p. 573.

7. Nakanoh, M., and Yoshida, F., Ind. Eng. Chem. Process Des. Develop., 19, 190 (1980).

8. El-Temtamy, S.A., Khalil, S.A., Nour-El-Din, A.A., and Gaber, A., Appl. Microbiol. Biotechnol., 19, 376 (1984).

9. McCabe, W.L. and Smith, J.C., Unit Operations of Chemical Engineering, McGraw-Hill, New York (1956), p. 151, 167.

10. Moo-Young, M. and Blanch, H.W., Adv. Biochem. Eng., 19, 1 (1981).

11. Sobotka, M., Prokop, A., Dunn, I.J., and Einsele, A., "Review of methods for the measurement of oxygen transfer in microbial systems," in Annual Reports on Fermentation Processes, vol. 5, Tsao, G.T., (Ed.), Academic Press, London, (1982), p. 127.

12. Fair, J.R., Lambright, A.J., and Andersen, J.W., Ind. Eng. Chem. Process Des. Develop., 1, 33(1962).

13. Kawagoe, K.,Inoue, T., Nakao, K., Otake, T., Int. Chem. Eng., 16 (1), 176 (1976).

14. Reuss, M., Debus, D., and Zoll, G., The Chemical Engineer (London), June (1982), p. 233.

BIOPROCESS MASS TRANSFER

The Oxygen Transfer Coefficient in Aerated Stirred Reactors and Its Correlation with Oxygen Diffusion Coefficients

CHESTER S. HO
MICHAEL J. STALKER
Department of Chemical Engineering
State University of New York at Buffalo
Buffalo, New York 14260

RAYMOND F. BADDOUR
Department of Chemical Engineering
Massachusetts Institute of Technology
Cambridge, Massachusetts 02139

Oxygen transfer coefficients were measured and correlated with oxygen diffusion coefficients and pertinent physical parameters for various single and mixed electrolyte solutions. The resulting correlation was

$$k_L \propto D_L^{0.67}$$

which was theorized for oxygen transfer from rigid spheres to bulk solutions by the boundary layer theory. Consequently, it appears that the small gas bubbles produced in electrolyte solutions behave as rigid spheres. The resulting correlation will allow for the prediction of oxygen transfer behavior by knowing the oxygen diffusion coefficients of aqueous solutions in aerated stirred reactors.

INTRODUCTION

Oxygen is known to be a growth limiting nutrient for submerged aerobic fermentations. Microorganisms typically obtain oxygen through mechanical aeration of fermentation media in which the oxygen is dissolved. Consequently, it is very important to understand the mechanism of oxygen transfer for aerobic fermentation systems, for it can lead to tremendous improvements in the design and scale-up of these industrially important fermentation processes.

Aerobic fermentation systems are a complex mixture of a number of different chemical components. It is thus important to understand oxygen transfer in the systems containing these individual components before the highly complicated fermentation systems can be studied. It is known that electrolyte solutions are a controlling parameter in oxygen diffusion for aerobic fermentation media (1). Therefore, electrolyte solutions will be studied in order to determine the relationship between the oxygen transfer coefficient and the corresponding diffusion coefficient. This relationship will further allow for the prediction of oxygen transfer coefficients through the manipulation of oxygen diffusion coefficients in electrolyte solutions.

EXPERIMENTAL METHODS

An overall flow diagram of the apparatus is shown in Figure 1. The agitated vessel consisted of a 14-liter tank with an inner diameter of 8 inches. The four baffles and the gas sparger were combined into one unit connected to the cover of the vessel. Two six-blade flat blade impellers of 3-inch diameter were used. The basic dimensions of the impeller used were D:L:W = 20:5:4. Four agitation speeds were used in this experiment, i.e., 400, 550, 700, 850 RPM. Air sparging was controlled by use of a rotameter. Three different aeration rates were used in this study, i.e., 2000, 4000 and 8000 cc/min.

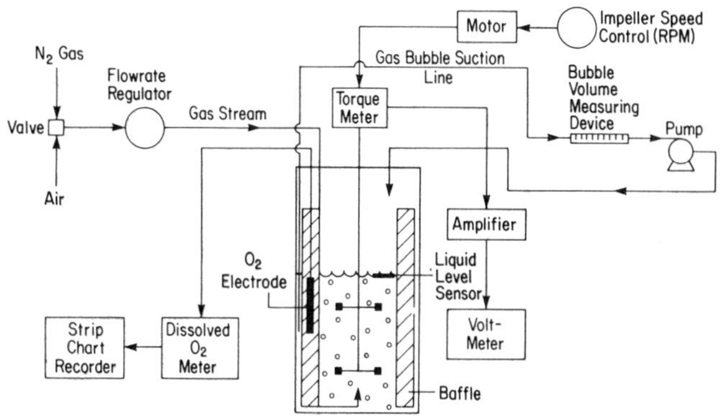

FIGURE 1. Schematic of the experimental system.

The dynamic method for the measurement of volumetric oxygen transfer coefficients was developed by Bandyopadhyay, Humphrey and Taguchi (2). This method involves a simple measurement of the change in dissolved oxygen concentrations with addition of aeration following a deaeration period using nitrogen gas. The following equation is resulted for a well-mixed bioreactor without microbial oxygen uptake:

$$\frac{dC_L}{dt} = k_L a(C^* - C_L) \qquad (1)$$

Bubble diameters were measured by the direct removal method described by Topiwala and Hamer (3). Bubbles were withdrawn from the dispersion through a small graduated glass capillary tube. Individual bubble volumes were measured in the capillary tube. Bubbles smaller than the tube diameter were treated as being spherical while those larger were treated as cylinders with hemispherical ends. A large quantity of samples were taken, and the Sauter mean diameter was calculated as:

$$d_m = \frac{\Sigma n d_i^3}{\Sigma n d_i^2} \qquad (2)$$

Gas holdup values are determined through measurements of the change in liquid height in dispersions with aeration in the agitated liquid:

$$H = \frac{h_a - h_o}{h_a} \qquad (3)$$

This method was used by Topiwala and Hamer (3), Foust et al. (4) and Lee and Meyrick (5).

These measurements allowed for the calculation of interfacial area by:

$$a = \frac{6H}{d_m} \qquad (4)$$

Interfacial area is then combined with the volumetric oxygen transfer coefficient to calculate the oxygen transfer coefficient:

$$k_L = \frac{(k_L a)}{a} \qquad (5)$$

Diffusion coefficient data were taken from Akita (6) and more recently Ho et al. (7) which compared very well with other literature values (8). These are the most widely ranged studies, using different cation and anion species.

RESULTS AND DISCUSSION

Agitation Power

Extensive work has been done by different authors to correlate power absorption in agitated tanks for different tank dimensions, impeller types and characteristics, and numerous other mechanical and physical properties (9 to 12). The study by Rushton et al. (10, 11) was generally regarded as the most complete and was thus used to compare with the present work.

The gassed power was measured in the same manner as the ungassed power. Three different aeration rates (0.29, 0.57, 1.14 VVM) were combined with the four impeller speeds (400, 550, 700, 850 RPM). The addition of gas to the agitation system significantly reduced the power requirement of the impeller. The gas decreased the fluid density thus making it easier to maintain the respective impeller speeds. In other words, the power required to mix the solution dispersions with air was less than the power required to mix those without air.

The following relationship was determined between the literature curve and the experimental values for the range of Reynolds numbers used in this study:

$$(N_{P_g})_{exp} = 0.1 + (N_{P_g})_{lit} \qquad (6)$$

The difference can be accounted for by the variation in the design of tank and impeller compared to that of Rushton et al. (10, 11). The small difference in power numbers is probably due to two variations: the different design of the impeller as well as the placement of baffles was not against the tank wall which could cause differences in power requirements. Figure 2 shows the comparison of the experimental results and the literature values (10, 11). The above correlation as shown in Equation (6) was determined for all electrolyte solutions.

It was also found that the gassed power measured fit the correlation of Michel and Miller (13):

$$P_g = 0.08(P_o^2 N D^3/Q^{0.56})^{0.45} \qquad (7)$$

where P_g and P_o are in horsepower, N is in RPM, D is the impeller diameter in feet, and Q is the gas flowrate in ft^3/min. The range of viscosity and surface tension for the electrolyte solutions used in the present study was slightly greater than those of Michel and Miller's study while the other parameters were held within the same range. The curve of Michel and Miller (13) is shown in Figure 3 along with data points from the present study.

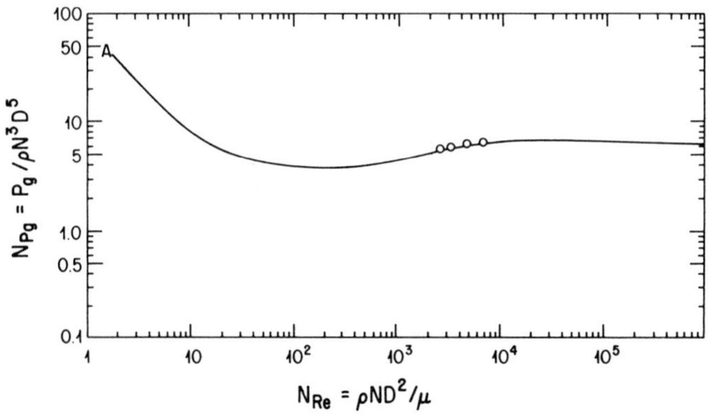

FIGURE 2. N_{P_g} vs. N_{Re} for agitation tank having six-blade turbines and four baffles. The curve is adapted from Rushton *et al.* (1950a,b). ○ represents experimental data for various electrolyte solutions.

Gas Holdup

The gas holdup is typically correlated (4, 14, 15) with (P_g/V) and V_s since both affect the amount of gas in the gas-liquid dispersion.

It was determined in this work that:

$$H \propto (P_g/V)^{0.45} (V_s)^{0.61} \quad (8)$$

This is shown in Figures 4 and 5. The superficial gas velocity ranged from 6.0 to 23.0 cm/min and the gassed power per unit volume ranged from 2.5×10^4 to 45.0×10^4 watts/cc. The gas holdup values did not vary from species to species for ionic strength ranging from 0 to approximately 10 mol/l, but did vary with increased ionic strength up to a value of 0.42 mol/l where the gas holdup began to level off.

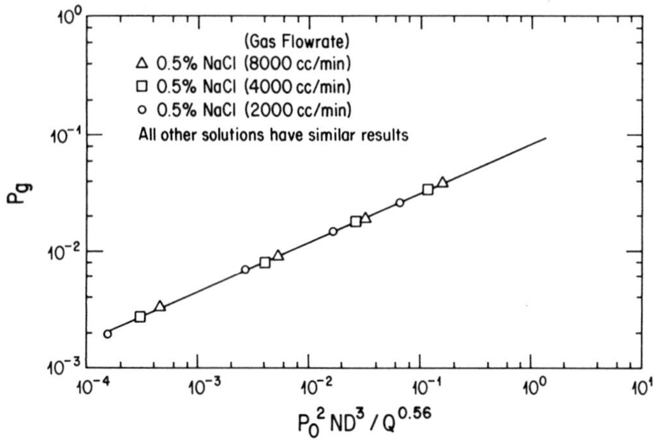

FIGURE 3. P_g as a function of $P_0^2 ND^3/Q^{0.56}$.

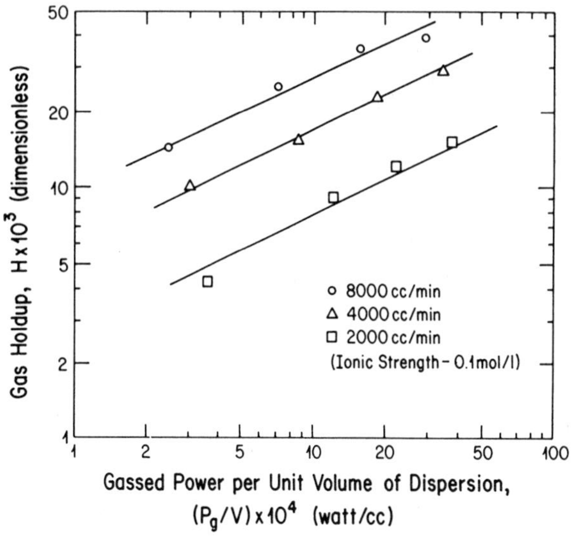

FIGURE 4. Correlation between gas holdup and P_g/V at constant superficial gas velocity (V_s).

It was expected that the gas holdup would be a strong function of (P_g/V) and V_s. If V_s was held constant and (P_g/V) was increased, the gas bubbles became more dispersed predicting a longer residence time and thus leading to a higher gas holdup. On the other hand, if (P_g/V) was held constant and V_s was increased, more gas was coming into the dispersion, and a higher gas holdup would result.

The most likely reason for the differences on gas holdup between the literature studies (4, 14, 15) and the present study is mechanical differences between the agitation systems. Slight discrepancies in the design of impeller, baffle placement, sparger placement, etc., will have an affect on the gassed power per unit volume and the superficial gas velocity with respect to gas holdups.

The effect of ionic strength on bubble diameter is also expected to cause changes in the gas holdup. It has been shown (3, 5, 16) that as ionic strength is increased, bubble size decreases up to a certain leveling point, i.e., 0.4 mol/l. The smaller gas bubbles are less buoyant and rise slower in the dispersions, thus leading to a longer residence time. This induces a greater gas holdup.

Bubble Diameter

The measurement of bubble diameter is extremely important for the determination of interfacial area and, hence, the oxygen transfer coefficient. In the present study, bubble diameter was measured according to the direct removal method as discussed by Topiwala and Hamer (3). It was found that the bubble diameter decreased with increased ionic strength. The bubble diameter decrease leveled off at the ionic strength of 0.42 mol/l. The interfacial area has been correlated with the gassed power per unit volume and the superficial gas velocity in the literature (16, 17, 18). A similar correlation was resulted in the present study as (Figures 6 and 7):

$$a \propto (P_g/V)^{0.62} \, (V_s)^{0.45} \qquad (9)$$

Interfacial area was also seen to increase with increased ionic strength up to 0.42 mol/l. Figure 8 shows interfacial area as a function of ionic strength at various agitation and aeration conditions.

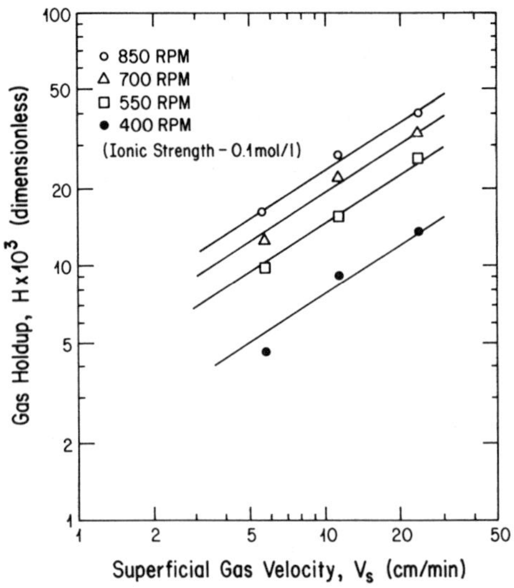

FIGURE 5. Correlation between gas holdup and V_s at constant gassed power per unit volume of dispersion (P_g/V).

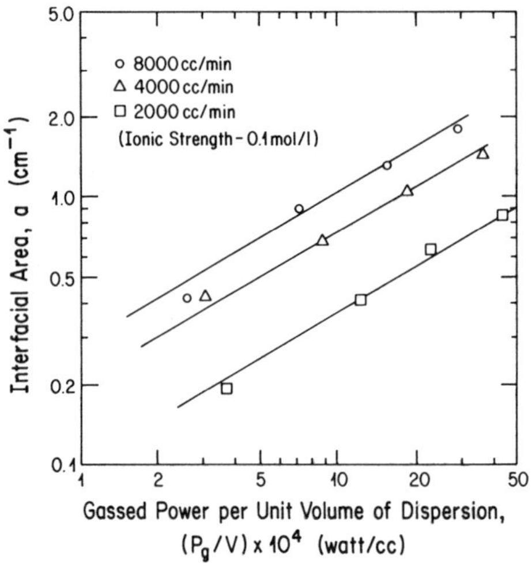

FIGURE 6. Correlation between interfacial area and P_g/V at constant superficial gas velocity (V_s).

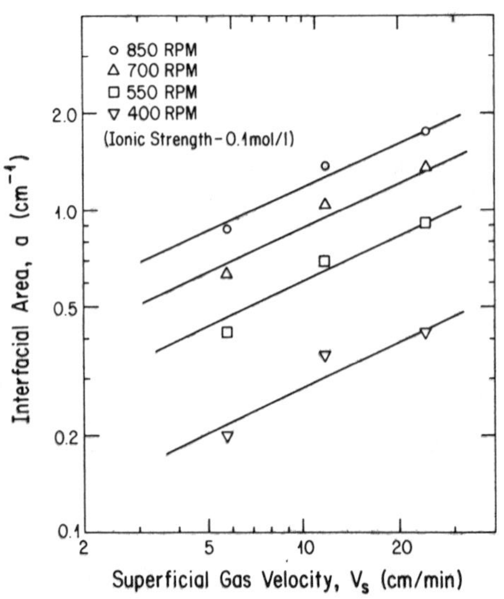

FIGURE 7. Correlation between interfacial area and V_s at constant gassed power per unit volume of dispersion (P_g/V).

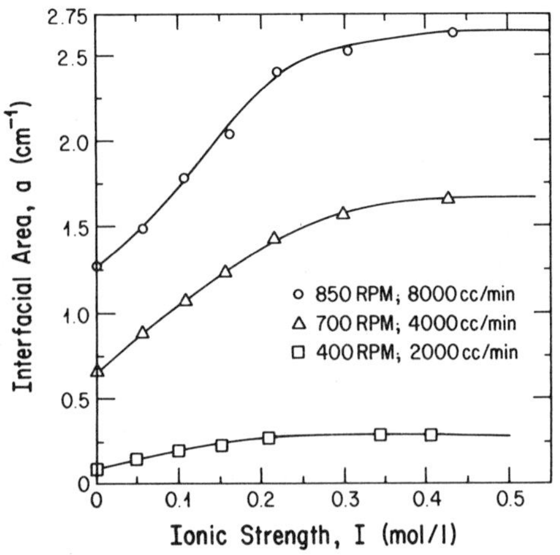

FIGURE 8. Correlation between interfacial area and ionic strength of electrolyte solutions.

The decrease of bubble diameters in electrolyte solutions has been shown by various authors (3, 5, 16, 19 to 23) to be directly related to a decrease in bubble coalescence and increased ionic strength. The bubble coalescence has also been measured and shown to be a function of ionic strength in a recent study by Oolman and Blanch (24). Interestingly, they reported that the bubble coalescence frequency leveled off when the ionic strength of the electrolyte solutions exceeded 0.4 mol/l.

A decrease in bubble coalescence would coincide with a decrease in bubble diameter (21, 22). The bubbles recombined at a lower rate with decreased bubble coalescence and thus tending to stay at their original sizes or at the size they were broken into by the dispersing device.

Marrucci (20) studied the bubble coalescence and theorized that the bubble coalescence was actually a thinning process of the liquid layer between two gas bubbles. This thinning process consisted of a rapid initial stretching followed by slower thinning and leaking of the gas through the liquid film layer until the bubbles joined together. He reported that the addition of surfactants created a rigid film surrounding the gas bubbles. The bubbles then acted as having rigid surfaces and the thinning process was greatly reduced. The rigid film increased with increased surfactant concentrations.

The differences between the literature correlations of interfacial areas (17, 18) and the present study are most likely due to the variations in the operating systems. The different fermentor types may take more (or less) power to disperse the gas phase in the liquid phase. The design of impeller and other tank dimensions will have an effect on ungassed power as reported by Rushton et al. (10, 11) and this effect should carry over for the ungassed power. This will further affect both (P_g/V) and V_s with respect to the relationship to interfacial area.

Volumetric Oxygen Transfer Coefficient

The volumetric oxygen transfer coefficient was measured by the dynamic method as presented by Bandyopadhyay et al. (2) and later improved by Dunn and Einsele (25), Dang et al. (26), Ruchti et al. (27) and Sobotka et al. (28). The values were correlated versus (P_g/V) and V_s. It was determined for electrolyte solutions that:

$$k_L a \propto (P_g/V)^{0.61} (V_s)^{0.45} \qquad (10)$$

as shown in Figures 9 and 10. The above proportionality held for electrolyte solutions of increased ionic strengths.

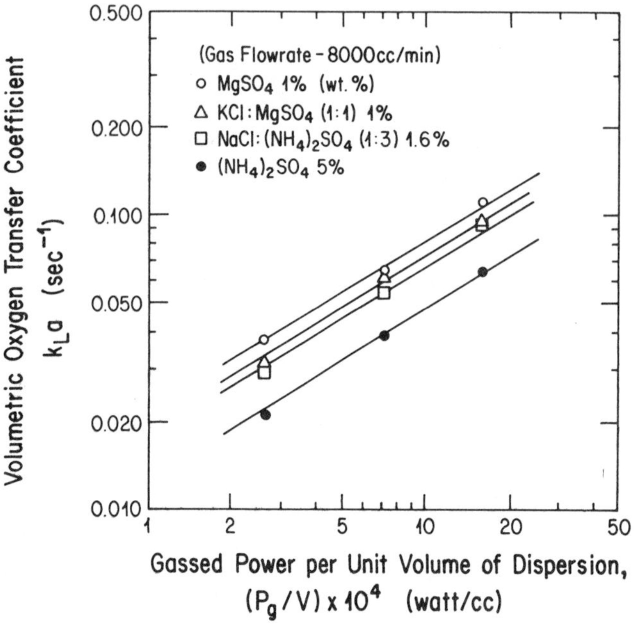

FIGURE 9. Correlation between volumetric oxygen transfer coefficient and P_g/V at constant superficial gas velocity (V_s).

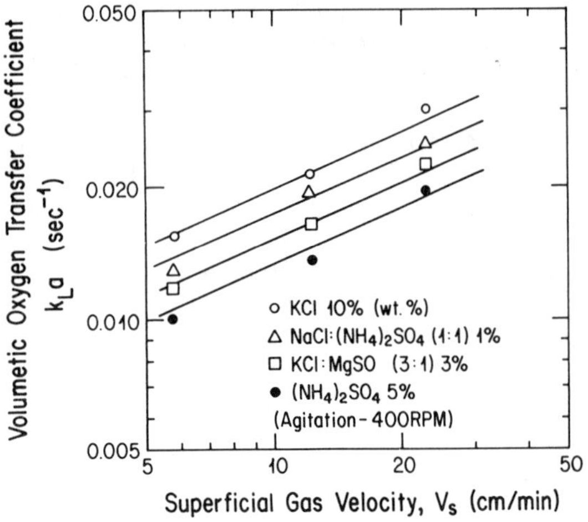

FIGURE 10. Correlation between volumetric oxygen transfer coefficient and V_s at constant gassed power per unit volume of dispersion (P_g/V).

Numerous correlations similar to Equation (10) appear in the literature (15 to 17, 28):

$$k_La \propto (P_g/V)^m (V_s)^n \qquad (11)$$

Cooper et al. (29) found for disk type impellers, m = 0.95, n = 0.67 and for paddle type impellers, m = 0.53, n = 0.67. The chemical method using aqueous sodium sulphite solutions was employed in their study. It is uncertain, however, as to the accuracy of this method due to the mechanical mixing effects on the chemical reaction (30). Moo-Young and Blanch (18) report for large mobile surface bubbles, m = 0.4, n = 0.5 and for small rigid surface bubbles, m = 0.7, n = 0.3. Robinson and Wilke (16) found that for water, m = 0.4, n = 0.35; for 0.1 N KCl, m = 0.63, n = 0.62; and for 0.22 N KCl, m = 0.71, n = 0.36. Hassan and Robinson (17) determined for water m = 0.42, n = 0.43 and for a solution with an ionic strength of 0.1 M, m = 0.52, n = 0.43.

No general correlation exists as shown above and it appears that the dependence of k_La on (P_g/V) and V_s varies from solution to solution for the above reported values from the literature. This was found not to be the case for the electrolyte solutions examined in this study. The proportionality in Equation (11) was determined to hold for all single and mixed electrolyte solutions that were studied. Though k_La appeared to increase with increased ionic strength for the solutions. Similar observations were also reported by other researchers (16, 17, 31).

The differences between the present study and the literature values reported on the effects of (P_g/V) and V_s on k_La can again be related to different mechanical properties varying from study to study. The differences seen by Moo-Young and Blanch (18) between the mobile-surface and rigid surface bubbles are due to the basic way in which the two types of bubbles react in dispersions. Mobile-surface bubbles have a constant internal circulation that helps to bring about greater mass transfer by regenerating the gas-liquid surface at the interface. Rigid surface bubbles have no internal circulation and are treated as rigid spheres in their behavior giving a lower mass transfer (18). Another possible difference between the equations derived by the chemical method and the dynamic method, is the inherent differences between the two methods. As discussed earlier, the extent to which mechanical mixing and other possible external interferences affecting k_La is unknown (30), therefore, the accuracy of the k_La results by the chemical method is quite debatable.

An interesting comparison exists between the equations describing the dependence of k_La and a on (P_g/V) and V_s. It was found that the oxygen transfer coefficient (k_La)/a has approximately no dependence on (P_g/V) and V_s. Robinson and Wilke (16) found similar results for the same range of (P_g/V). This type of observation is expected since the oxygen transfer coefficient has been shown experimentally (14, 32, 33) to be a function of solution properties in particular the diffusion coefficients and not the amount of agitation or aeration. The oxygen transfer coefficient can be viewed, in effect, as a physical property since it is dependent on different species in presence and not on the mechanical properties as the amount of aeration or agitation in the system.

The Relation Between k_L and D_L

It was desired to determine the correlation between the oxygen transfer coefficient and the diffusion coefficient for the electrolyte solutions studied. The range of diffusivities used in this study were 0.95×10^{-5} cm^2/sec - 2.34×10^{-5} cm^2/sec. Oxygen transfer coefficients ranged from 0.028 cm/sec - 0.051 cm/sec. As shown in Figure 11, the correlation was determined to be:

$$k_L \propto D_L^{0.67} \qquad (12)$$

This correlation was found to hold for both the single and mixed electrolyte solutions studied.

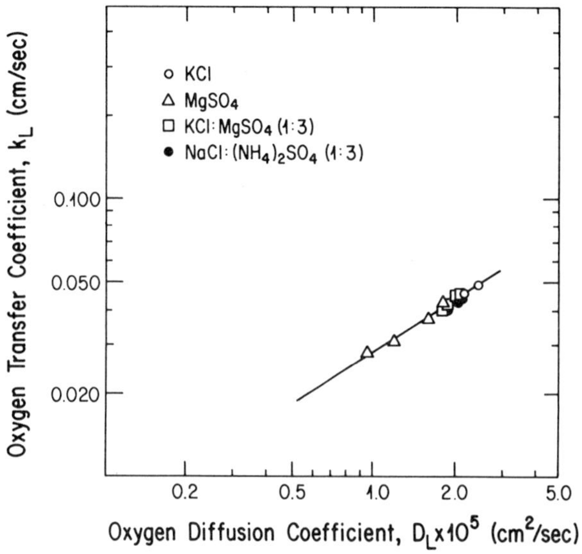

FIGURE 11. Correlation between oxygen transfer coefficients and oxygen diffusion coefficients.

Various correlations of k_L and D_L in the literature are shown in Table 1. It is interesting to note that the proportionality was the same for small gas bubbles, small liquid drops and solid spheres. The correlations shown in Table 1 follow the boundary layer theory developed by Froessling (34) for rigid non-slip interfaces which can be generally summarized as:

$$N_{Sh} \propto N_{Pe}^{1/3} \tag{13}$$

or

$$k_L \propto D_L^{2/3} \tag{12}$$

Therefore, it appears that for the electrolyte solutions studied, gas-liquid dispersions will act in a similar fashion to solid-liquid dispersions. In other words, the gas bubbles act like rigid spheres in these electrolyte solutions. Garner and Hammerton (35) determined a critical bubble diameter of 0.2 cm which differentiated between mobile-surface (larger) bubbles and rigid-surface (smaller) bubbles. This was also determined by others (32, 36). The largest bubble diameter measured in the current study was 0.2 cm and, therefore, can be viewed as small bubbles according to the above citations.

The observations can be explained by combining the results for interfacial area and the oxygen transfer coefficient. It was determined that the bubble diameter decreased with increased ionic strengths up to 0.42 mol/l. This phenomenon can be interpreted as the cause of ion-water interactions. Marrucci (20) noted that surfactants would create a rigid film surrounding the gas bubbles which would decrease bubble coalescence. This was determined experimentally (5, 21, 22, 24) for electrolyte solutions and was shown in this study by the decrease in bubble diameter with the increased ionic strength. Consequently, the small gas bubbles surrounded by a rigid liquid film had a tendency not to coalesce. This rigid film prevented the bulk liquid that passed by the bubble from creating internal motion in the bubble by moving the gas near the gas-liquid interface at a similar rate as the bulk fluid. This created the internal motion in mobile-surface bubbles since the gas continually circulated. The small bubble was in effect acting as a rigid sphere allowing the liquid to flow by the bubble without creating any internal motion. As a result, the bubble should behave as a rigid sphere for mass transfer purposes. The electrolyte solutions produced small gas bubbles (< = 0.2 cm) which behaved as solid spheres due to a rigid film surrounding them. This would justify a mass transfer relationship similar to that of solid spheres as shown in Equation (12).

CONCLUSIONS

The measurement of the oxygen transfer coefficient in aqueous electrolyte solutions has been studied. This is important for a number of reasons. The most important is that accurate knowledge of the oxygen transfer coefficient and its relation to system parameters (i.e. diffusion coefficient) must be determined before rational scale-up of aerobic fermentation systems can be achieved.

Values of the volumetric oxygen transfer coefficient were obtained by use of a membrane-covered oxygen electrode using the dynamic method. Interfacial area values were calculated from the measurements of gas holdup and mean bubble diameters according to:

$$a = 6H/d_m.$$

It was determined that interfacial area increased with ionic strength and then plateaued at a value of 0.42 mol/l. This was not species dependent. This was explained (through bubble diameter relations) by the theory of ion-water interactions. The

TABLE 1
Literature References for Mass Transfer and Diffusion Coefficients for Small Gas Bubbles, Small Liquid Drops, and Solid Spheres in Liquid Dispersions

	System Description	Reference
$k_L \propto D_L^{2/3}$	small liquid drops in liquid dispersions	Froessling (34)
$k_L \propto D_L^{2/3}$	O_2 gas in water (small bubbles) (Absorption in bubble columns)	Coppock and Meilkejohn (37)
$k_L \propto D_L^{2/3}$	small bubbles (less than 0.2 cm) in pure liquids	Garner and Hammerton (35)
$k_L \propto D_L^{2/3}$	brine particles in NaCl solution (Stirred tank - dissolution)	Piret et al. (38)
$k_L \propto D_L^{2/3}$	solid spheres in liquids (pure)	Garner and Suckling (39)
$k_L \propto D_L^{2/3}$	small bubbles (less than 0.2 cm) and solid spheres in pure liquids	Calderbank (36)
$k_L \propto D_L^{2/3}$	brine particles in NaCl solution (Fluidized bed crystallization)	Rumford and Bain (40)
$k_L \propto D_L^{2/3}$	small gas bubbles (less than 0.2 cm) and solid spheres in pure liquids	Calderbank and Moo-Young (41)
$k_L \propto D_L^{2/3}$	solid spheres in forced convection	Grafton (42)
$k_L \propto D_L^{2/3}$	liq.-liq. dispersions (rigid sphere behavior for small liquid drops)	Calderbank (36) Boyadzhieu and Elenkov (43)
$k_L \propto D_L^{2/3}$	solid benzoic acid pellets in water	Miller (44)

increased rigidity caused a decrease in bubble diameter up to a characteristic ionic strength, determined as 0.42 mol/l in this study.

It was generally observed that k_La increased with increasing ionic strengths. The correlation of gassed power per unit volume and superficial gas velocity with the volumetric oxygen transfer coefficient and interfacial area was shown to be proportionally similar. This would suggest that the oxygen transfer coefficient, K_L, was not a function of the gassed power per unit volume or superficial gas velocity for the range of parameters studied, but rather a function of the solution properties and the type of species involved.

The oxygen transfer coefficient was correlated with oxygen diffusion coefficients and it was found that:

$$k_L \propto D_L^{0.67}$$

where k_L is the oxygen transfer coefficient and D_L is the oxygen diffusion coefficient. This corresponds with the boundary layer theory predictions of oxygen transfer for rigid spheres.

This study shows that small gas bubbles are produced in aqueous electrolyte solution dispersions. The oxygen transfer coefficient for these electrolyte solutions can be correlated with the diffusion coefficient according to the boundary layer theory prediction for the rigid spheres. These small gas bubbles as produced in aqueous electrolyte dispersions can, therefore, be treated as rigid spheres for mass transfer purposes.

NOTATION

a	interfacial area per unit dispersion, cm^2/cm^3.
C	dissolved oxygen concentration, mg/l.
C^*	oxygen concentration at saturation, mg/l.
D	impeller diameter, ft.
D_L	oxygen diffusion coefficient, cm^2/sec.
d_i	equivalent spherical bubble diameter, cm.
d_m	mean bubble diameter, cm.
H	gas holdup.
h_a	liquid height with aeration, cm.
h_o	liquid height without aeration, cm.
k_L	oxygen transfer coefficient, cm/sec.
k_La	volumetric oxygen transfer coefficient, 1/sec.
L	impeller length, ft.
N	agitation speed, rev/min (RPM).
N_{P_g}	power number ($= P_g/\rho N^3 D^5$).
N_{Pe}	Peclet number ($= d_i\ U/D_L$).
N_{Sh}	Sherwood number ($= k_L d_i/D_L$).
n	number of gas bubbles.
P_o	ungassed power, horsepower.
P_g	gassed power, horsepower.
Q	volumetric gas flow rate, ft^3/min.
t	time, sec.
U	bubble free rising velocity, cm/sec.
V	volume of gas-liquid dispersion, liter.
V_s	superficial gas velocity, cm/sec.
W	impeller width, ft.

LITERATURE CITED

1. Ho, C.S., Ju, L. and C. Ho, "Measuring Oxygen Diffusion Coefficients with Polarographic Oxygen Electrodes. II. Fermentation Media", Biotechnol. Bioeng., 28, 1086-1093 (1986).

2. Bandyopadhyay, B., Humphrey, A.E. and H. Taguchi, "Dynamic Measurement of the Volumetric Oxygen Transfer Coefficient in Fermentation System", Biotechnol. Bioeng., 9, 533-544 (1967).

3. Topiwala, H.H. and G. Hamer, "Mass Transfer and Dispersion Properties in a Fermenter with a Gas-Inducing Impeller", Trans. Instn. Chem. Engrs., 52, 113-120 (1974).

4. Foust, H.C., Mack, D.E. and J.H. Rushton, "Mixing", Ind. Eng. Chem., 48, 552-555 (1956).

5. Lee, J.C. and D.L. Meyrick, "Gas-Liquid Interfacial Areas in Salt Solution in an Agitated Tank", Trans. Instn. Chem. Engrs., 48, 737-745 (1970).

6. Akita, K., "Diffusivities of Gases in Aqueous Electrolyte Solutions", Ind. Eng. Chem. Fund., 20, 89-94 (1981).

7. Ho, C.S., Ju, L., Baddour, R.F. and D.I.C. Wang, "Simultaneous Determination of Oxygen Diffusion Coefficients and Solubilities in Electrolyte Solutions with Polarographic Oxygen Electrodes", submitted to Chem. Eng. Sci. for publication (1986).

8. Ratcliff, G.A. and J.G. Holdcroft, "Diffusivities of Gases in Aqueous Electrolyte Solutions", Trans. Instn. Chem. Engrs., 41, 315-319 (1963).

9. Olney, R.B. and G.J. Carlson, "Power Absorption in Mixers", Chem. Eng. Prog., 43, 473-480 (1947).

10. Rushton, J.H., Costich, E.W. and H.J. Everett, "Power Characteristics of Mixing Impeller, (Part I)", Chem. Eng. Prog., 46, 395-404 (1950a).

11. Rushton, J.H., Costich, E.W. and H.J. Everett, "Power Characteristics of Mixing Impeller, (Part II)", Chem. Eng. Prog., 467-476 (1950b).

12. Metzner, A.B. and R.E. Otto, "Agitation of Non-Newtonian Fluids", AIChE Journal, 3, 3-10 (1957).

13. Michel, B.J. and S.A. Miller, "Power Requirements of Gas-Liquid Agitated Systems", AIChE Journal, 8, 262-266 (1962).

14. Calderbank, P.H., "Physical Rate Processes in Industrial Fermentation. Part 1: The Interfacial Area in Gas-Liquid Contacting with Mechanical Agitation", Trans. Instn. Chem. Engrs., 36, 443-463 (1958).

15. Ade Bello, R., Robinson, C.W. and M. Moo-Young, "Diffusion of the Volumetric Mass Transfer Coefficient in Pneumatic Contactors", Chem. Eng. Sci., 40, 53-58 (1985).

16. Robinson, C.W. and C.R. Wilke, "Simultaneous Measurement of Interfacial Area and Mass Transfer Coefficients for a Well-Mixed Gas Dispersion in Aqueous Electrolyte Solutions", AIChE Jnl.,20,285-294(1974).

17. Hassan, I.T.M. and C.W. Robinson, "Mass-Transfer-Effective Bubble Coalescence Frequency and Specific Interfacial Area in a Mechanically Agitated Gas-Liquid Contactor", Chem. Eng. Sci., 35, 1277-1289 (1980).

18. Moo-Young, M. and H.W. Blanch, "Design of Biochemical Reactors. Mass Transfer Criteria for Simple and Complex Systems", Advances in Biochemical Eng., 19, 1-69 (1981).

19. Marrucci, G. and L. Nicodema, "Coalescence of Gas Bubbles in Aqueous Solution of Inorganic Electrolytes", Chem. Eng. Sci., 22, 1257-1265 (1967).

20. Marrucci, G., "A Theory of Coalescence", Chem. Eng. Sci., 24, 975-985 (1969).

21. Lessard, R.R. and S.A. Zieminski, "Bubble Coalescence and Gas Transfer in Aqueous Electrolytic Solutions", Ind. Eng. Chem. Fundam., 10, 260-269 (1971).

22. Zieminski, S.A. and R.C. Whittemore, "Behavior of Gas Bubbles in Aqueous Electrolyte Solutions", Chem. Eng. Sci., 26, 509-520 (1971).

23. Keitel, G. and U. Onken, "The Effect of Solutes on Bubble Size in Air-Water Dispersions", Chem. Eng. Commun., 17, 85-98 (1982).

24. Oolman, T.O. and H.W. Blanch, "Bubble Coalescence in Air-Sparged Bioreactors", Biotechnol. Bioeng., 28, 578-584 (1986).

25. Dunn, I.J. and A. Einsele, "Oxygen Transfer Coefficients by the Dynamic Method", J. Appl. Chem. Biotech., 25, 707-720 (1975).

26. Dang, N.D.P., Karrer, D.A. and I.J. Dunn, "Oxygen Transfer Coefficients by Dynamic Model Moment Analysis", Biotechnol. Bioeng., 19, 853-865 (1977).

27. Ruchti, G., Dunn, I.J. and J.R. Bourne, "Comparison of Dynamic Oxygen Electrode Methods for the Measurement of K_La", Biotechnol. Bioeng., 23, 277-290 (1981).

28. Sobotka, M., Prokop, A., Dunn, I.J. and A. Einsele, "Review of Methods for the Measurement of Oxygen Transfer in Microbial Systems", Ann.Rpts on Ferment.Processes, 5, 127-210 (1982).

29. Cooper, C.M., Fernstrom, G.A. and S.A. Miller, "Gas-Liquid Contactors", Ind. Eng. Chem., 36, 504-509 (1944).

30. Sridhar, T. and O.E. Potter, "Interfacial Area Measurements in Gas-Liquid Agitated Vessels", Chem. Eng. Sci., 33, 1347-1353 (1978).

31. Robinson, C.W. and C.R. Wilke, "Oxygen Absorption in Stirred Tanks: A Correlation for Ionic Strength Effects", Biotechnol. Bioeng., 15, 755-782 (1973).

32. Calderbank, P.H. and M.B. Moo-Young, "The Prediction of Power Consumption in the Agitation of Non-Newtonia Fluids", Trans. Instn. Chem. Engrs., 37, 26-33 (1959).

33. Akita, K. and F. Yoshida, "Bubble Size Interfacial Area, and Liquid-Phase Mass-Transfer Coefficient in Bubble Columns", Ind. Eng. Chem. Process Des. Develop., 13, 84-90 (1974).

34. Froessling, N., "Uber die Verdunstung fallender Tropfen", Beitr. Geophys., 52, 170-216 (1938).

35. Garner, F.H. and D. Hammerton, "Gas Absorption from Single Bubbles", Trans. Instn. Chem. Engrs., 32, S18-S24 (1954).

36. Calderbank, P.H., "Physical Rate Processes in Industrial Fermentation. Part II: Mass Transfer Coefficients in Gas-Liquid Contacting With and Without Mechanical Agitation", Trans. Instn. Chem. Engrs., 37, 173-185 (1959).

37. Coppock, P.D. and G.T. Meiklejohn, "The Behaviour of Gas Bubbles in Relation to Mass Transfer", Trans. Instn. Chem. Engrs., 29, 75-86 (1951).

38. Piret, E.L., Mattern, R.U. and O. Bilous, "Continuous-Flow Stirred Tank Reactors: Solid-Liquid Systems", AIChE Journal, 3, 497-505 (1957).

39. Garner, F.H. and R.D. Suckling, "Mass Transfer from a Soluble Solid Sphere", AIChE Journal, 4, 114-124 (1958).

40. Rumford, F. and J. Bain, "The Controlled Crystallization of Sodium Chloride", Trans. Instn. Chem. Engrs., 38, 10-20 (1960).

41. Calderbank, P.H. and M.B. Moo-Young, "The Continuous Phase Heat and Mass-Transfer Properties of Dispersions", Chem. Eng. Sci., 16, 39-54 (1961).

42. Grafton, R.W., "Prediction of Mass Transfer from Spheres and Cylinders in Forced Convection", Chem. Eng. Sci., 18, 457-466 (1963).

43. Boyadzhieu, L. and D. Elenkov, "On The Mechanism of Liquid-Liquid Mass Transfer in a Turbulent Flow Field", Chem. Eng. Sci., 21, 955-959 (1966).

44. Miller, D.N., "Scale up of Agitated Vessels - Mass Transfer From Fixed Solute Surfaces", Chem. Eng. Sci., 22, 1617-1626 (1967).

Improvements in Multi-Turbine Mass Transfer Models

FREDRIC G. BADER
Industrial Division
Bristol-Myers Co.
Syracuse, New York 13221-4755

Mathematical modeling approaches are being developed to better understand and describe the overall performance of large-scale agitated fermentors. These new approaches are based upon subdividing the fermentor into various mixing cells or stages. The further development of these models requires information on the power input per stage, its effect on mass transfer within the stage, fluid mixing patterns between stages, and the distribution of gas in the fermentor stages. The following discussion presents data and comments related to these topics for aerobic fermentations of low to medium viscosity.

A mathematical approach for simulating the performance of large industrial multi-turbine fermentors has been proposed by Manfredini et al. (1), Oosterhuis and Kossen (2), Bader (3,4) and Cohen and Gaden (5). Each of these reports subdivide the fermentor into a number of discrete mixing cells or stages which can be analyzed as interconnected but separate reactors. With the exception of Oosterhuis and Kossen (2), all assume that each stage is well mixed, an assumption that is generally appropriate for fermentation broths of low to medium viscosity.

One of the major strengths of the stage-wise approach is that it provides for optimization on a per turbine basis rather than by combining the entire agitation system together and calculating an overall k_La. It also provides a system that accounts for the hydrostatic head variation of large fermentors and permits the estimation of the axial dissolved oxygen profile. The major weakness of this approach has been the rather limited level of knowledge that exists on the mixing patterns, power distribution, and mass transfer of multi-turbine agitation systems. The following data provide additional information in these areas.

MODELLING STRATEGY

The stage-wise approach to modelling large fermentors subdivides the vessel shown in Figure 1 into a series of separate mixing stages as shown in Figure 2. The mixing cells are interconnected by liquid flowing between stages (F) and by gas flowing from the bottom stage towards and out of the top stage. Backmixing of the gas phase is ignored for simplicity. A summary of the modelling strategy (3) is presented as follows:

1. Define the system geometry.
2. Perform pressure-volume hold up calculations on each stage.
3. Estimate the liquid flow between stages.
4. Estimate the power input per turbine.
5. Correlate the mass transfer rate to power input for each stage.
6. Calculate the gas composition between cells from oxygen uptake rate data.
7. Calculate the axial dissolved oxygen profile from oxygen balances on each cell.

Figure 3 shows the predicted stage by stage dissolved oxygen concentration for a representative set of fermentation conditions as a function of agitator RPM as calculated from a stage-wise model. (4)

The development of this modelling strategy raises questions as to how to perform some of the calculations outlined above. This paper will present comments and/or data relevant to estimating percent hold-up, interstage liquid flow rate, power input for multiple agitators and the relationship of k_La to operating variables.

COMMENTS ON PERCENT HOLD-UP

Many attempts have been made to correlate the percent gas hold-up in gas-liquid agitated systems

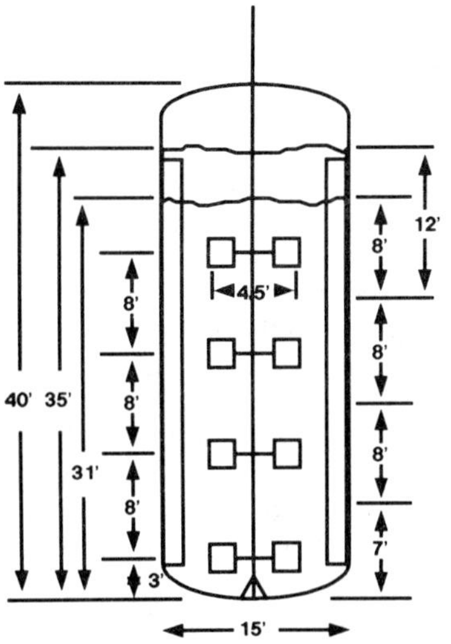

FIGURE 1. Typical production scale antibiotic fermentor.

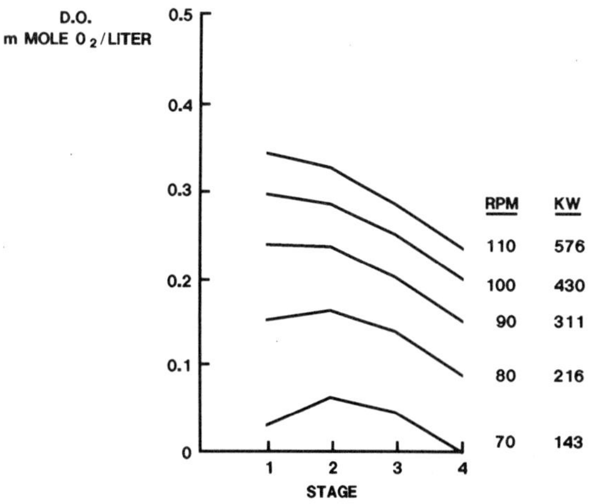

FIGURE 3. The effect of agitator speed (rpm) on the predicted D.O. profile for a hypothetical fermentor (4).

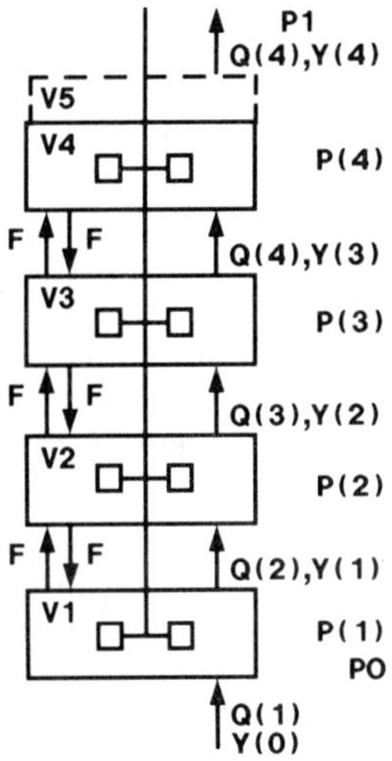

FIGURE 2. The breakdown of a multi-turbine fermentor into mixing cells. Volume V5 represents expansion due to aeration.

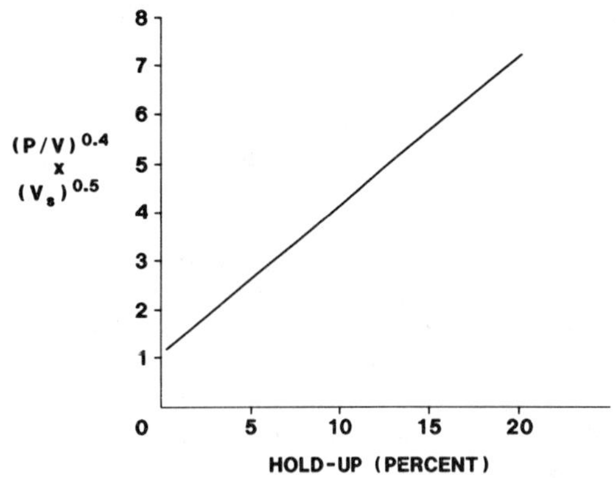

FIGURE 4. Relationship between percent hold-up, power per unit volume, and superficial air velocity as presented by Richards (7).

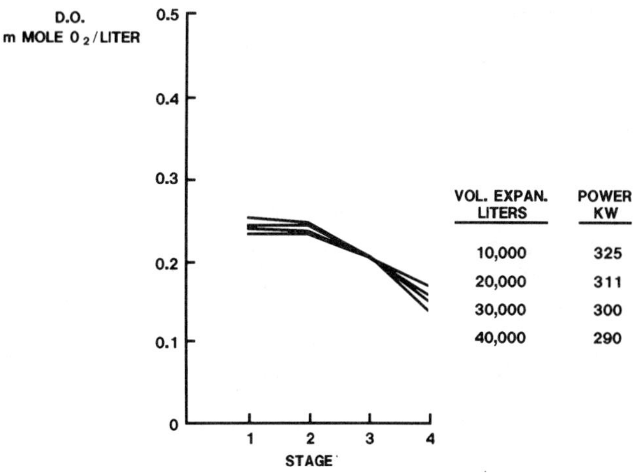

FIGURE 5. The effect of volume expansion or percent hold-up on the D.O. profile for a hypothetical fermentor (4).

with variables such as power input per unit volume (P/V) and superficial gas velocity (V_S). Calderbank (6) experimentally found the following equation to apply in predicting percent hold-up:

$$H = \left(\frac{V_S\,H}{V_t}\right)^{0.5} + 0.0216\left[\frac{(P/V)^{0.4}\;\rho^{0.2}}{\sigma^{0.6}}\right]\left[\frac{V_S}{V_t}\right]^{0.5} \quad (1)$$

where:
- σ = liquid surface tension (dyn/cm)
- P/V = power input per unaerated volume of liquid (hp per ft ^{3})
- ρ = liquid density (gm/ml)
- V_S = superficial gas velocity (ft/sec)
- V_t = terminal rising velocity of bubbles (ft/sec)

Calderbank's work was conducted primarily in clean liquids. It is not well documented whether this equation can be applied generally to fermentation broths where solids may affect V_t and where the surface tension can vary over a substantial range. Figure 4 shows this correlation as plotted for water by Richards. (7) This relationship is useful in that it shows that gas hold-up is increased by either increasing agitator power or airflow rate. However, the Richards relationship has little use alone in predicting percent hold-up in real fermentation broths. Steel and Maxon (8) showed that the addition of antifoam agent could reduce gas retention from as high as twenty percent down to three percent in the novobiocin fermentation under identical conditions of aeration and agitation. The percent hold-up can reach levels as high as 90% under severe foaming conditions. It is possible that these observations can be explained by the surface tension term in Equation 1. However, this has not been done for real broths. It is likely that these observations fall outside of the range tested in establishing Equation 1. In practice, it is difficult to distinguish between percent hold-up and actual foaming. It is likely that hold-up and foaming are similar phenomena, but it is also likely that they do not fit a single correlation such as Equation 1.

During normal fermentor operation, it is not necessary to predict percent hold-up as it is controlled by the addition of an antifoam agent. This requires some method of measuring percent hold-up, which is difficult in large equipment. At best, it can be estimated from physical observation of the fermentor and is limited in the extreme by the use of antifoam probes. The typical range observed for percent hold-up is between 5 and 15 percent.

It is difficult to determine the effect of percent hold-up on the mass transfer performance of a fermentor. Low levels of entrained gas may indicate poor gas dispersion and low bubble surface area for mass transfer. Very high levels of gas hold-up lead to decreased agitator efficiency, wasted tank volume, and may or may not indicate high bubble area. In the modelling approach described by Bader (4), percent hold-up has been an assumed variable and has only been used to determine liquid densities and levels in the fermentor. Any effect of hold-up on bubble surface area is taken care of separately in the k_la relationship.

Figure 5 shows a model prediction of the effect of increasing the volume expansion on a 150,000 L fermentor from 10,000 to 40,000 liters (6.25 to 21.05% hold-up). Only minor changes in the axial D.O. profile were predicted. If, however, the increase in entrainment substantially increased bubble area and k_la, the effect would be larger. This will be discussed in more detail later.

In light of the above comments, certain recommendations can be made related to future work on percent hold-up. First, a simple on-line method of measuring it would be useful both for control and future modelling work. Second, more work is needed to correlate H with variables such as percent solids, surface tension, and viscosity. Third, studies are needed to better understand the effect of gas hold-up on agitator and mass transfer performance. Fourth, the effect of antifoam on hold-up needs to be studied. Finally, the potential of using surface tension reducing agents to increase hold-up and improve mass transfer may be worth investigating.

LIQUID FLOW RATE BETWEEN STAGES

Manfredini et al. (1) estimated the liquid flow rate (F) between mixing stages with the equation:

$$F = Q'N_Q\;S\,D^3 \quad (2)$$

where
- Q' = fraction of impeller flow in the axial direction
- N_Q = impeller discharge coefficient
- S = RPM
- D = turbine diameter

In their estimate, N_Q had a value of 0.5 to 1.0 and Q' was estimated at 0.2 to 0.3, giving $Q'N_Q$ a range of 0.1 to 0.3. However, no experimental data was presented to verify the estimates.

Experiments were conducted on 100, 1,000, and 30,000 gallon fermentors to determine the mixing time constants of the tanks. Acid was added to the top stage of the fermentors and the pH profile at the bottom stage was monitored as a function of time. In the smaller two tanks, the manhole was opened and a large single addition of acid was added. Figure 6 shows the response of the pH probe with time after addition to the 1,000 gallon tank. A computer flow simulation model (solid line) predicted that an average holding time per stage of ten seconds, was required to produce the response shown for the five stage fermentor.

Figure 7 shows data for a 30,000 gallon (9) tank with five mixing stages. In this experiment, acid was pumped into the top stage at a fixed rate for 210 seconds and the pH was monitored at the bottom stage. Computer simulation predicted a holding time per stage of 15 seconds for this tank. Figure 8 shows a computer simulation of the concentration profile in the top, middle, and bottom stage for the 30,000 gallon tank. The simulation is for a single addition and is, therefore, comparable to the experimental results in Figure 6.

Table 1 tabulates the data for the mixing time studies for the three tank sizes. In this table, the volume of each stage is calculated as;

$$V(N) = \pi D_T^2 L_S/4 \qquad (3)$$

and the flow rate between stages is calculated from;

$$F = V(N)/\Theta \qquad (4)$$

where L_S is the turbine spacing, D_T is the tank diameter and Θ is the measured holding time per stage. Based upon these results, the product $Q'N_Q$ was calculated for each tank as shown in Table 1. The value of $Q'N_Q$ varied substantially and rather randomly between the three scales of fermentor but agrees with the results of Singh et. al. (10)

Figure 9 shows a model simulation of the D.O. profile in a four stage fermentor where the holding time per stage was varied over a 25 fold range. The curves were generated by multiplying the actual holding time by the factors shown. As would be expected, a fermentor with a very short holding time per stage (high mixing between stages) shows a much flatter D.O. profile than for a poorly mixed fermentor. However, the overall D.O. profile does not appear to be particularly sensitive to the holding time per stage. This would be expected if mass transfer is occurring rather uniformly throughout the fermentor as was assumed in the model. If this is not the case, then the liquid flow rate between stages would be expected to more severely affect the D.O. profile.

AGITATOR POWER DISTRIBUTION

Rushton et. al. (11) determined that the power delivered by an agitator fits the equation:

$$P_o = \frac{N_P \, \rho \, S^3 \, D^5}{5.118 \times 10^9} \qquad (5)$$

where P_o is the ungassed power in Kwatts, N_P is the power number for the turbine and ρ is the density of the liquid. This equation fits as long as the agitator is operating in the turbulent region (Reynolds numbers greater than 1,000). In applying this equation to numerous fermentation broths, the relationships between power and agitation speed S and power and turbine diameter D have held to be true. The main difficulty that occurs in using Equation 5 lies in knowing how to predict the power number N_P.

In order to explore better ways of predicting N_P in fermentation equipment, a study was conducted on a 100 gallon fermentor equipped with up to four 9.5 inch diameter eight bladed Rushton turbines. (12) Figure 10 shows the effect of liquid level above (L_A) and below (L_B) the turbine on the power number N_P of a single turbine. Figure 11 shows the power number for multiple turbines spaced one diameter apart as a function of the liquid level above the top turbine. All data are based upon delivered power measurements. These figures illustrate that the power input of a turbine is highly dependent upon its spacial location within the vessel, and that equally spaced turbines draw equal amounts of power.

When sparging turbines, the air flow rate is generally expressed in the dimensionless form of the aeration number ($N_A = Q/S\,D^3$) where Q is the air flow rate at reactor temperature and pressure. (6,13,14) Figure 12 shows power measurements as a function of N_A for various liquid levels above (L_A) and below (L_B) a single turbine. It is interesting to observe that increasing L_B increases the power at all values of aeration. However, increasing L_A only increased power at low air flow rates. At high air rates, the liquid level above the turbine no longer affected the turbine power.

Figure 13 shows power curves for a number of turbines spaced equally at one diameter apart. The most striking feature of this data is the relatively equal distribution of power between the four turbines. This is different from what would be predicted by simply correcting the upper turbines for percent hold-up. (15,16)

The solid lines that are shown in Figures 10 to 13 were generated by a curve fitting correlation. (12) The correlation was used to predict the power draw on a 1,000 gallon and a 30,000 gallon fermentor to see how it compared with actual data. The 1,000 gallon fermentor has five Rushton turbines spaced one diameter apart with six blades per turbine. Delivered power measurements for a number of identical fermentations are shown in Figure 14, along with the prediction of the model. Only three turbines are submerged at the beginning

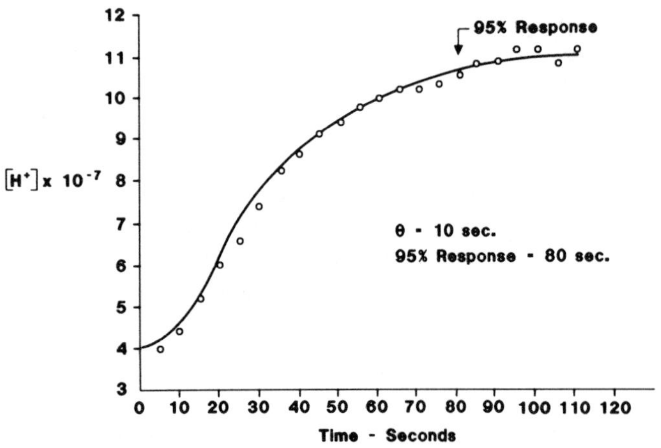

FIGURE 6. Acid concentration response profile of a 1000-gallon five-stage fermentor. An acid shot was added to the top stage of the fermentor and the response as monitored at the bottom stage with a pH probe. The points represent data and the solid line represents the model simulation.

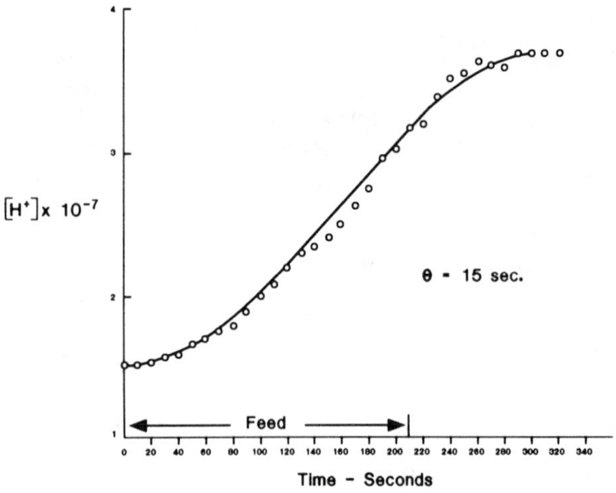

FIGURE 7. Acid concentration response profile in a 30,000-gallon five-stage fermentor. Acid was fed to the top stage for 210 s and the pH response was monitored at the bottom stage. The points represent data and the solid line represents the model simulation.

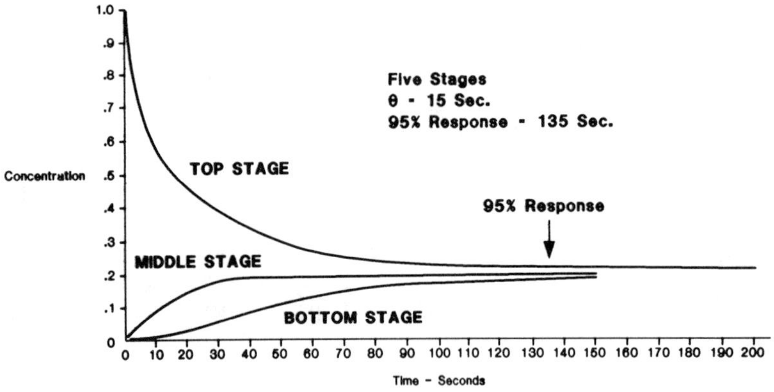

FIGURE 8. A computer simulation of the concentration profiles at the top, middle, and bottom stages of a five-stage 30,000-gallon fermentor to a hypothetical addition of material. The holding time per stage is 15 s as measured from the data in Figure 7.

TABLE 1
Tank Dimensions and Parameters for Mixing Studies

Tank Volume Gal.	Number of Turbines	Pressure PSIG	Air VVM	Agitator Speed RPM	Turbine Diameter D_i, in	Tank Diameter D_T, in	Turbine Spacing L_S, in	Volume Per Stage V(N), ft^3
100	4	0	.42	280	9.5	23.5	9.5	2.38
1,000	5	0	.42	155	22	48	22.0	23.04
30,000	5	15	.47	100.6	48	142.75	70.0	648.33

Tank Volume Gal.	SD^3 ft^3/sec.	V(N)/SD^3 sec.	95% Mixing Response Time sec.	O(N) sec.	F(N) ft^3/sec.	$Q'N_Q$
100	2.32	1.03	17	3	0.79	.3426
1,000	15.92	1.45	85	10	2.30	.1447
30,000	107.31	6.04	135	15	43.22	.4028

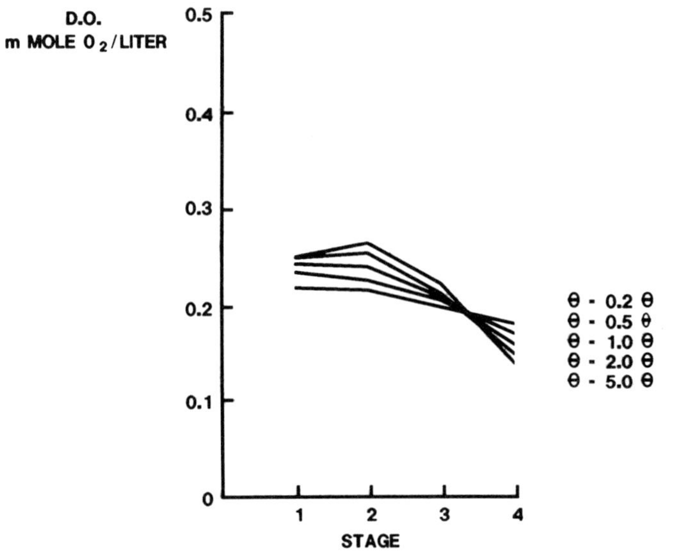

FIGURE 9. The effect of varying the holding time per stage on the predicted D.O. profile (4). In the model simulation, the original holding time was varied by the factors indicated. At θ = 0.2θ, the interstage flow rate is increased by a factor of five which generates the flattest D.O. profile.

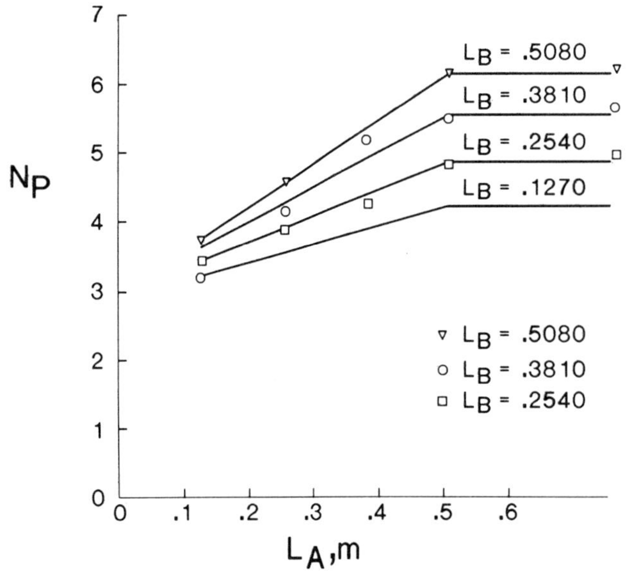

FIGURE 10. The ungassed power number for a single 9.5-inch, 8-blade Rushton turbine as a function of liquid levels above (L_A) and below (L_B) the turbine. The solid lines are linear fits to the data presented. The line $L_B = 0.1270$ is an extrapolation from the other data.

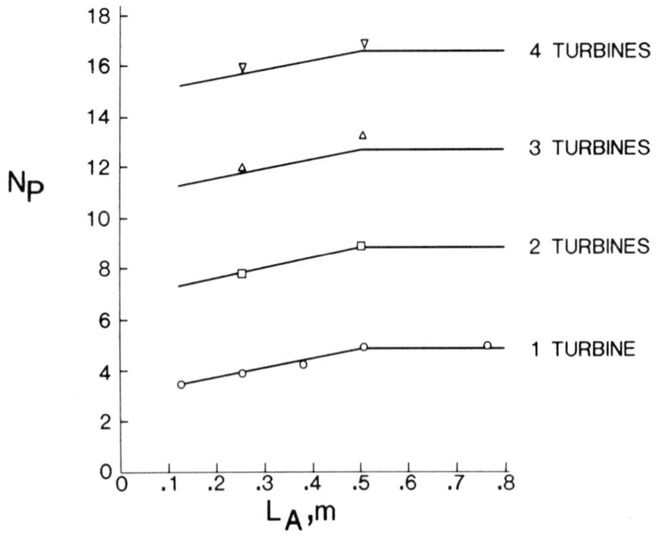

FIGURE 11. The ungassed power number as a function of liquid level above the top turbine L_A for various numbers of equally spaced turbines.

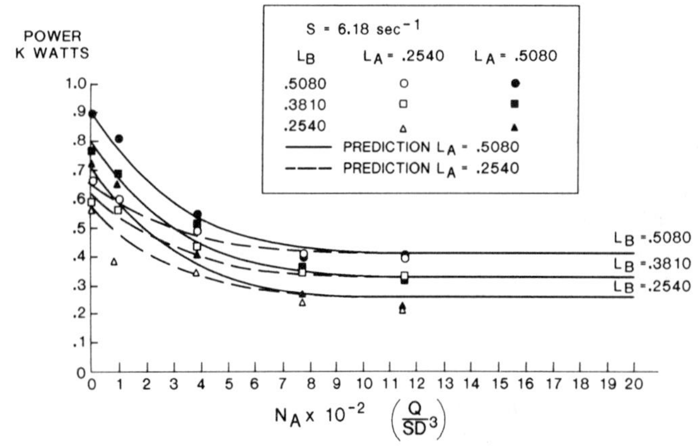

FIGURE 12. Agitator power demand for a single turbine as a function of aeration number for three values of L_B and two values of L_A.

of the run, but all five are submerged by the end. The air flow rate to the fermentor is changed at 20 hours and again at 30 hours. Pressure was held constant throughout the runs.

The prediction of the correlation was close to the actual data in this case. There is, however, substantial variation in the power draw between the actual runs with a standard deviation of ± 24% at time zero and 12% at 180 hours. Although it is not clear why such variation occurs, the most logical hypothesis is that the percent hold-up is different due to random differences in the foaminess of the broth. For example, if one assumes that 50 SCFM of air is sparged to the bottom trubine and the bottom stage is at 25°C and 15 PSIG, then this represents a value of Q of 27 ft^3/min. At 155 RPM, the full pumping capacity of a 22 inch turbine is 955. ft^3/min (See Table 1). At 10% hold-up, the total air (sparged and recirculated) going to the bottom turbine would be 122.5 ft^3/min. At 20% hold-up, this number would be 218 ft^3/min. Recirculated gas is not used in predicting power and it is possible that recirculated gas can be ignored if the surface tension in Equation 1 is held constant, (e.g., the case for studies in water). In fermentation broths, where H is likely to vary significantly, some method may be needed to account for large changes in recirculated gas flow.

A similar comparison was made between the correlation and power data for a 30,000 gallon production fermentor where the correlation predicted results that were consistently low. Three potential reasons for the low prediction are: (1) The production fermentor has a mixture of blades per turbine which makes it difficult to predict N_P from the eight bladed turbine data; (2) The ratio of D_i/D_T for the production tank was 0.3363 versus 0.4043 and 0.4583 for the 100 and 1,000 gallon tanks, respectively; and (3) the correlation is not sufficiently general to cover such a broad range. It is also likely that the effect of turbine spacing on N_P is also affected by the ratio D_i/D_T.

The conclusion that must be drawn from the above is that the correlation that was used to fit the data in Figures 10 through 13 cannot be used for general scale-up. That leaves the problem of determining the power distribution between turbines unanswered. There are certain rules that may be the most accurate method of predicting power distribution. The rules for Rushton turbines in a large fermentor are:

(1) Measure the total motor power reading

$$\text{Power (Kwatts)} = \frac{(\text{Volts X Amps X 1.732 X Power Factor})}{1{,}000} \quad (6)$$

(2) Subtract 18% for motor and drive losses to determine the power delivered to the broth.

(3) Divide the delivered power by the number of turbine blades that are fully immersed in the broth. This gives the power per blade.

(4) To determine the power for a particular turbine, multiply the power per blade by the number of blades on the turbine.

(5) In the above calculations, if the top turbine is not fully immersed or is "splashing", ignore the turbine in the calculation as it draws very little power in that mode.

These rules may not be theoretically elegant, but are closer to reality than any of the theories proposed thus far. These rules are easy to use and are consistent with the data shown in Figures 11 and 13.

MASS TRANSFER

The mass transfer coefficient k_la for large fermentors has generally been determined from the equation:

$$k_la = \frac{OUR}{\ln\left[\dfrac{C_{in}^* - C_{out}^*}{\dfrac{C_{in}^* - C}{C_{out}^* - C}}\right]} \quad (7)$$

where OUR is the oxygen uptake rate in mmole O_2/L/hr, C^*_{in} and C^*_{out} are the D.O. concentrations in equilibrium with the inlet and exit partial pressure of oxygen in the air, C is the actual fermentor dissolved oxygen concentration in mmole O_2/L, and k_la has units of hr^{-1}. Equation 7 does not provide information on the mass transfer efficiency of each turbine in a multiturbine fermentor, and leaves open the question of what value of C to use, since C varies throughout the fermentor.

Models which subdivide the fermentor into multiple mixing stages overcome many of the limitations of Equation 7, but raise a new question of how to determine the efficiency of each turbine as a mass transfer device. Most correlations which relate mass transfer to agitation-aeration parameters have involved single turbines which provide both initial bubble breakup and bulk motion. Rushton (17) used dimensional analysis to determine a general relationship for k_l;

$$\left(\frac{k_l D}{d}\right)\left(\frac{\mu}{\rho d}\right)^a = K\left(\frac{D^2 S \rho}{\mu}\right)^{0.5} \quad (8)$$

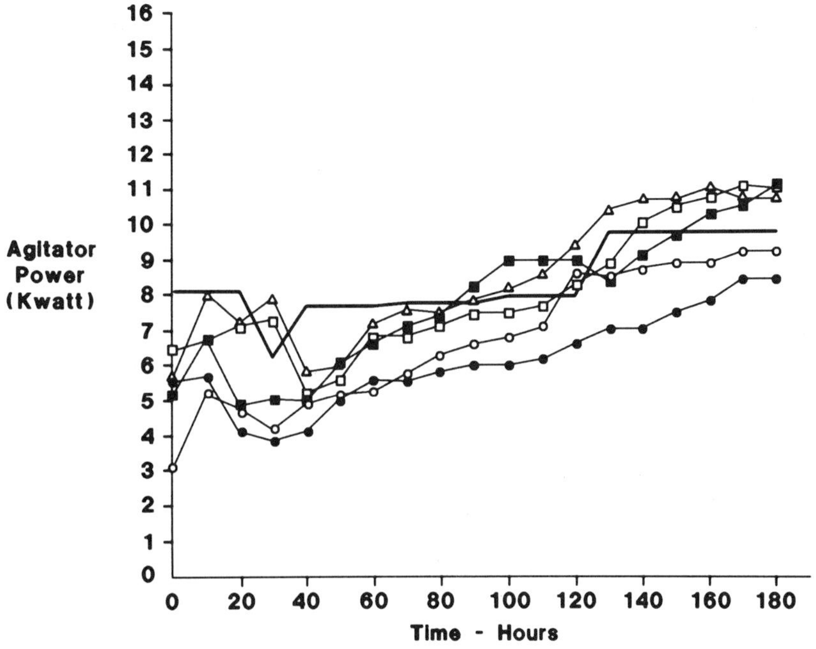

FIGURE 13. Agitator power demand versus aeration number for one to four equally spaced turbines at two values of liquid level above the top turbine.

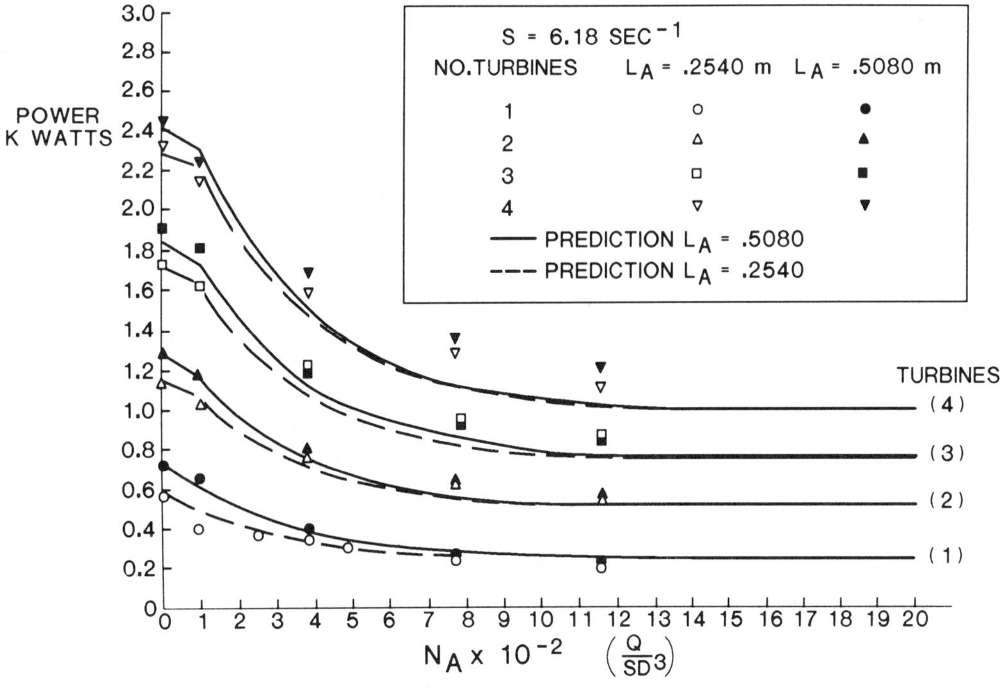

FIGURE 14. Power data for five identical fermentations at the 1000-gallon scale as a function of run time. The heavy solid line represents the prediction of the model (*12*).

and Calderbank (6) experimentally determined a relationship for the bubble area a:

$$a = c \left[\frac{P/V^{0.4} \ \rho^{0.2}}{\sigma^{0.6}}\right] \left(\frac{V_s}{V_t}\right)^{0.5} \quad (9)$$

From these two equations, it can be seen that bubble area is increased by high power input, high superficial gas velocity and low surface tension. On the other hand, k_l is increased by high Reynolds number and low viscosity. This means that a high shear turbine is needed for production of bubble surface area and a high pumping capacity turbine for mass transfer. When a single turbine is used, it must be capable of both, making the Rushton turbine ideal for mass transfer operations. What is not known is whether mass transfer correlates with agitation-aeration parameters in the same way for both upper and lower turbines.

In a multiple turbine system, if bubble coalescence occurs, each turbine may need to provide sufficient shear to maintain the bubble area. However, if coalescence does not occur, overall mass transfer might be enhanced by the use of a high pumping capacity agitator at significantly lower shear and power per unit volume. Studies of this question (see Gbewonyo et. al. (18)) are beginning to occur in the fermentation industry.

Equations 8 and 9 can be combined to give the following expression:

$$k_l a = C \left(\frac{P}{V}\right)^{0.4} V_S^{0.5} \ S^{0.5} \ X \quad (10)$$

$$\left[\left(\frac{D^2\rho}{\mu}\right)^{0.5} \left(\frac{\rho\, d}{\mu}\right)^{a} \left(\frac{d}{D}\right) \frac{\rho^{0.2}}{\sigma^{0.6}} \ \frac{1}{V_t^{0.5}}\right]$$

The first three terms of equation 10 are the terms most generally used in correlating $k_l a$ data. However, the exponents for P/V, V_s, and S vary widely in the literature making it difficult to arrive at any predictive relationship. In reality, these terms are of little use to the production engineer since none of the three terms can be varied over any significant range on large scale fermentors. The second set of variables has generally been ignored but should be studied in more detail.

Two terms in the second set are likely to have a major effect on mass transfer in real fermentations. The first term is the surface tension. As mentioned earlier, the percent hold-up is generally controlled by the addition of antifoam agent which increases surface tension, increases bubble coalescence, decreases bubble area and lowers $k_l a$. On the other hand, constituents of the medium, additions to the medium, and products of metabolism often decrease surface tension and increase bubble area and raise $k_l a$.

The second term of interest is viscosity. It has been shown (19) that viscosity can be correlated to the percent solids in a fluid by the relationship;

$$\mu_p = \mu_o \left[1 - \frac{\varepsilon_p}{0.4}\right]^{-1} \quad (11)$$

where μ_o is the liquid viscosity and ε_p is the percent solids in the liquid. When ε_p reaches 36%, the viscosity will start to rise rapidly and approach infinity at 40% solids. Such a rapid rise in viscosity would severely decrease mass transfer as indicated by equation 10. In addition to the effect of solids on viscosity, Frijlink and Smith (20) have shown that solid particles of 0.1 to 3mm can induce bubble coalescence and reduce bubble rise velocity.

Many industrial antibiotic fermentations produce mycelial aggregates or pellets in this size range and the percent solids is very likely to approach the 40% level. A sudden rise in viscosity, as would be suggested by Equation 11, is frequently observed. These phenomena, along with the presence of antifoam agents would indicate that the drop in $k_l a$ that occurs in many industrial fermentations may be due to the production of large quantities of cellular solids. The only way to counteract this is by diluting the broth with water or decreasing the strength of the medium.

For fermentations which have a high percent solids and high viscosity, it is likely that high shear impellers will continue to be required throughout the fermentor. The one exception might be the top impeller where low shear may permit bubble coalescence, reduce the need for antifoam and promote overall better $k_l a$ values. In low solids fermentations, lower power number turbines may be useful in saving power and promoting better mixing. Underlying these comments remains the fundamental question concerning the relationship between mass transfer and agitation-aeration parameters for multi-turbine tanks. It is likely that these questions will only be solved by the use of a modelling approach which subdivides the fermentor into multiple mixing stages so that the performance of each turbine can be more individually analyzed.

CONCLUSIONS

The development of a stage-wise modelling approach for analyzing multi-turbine fermentors has increased interest in understanding the mixing patterns, power distribution, and mass transfer relationships in large vessels. In the above discussion, it was pointed out that percent gas hold-up is not likely to be predictable for fermentation broths. Since percent hold-up is generally limited or controlled, it would be a useful parameter to measure if a simple measuring device could be developed. Routine measurements might provide new insights on its effect on other parameters.

The liquid flow rate between stages is related to impeller pumping, but no general method for predicting the exact relationship exists. The use of a simple acid addition method appears to be adequate for measuring the flow rate between stages and provides good agreement with published values. The liquid flow rate between stages only mildly affects the axial dissolved oxygen profile in a fermentor.

Many attempts have been made to predict power input for gassed and ungassed Rushton turbines. A correlation from a previous study (12) did not scale-up to production scale equipment. A set of rules are proposed which fit the earlier results. These have generally proven to hold true on large equipment.

Finally, studies need to be conducted to determine whether upper turbines affect mass transfer in the same way as bottom turbines. In addition, more work relating surface tension, percent solids, and viscosity might prove useful in providing the production engineer with a better method for on-line control of mass transfer in the fermentor.

NOMENCLATURE

a	bubble interfacial area, L^{-1}
C	dissolved oxygen concentration, mmole O_2 L^{-3}
C^*_{in}	D.O. in equilibrium with inlet O_2 partial pressure in gas, mmole O_2 L^{-3}
C^*_{out}	D.O. in equilibrium with outlet O_2 partial pressure in gas, mmole O_2 L^{-3}
d	diffusivity, L^2 T^{-1}
D, D_i	turbine diameter, L
D_T	tank diameter, L
F	liquid flow rate between stage L^3 T^{-1}
H	percent hold-up
k_l	liquid phase mass transfer coefficient, L T^{-1}
k_la	overall mass transfer coefficient, T^{-1}
K	constant
L_A	liquid level above turbine
L_B	liquid level below turbine
L_S	turbine spacing
N	stage number from bottom of vessel
N_A	aeration number
N_P	power number
N_Q	impeller discharge coefficient
OUR	oxygen uptake rate, mmole O_2 L^{-3} T^{-1}
P	Power, Kwatt
P_g	gassed power, Kwatt
P_o	ungassed power, Kwatt
Q	air flow rate, L^3 T^{-1}
Q'	fraction of impeller flow in axial direction
S	agitator speed, T^{-1}
t	time, T^{-1}
V(N)	volume of Nth mixing stage
V_S	superficial gas velocity, LT^{-1}
V_t	terminal bubble rise velocity LT^{-1}
ε_p	percent solids in liquid
Θ	holding time, T
μ	viscosity, ML^{-1} T^{-1}
μ_o	liquid viscosity, ML^{-1} T^{-1}
μ_p	suspension viscosity, ML^{-1} T^{-1}
ρ	density, ML^{-3}
σ	surface tension MT^{-2}

Literature Cited

1. Manfredini, R., V. Cavallera, L. Marini, and G. Donati, Biotechnol. Bioengr., 25, 3115 (1983).

2. Oosterhuis, N.M.G. and N.W.F. Kossen, Biotechnol. Bioengr., 26, 546 (1984).

3. Bader, F.G., "A Multi-stage model for oxygen transfer in large fermentors." Presented at the 188th National Meeting, ACS, Philadelphia (1984).

4. Bader, F.G., "Modelling mass transfer and agitator performance in multi-turbine fermentors." Biotechnol. Bioengr. In Press.

5. Cohen, J.D. and E.L. Gaden Jr., "Oxygen Transfer in Production Scale Bioreactor." Presented at the 190th National Meeting, ACS, Chicago (1985).

6. Calderbank, P.H., Transactions of the Institution of Chem. Engr., 36 443, (1958).

7. Richards, J.W., Progress in Ind. Microbiol. 3, 143 (1961).

8. Steel, R. and W.D. Maxon, Ind. & Eng. Chem. 53, 739 (1961).

9. Kovac L., and J.D. Gregory. Personal Communication.

10. Singh, V., W. Hensler, R. Fuchs, and A. Constantinides. On-line determination of mixing parameters in fermentors using pH transients. Paper 18. International Conference on Bioreactor Fluid Dynamics, BHRA, Cranfield, U.K. (1986).

11. Rushton, J.H., E.W. Costich, and H.J. Everett. Chem. Eng. Prog., 46 (9), 467 (1950).

12. Bader, F.G. "Mass transfer in a multiturbine fermentor. I. Gassed and ungassed power inputs." Paper No. 20 International Conference on Bioreactor Fluid Dynamics. BHRA, Cranfield, U.K. (1986).

13. Dickey, D.S. "Turbine agitated gas dispersion-power, flooding, and hold-up." Paper No. 116d. 72nd National Meeting, AIChE, San Francisco, (1979).

14. Nienow, A.W., D.J. Wisdom, and J.C. Middleton. "The effect of scale and geometry on flooding, recirculation, and power in gassed stirred vessels," Paper F1, Proceedings 2nd European Conference on Mixing, BHRA, Cranfield, U.K. (1977).

15. Nienow, A.W. and M.D. Lilly, Biotech. and Bioengr., 21, 2341 (1979).

16. Hicks, R.W. and L.E. Gates, Chem. Eng. July 19, 141 (1976).

17. Rushton, J.H., Chem. Eng. Prog. 47, 485 (1951).

18. Gbewonyo, K., D. DiMasi and B.C. Buckland. "The use of hydrofoil impellers to improve oxygen transfer efficiency in viscous mycelial fermentations." Int'l Conf. Bioreactor Fluid Dynamics, BHRA, Cranfield, UK, (1986).

19. Loh, V.Y., S.R. Richards, and P. Richmond. "Fluid dynamics and mass transfer in a three-phase circulating bed fermentor. International Conference on Bioreactor Fluid Dynamics. BHRA. Cranfield, U.K. (1986).

20. Frijlink, J.J. and J.M. Smith. Coalescence in three phase systems. International Conference on Bioreactor Fluid Dynamics, BHRA, Cranfield, U.K. (1986).

Flow Conditions in Vessels Dispersing Gases in Liquids with Multiple Impellers

JOHN M. SMITH*
MARIJN M.C.G. WARMOESKERKEN
ERIK ZEEF
Delft University of Technology
Delft, The Netherlands

This paper considers gas dispersion in mechanically agitated tall vessels.

Two phase hydrodynamic conditions have been studied in a 3:1 (H/T) tank of 0.64 m diameter with triple Rushton turbines dispersing air in water. A unified treatment links the operating regimes in vessels with multiple disc turbines to those in "standard" stirred tanks. The lowest impeller of a multiturbine agitator behaves as if it were operating alone. Radial gas distribution reduces the gas loading of the upper ones to about half that of the bottom one. The data can be presented in a simple generalised flow map based only on the Froude and Gas Flow Numbers. This is scale independent and provides a basis for the design of most forms of turbine agitated vessels dispersing gases in low viscosity liquids.

Gas-liquid technology has expanded in recent years as a result of the growth of interest in biotechnology in general and fermentation processes in particular. The need to ensure efficient utilisation of the gas supplied to a fermentor has led to a preference for taller vessels than are usual in more conventional chemical reactors.

The operation of individual Rushton turbine impellers in gas dispersion applications is now reasonably well understood, at least in vessels of standard geometry. In operation, ventilated cavities develop behind the blades. The form and configuration of these depends on the operating conditions, i.e. for a particular scale and geometry on gas flow rate and impeller speed. The hydrodynamic regime of the impeller controls the mixing performance of the whole reactor, as is reflected in liquid phase homogenization, retained gas fraction, bubble size distribution, gas-liquid mass transfer, heat transfer and the suspension of solids.

Fermentors are often designed with a height about three times the vessel diameter. In order to produce and maintain good gas dispersion, nutrient distribution and mass transfer conditions in a fermentor three impellers mounted on a single shaft are commonly used. This paper is concerned with the operation and performance of a 3:1 (H/T) tank with triple Rushton turbines dispersing air in water. New measurements are presented and placed in the context of the relatively limited information available in the literature, notably the papers of Hicks and Gates, (1976), Nienow and Lilley (1979), Kuboi and Nienow (1982), Kurpiers et al. (1984), Machon et al. (1985) and Roustan (1985).

The objective of the research reported here is to relate the flow regimes which develop in multi-impeller equipment to the operating and performance variables.

EQUIPMENT.

The investigation has been carried out in a 0.64m diameter tank with a total height of 2.2m, so that a liquid level at a 3:1 H/T ratio can be maintained. The equipment is shown in Fig.1. The tank is fitted with four symmetrical wall baffles of width 0.1T. Most of the results reported here refer to an impeller assembly of three six-blade Rushton disc turbines, with diameter 0.256m, i.e. D = 0.4T. For each turbine the blade height W = 0.2D, the length L = 0.25D and the disc diameter 0.75D. The blade thickness is 3mm and the shaft diameter 30mm. Power demand is determined with a Vibro-torque inductive transducer mounted in the shaft, between the DC drive motor and the impeller assembly. The lower end of the shaft is

* J.M. Smith is now at University of Surrey, Guildford, England

centred by a non-load carrying PTFE bearing. Liquid level is maintained constant by a continuous draw-off and recirculation system. This maintains the upper level of the dispersion constant, as is usual in industrial practice.

POWER DEMAND IN UNGASSED CONDITIONS.

There is some information in the literature concerning the power demand of multiple impellers mounted on a single shaft. Various authors are quoted by Kuboi and Nienow (1982) as suporting the concept that the power demanded by two impellers is twice that of a single impeller providing that the separation between the impellers is at least as great as their diameter. Fig.2 shows how the flow fields in tanks agitated by multiple impellers depend on their spacing. When they are close the fluid mass between the impellers tends to rotate as a forced vortex, and the energy of the single discharge stream is not much greater than that from an isolated impeller. When spaced further apart than their diameter separate circulation cells are established, and the total pumped capacity approaches that of twice that of a single impeller.

With the present equipment determining the power demand with various impeller spacings is straightforward. The results are given in Table 1 which sets out the number, n, of turbines on the shaft, the separation C_i between them expressed as a multiple of the impeller diameter, and the factor between the multiple and single Power Numbers as measured at Reynolds Numbers in excess of 300,000

TABLE 1

n	C_i/D	Po_n/Po_1
2	1	1.76
2	1.5	1.97
2	2.0	2.00
3	0.2	1.16
3	0.5	1.25
3	0.75	1.43
3	1.0	2.46
3	1.2	2.68
3	1.5	2.86
3	2.0	2.90
3	2.5	2.98

It is clear that if the impellers are closer than about 0.75D the additional power demand of the extra impellers is very modest indeed - always less than half that of separate impellers. Very similar conclusions were reached by Kurpiers et al. (1984). At about 1.0*D the loading of each impeller becomes much more efficient and the power demand rises steeply until at spacings of about 1.5*D the additional power demand is over 90% of the value in isolation. In view of the sensitivity of power to spacing in the range 0.5 to 1.25 impeller diameters, it is clearly desirable to avoid these.

A separation of 2.5*D (=T) has been used in the present work with dispersions of air in water.

GASSED IMPELLERS

Turbine hydrodynamics and dispersion performance are controlled by the ventilated cavities which develop behind the blades of the impeller, (Bruin et al., 1974). The various shapes of these cavities are determined either by the dynamics of the trailing vortices from the blade tips or by the characteristic flows behind bluff bodies. The general forms of the two main types, ´vortex´ and ´large´ cavities, are shown in Figs 3 and 4. With disc turbines there is a further unexpected phenomenon in that the large ventilated cavities are not all alike, but stable larger and smaller cavities are maintained behind alternate blades of an impeller with an even number of blades. A six blade Rushton turbine produces a 3-3 structure therefore. Disc turbines with an odd number of blades are unstably and asymmetrically loaded even at moderate gas rates and should be avoided, (Warmoeskerken et al. 1981).

A coherent classification of the hydrodynamic regimes developed by a single disc turbine can be given in terms of the Gas Flow Number, $Fl = Q_g/ND^3$ and the Froude Number, $Fr = N2D/g$. Generalised flow regime charts which predict the conditions in agitated vessels of H/T=1 have been successfully developed, Fig.5, (Warmoeskerken and Smith, 1986)

Flow Map Equations for Single Impellers

The various transitions represented on the flow map are well described by the following dimensionless equations:

a) the minimum stirrer speed for the impeller to hold any stable cavity structure, (van Dierendonk et al. 1968) which is given by (Warmoeskerken et al, 1981):

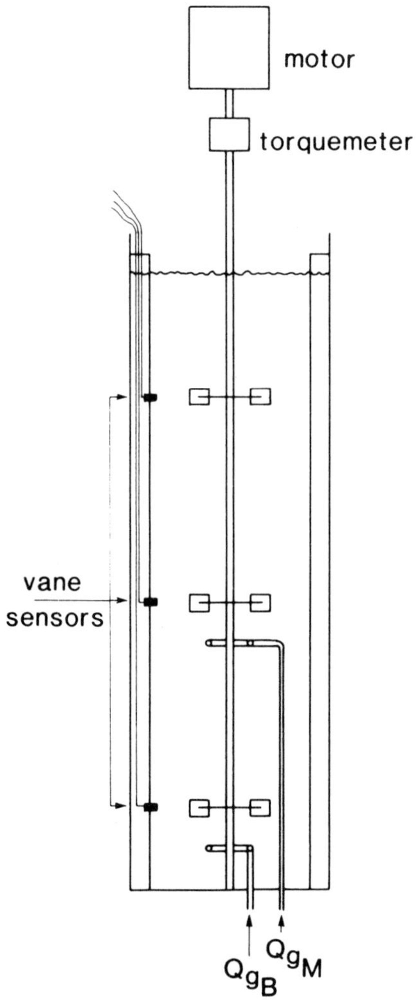

FIGURE 1. Experimental arrangement.

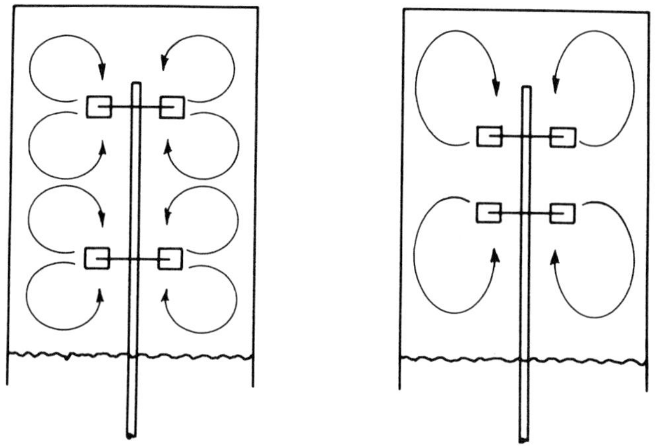

FIGURE 2. Flow fields in tanks with multiple impellers at various spacings.

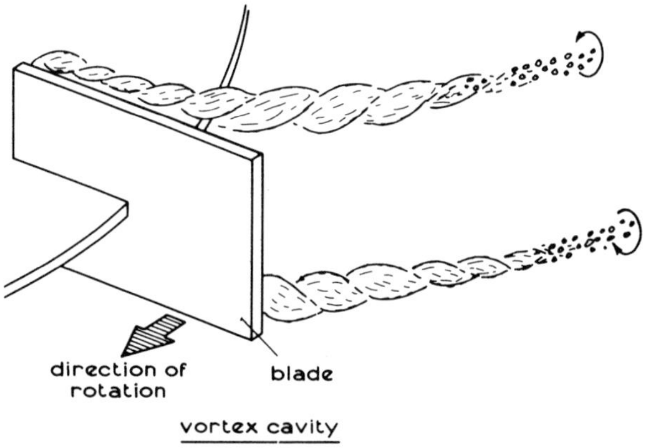

FIGURE 3. Vortex cavities behind a turbine blade.

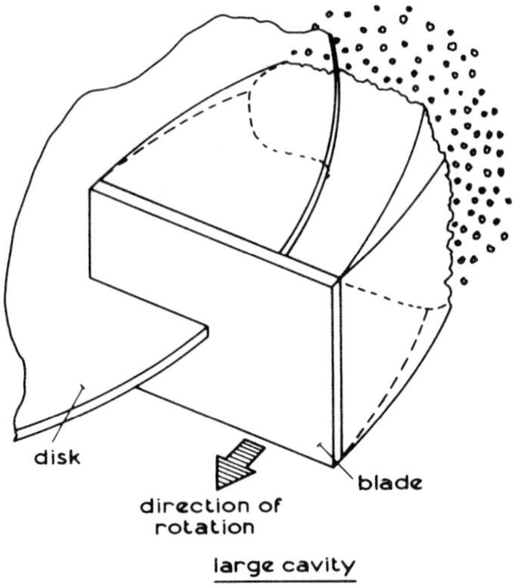

FIGURE 4. A large cavity behind a flat turbine blade.

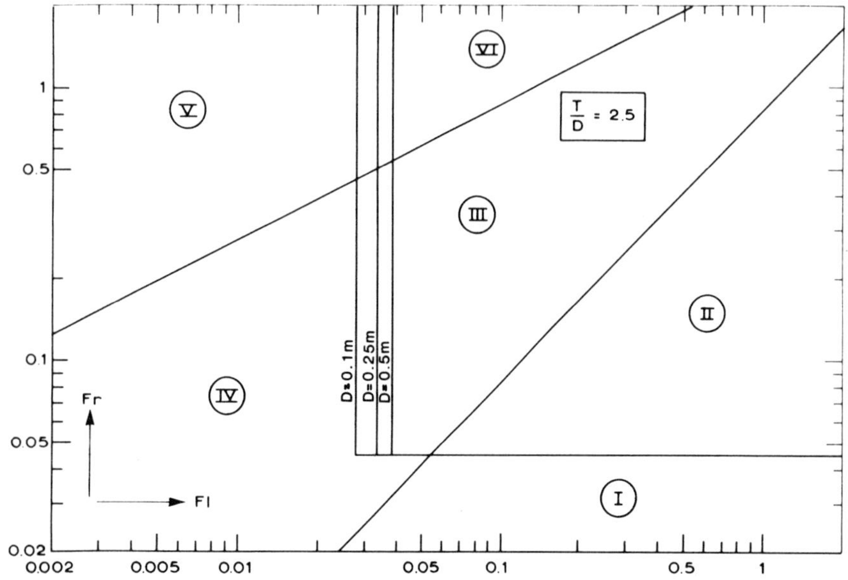

FIGURE 5. The flow map for 6-blade Rushton turbines. I, Bubble column; II, Flooded (ragged cavities); III, "3–3" structure; IV, Vortex-clinging cavities (I–IV with limited recirculation); V, Vortex-clinging cavities; VI, "3–3" structure (V–VI with fully established recirculation).

$$Fr = 0.045 \quad (1)$$

b) the transition from vortex/clinging cavities to the first large cavities (Warmoeskerken et al, 1981):

$$Fl = 3.8*10^{-3}\ (Re^2/Fr)^{1/15} * (T/D)^{0.5} \quad (2)$$

(The Re^2/Fr group here is a tentative attempt to describe the $D^{0.2}$ scale dependence of this phenomenon)

c) the maximum gas flow rate at which a given impeller speed will recirculate gas throughout a standard (T/D=1) vessel, (after Nienow et al. 1977):

$$Fl = 13\ Fr^2 * (T/D)^{-5} \quad (3)$$

d) the maximum gas rate that can be handled by a given impeller without flooding, (Warmoeskerken and Smith, (1985) and Nienow et al, (1985)):

$$Fl = 30\ Fr * (T/D)^{-3.5} \quad (4)$$

The gas dispersion regimes associated with the smaller vortex and clinging cavities and those of the more highly aerated systems are so different that it is reasonable to use different correlations for the prediction of mass transfer rates and gas fraction. The formation of the 3-3 cavity configuration of alternate larger and smaller cavities behind successive blades is the most crucial change-over affecting performance.

A flexible vane to which strain gauges are attached (Fig. 6) will easily detect the blade passage frequency if it is mounted in the outflow from a turbine. The presence of large gas cavities behind alternate turbine blades alters the nature of the discharge from the impeller. When the impeller is operating in the 3-3 regime unequal pulses are detected by the vane as the differently loaded blades pass by. A spectrum analyser can be used to give an unequivocal indication of the transition from 6-symmetric vortex or clinging cavities to the 3-3 alternate large - small cavities.

When the gas loading is excessive the impeller is said to be flooded. In these circumstances the buoyancy forces of the bubble column dominate the liquid circulation and the impeller ceases to pump effectively. When this occurs there is a change in the flow regime. The ventilated cavities become irregular, or 'ragged', and the stability of the 3-3 structure is lost.

Transitions with Multiple Turbines

The two phase hydrodynamics of this triple impeller geometry has been studied for the air-water system using this technique. Ring spargers below each turbine could be supplied independently with metered air flows. The mixing vessel was equipped with vane sensors at three points so that the hydrodynamic loading condition of the individual turbines could be examined.

As evident from equation (2) above, in single impeller installations the onset of the 3-3 configuration occurs at a given value of the gas Flow Number, with a slight dependence on the equipment scale. It is convenient to display data concerning the transition in terms of a graph of Q_g/D^2 versus ND, essentially the superfical gas velocity against impeller tip speed, Fig. 7. A straight line through the origin on this graph corresponds to a constant Gas Flow Number. The broken line which has been drawn on the figure is the established point of transition to the 3-3 structure in a standard vessel of this scale and geometry.

The conditions leading to the development of large cavities on the various impellers have first been studied introducing gas only through the lowest sparger, i.e. below the bottom stirrer.

Cavity development on the lowest impeller. In fig. 7. the points for the lowest impeller are seen to lie very slightly above the established correlating line for an isolated impeller. The difference, which with the transition being reached at Fl = 0.037, amounts to an increase of about 10% above that predicted by eq. (2), is not thought to be significant. It may reflect marginally increased liquid pumping by the lowest impeller in the multi-turbine system compared to that in a standard vessel with a free surface.

Cavity formation on the upper impellers. For a given impeller speed it can be seen from Fig. 7 that the development of large cavities behind the blades of the upper impellers occurs only at much higher gas superficial velocities than those which lead to the formation of the 3-3 structure on the lowest impeller. It is also evident that the 3-3

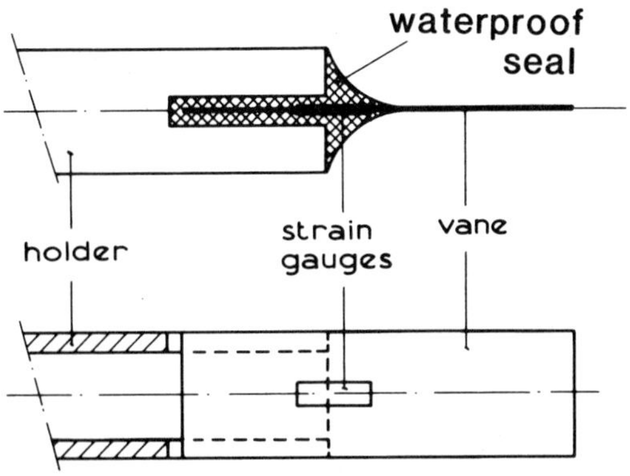

FIGURE 6. The vane sensor for detecting large cavity development.

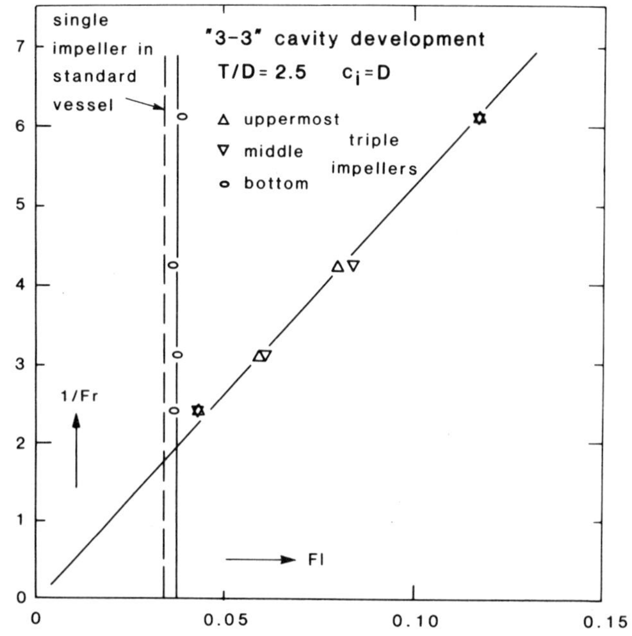

FIGURE 8. Large cavity development lines for lowest and upper turbines, expressed as Gas Flow Number against the reciprocal of the agitator Froude Number.

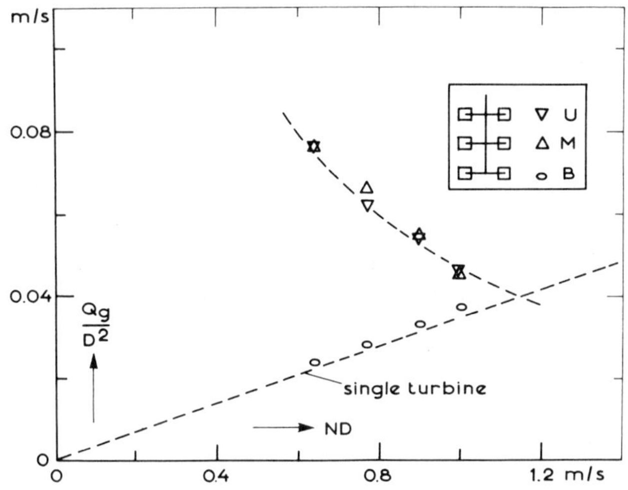

FIGURE 7. Large cavity development lines for lowest and upper turbines. (Essentially gas phase superficial velocity against impeller tip speed).

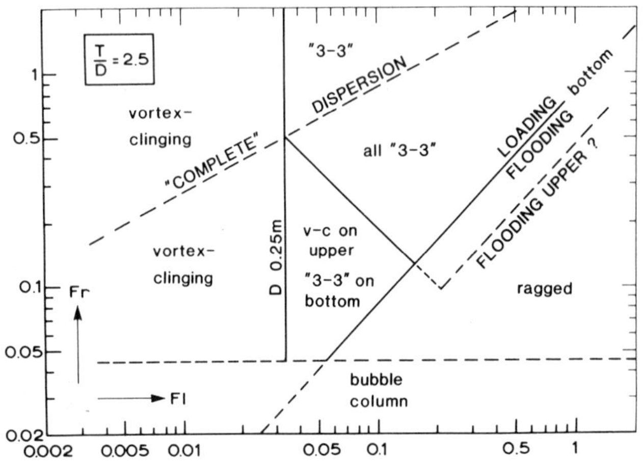

FIGURE 9. The flow map modified for multi-turbine vessels of $(T/D) = 2.5$

configuration develops on both the upper impellers at the same time.

The gas rising to the upper impellers is naturally more uniformly distributed over the tank cross section than is the case with the lowest one, at least at modest stirrer speeds. With these conditions the upper impellers actually pump less gas and more liquid and so can be expected to have a greater proportional power draw.

At higher stirrer speeds the transitions for the lowest and the upper two impellers come closer together. The fall in the Flow Number at which cavities form on the upper impellers is a result of the steadily increasing circulation in the upper cells, which serves to concentrate the gas stream arriving at the impeller and which would otherwise be distributed across most of the tank area.

It is evident from Fig. 7 that as the stirrer speed is raised the increasing gas recirculation allows large cavities to develop at reduced superficial gas velocities. The intersection of the curves describing the behaviour of the lowest and the upper turbines becomes even more precisely defined if the Gas Flow Number is plotted against the reciprocal of the Froude Number, Fig. 8.

As noted above, the cavity development transition line for the lowest turbine is at Fl = 0.037. The straight line defining cavity development on both the upper turbines is given by the equation:

$$Fr^{*}Fl = 0.019 \qquad (5)$$

The point of intersection of these lines, at Fr = 0.50, corresponds to an impeller tip speed of 4.38m/s

It is particularly interesting to note that eq (3), due to Nienow et al., predicts that in order to attain complete recirculation at this Gas Flow Number in a standard vessel, the Froude Number would have to exceed 0.52. This excellent agreement is a useful confirmation of the validity both of the equivalence of the normal operating conditions in the upper cells to full recirculation and of the concept of considering the tall fermentor geometry to be in many ways equivalent to a simple addition of standard vessels.

Using this information it is now straightforward to modify the flow map presented in Fig.5 to show conditions in a tall multiple impeller vessel, Fig. 9. This modified map should be used with caution for vessels with a T/D ratio other than 2.5 since there is no present information on the (probable) relevance of the geometry on this transition on the upper turbines. However we believe that using the lower cavity formation and recirculation intersection to locate the line describing the upper impeller transitions provides a sound basis. On this assumption eq.(5) for cavity development on all the upper turbines can be reformulated :

$$Fr^{*}Fl = 1.85\ (T/D)^{-5} \qquad (6)$$

Gas supply to the middle impeller. Determinations of the cavity transition were made with gas supply to the mid turbine both instead of and as well as to the lowest. The gas supply to the lowest impeller is held constant in each of the three cases shown in Fig. 10 with values of Q_g/D^2 of 0, 0.018 m/s and 0.037 m/s respectively. Cavity development is influenced by both flow rates.

With no gas from the bottom sparger, the points of transition lie slightly higher again than those for a single impeller or when gas supply is only through the lowest sparger. Once more this can be regarded as evidence for a small degree of assistance for the liquid pumping action from adjacent impellers. The vertical displacement corresponds to an increase of about 20% in terms of gas handling capacity at a given impeller speed.

Fig. 10 also allows us to estimate the fraction of ´primary´ gas which reaches the second (and by implication third) impeller. When gas is supplied to both lowest and mid impeller positions the curves are displaced. The extent of this shift corresponds to about 40% of the gas from the lowest cell actually entering the immediate impeller regions of those above. The remaining 60% apparently bypasses the mixed region and escapes upwards.

The picture that emerges is generally consistent with that of Hicks and Gates, (1976) and Nienow and Lilly (1979), though both these considered the operation of the upper impellers could be described in terms of a reduced local fluid density. A two region model of this form has also been considered by Nishikawa et al. (1984) as a basis for predicting mass transfer.

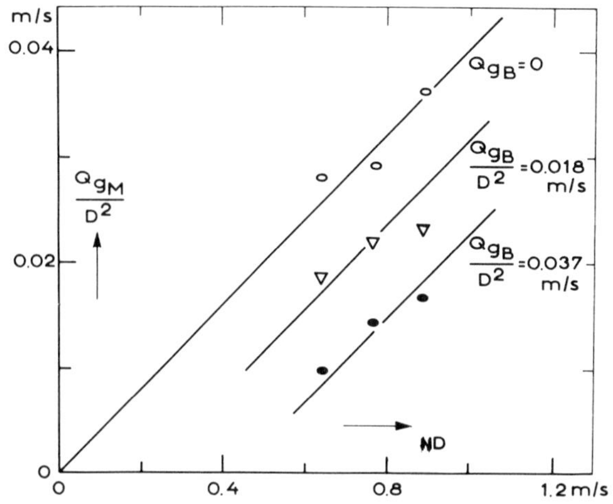

FIGURE 10. Cavity development lines for middle impeller with three given gas rates to the lowest impeller. (Essentially the total superficial gas velocity at which large cavities form on the middle impeller when some gas is supplied directly to it.)

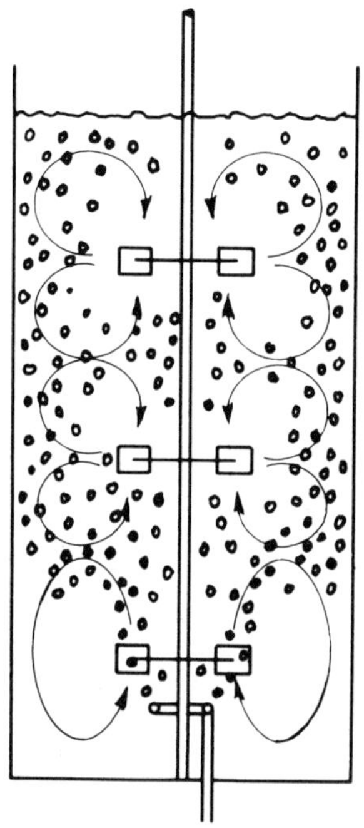

FIGURE 11. Flow field in vessel when the lowest impeller is flooded.

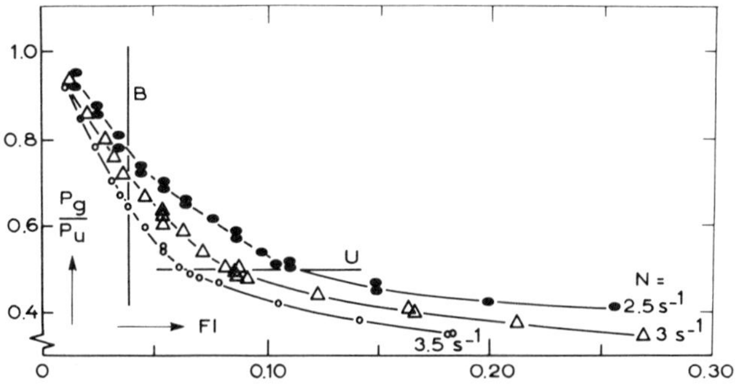

FIGURE 12. Power demand curves, all gas supplied to lowest turbine. Lines (B) and (U) are the points of cavity formation on lowest and upper impellers.

The situation is a little more complex when the recirculation is fully established below the lowest impeller since the hydrodynamic loadings of all the turbines are then similar.

It is very probable that a similar reduction of the gas loading of the upper impellers occurs under ´normal´ operating conditions, when gas is supplied through a single sparger. This explains the well known fact that when three similar disc turbines are mounted on a single shaft, the lowest impeller will flood long before the upper ones have ceased to function properly. The onset of flooding will completely change the flow field around the lowest impeller. Fig. 11 illustrates how the large scale circulation in the lowest part of the vessel will change when the bottom impeller is flooded. Although there is insufficient data to quantify the effect precisely, even a flooded impeller redistributes the gas somewhat and it is reasonable to expect that the loading/flooding transition of the upper turbines will occur at a Gas Flow Number about twice that which floods the lowest. This would be consistent with the flooding relationship suggested by Roustan et al. (1985) on the basis of measurements done in equipment in which the gas supply was distributed over the whole of the base of the vessel, and also supports the conclusion of Rushton et al. (1956) that there are advantages in using multiple impellers when a large gas handling capacity is required. The implications of this for the flow map can be judged from the dashed line which is included on figure 14 to indicate approximately where the upper impeller flooding transitions might reasonably be expected to lie.

The redistribution effect can be exploited by arranging for the gas to be introduced at individual spargers below two or more impellers - as is done in the present experimental rig. A simpler arrangement is to use a somewhat larger impeller as the lowest. Gas handling capacity increases rapidly with impeller diameter; the Fr/Fl relationships imply a 4.5 power dependence for the flooding transition. If the lowest impeller is about 15% larger than the upper ones, the design will be reasonably in balance.

POWER DEMAND IN MULTIPLE IMPELLER SYSTEMS.

The correspondence between power draw and hydrodynamic condition of a single Rushton Turbine is well established, (Warmoeskerken et al. 1981). The fall in gassed power demand is more complex when multiple impellers are involved since the transitions do not occur at the same operating pointq on the lowest and the upper impellers. Fig. 12 shows the power demand for the present equipment at three different shaft speeds.

The successive stages in the loading of the various impellers as the Gas Flow Number is increased in this tank have been described above. The vertical line, (B), on Fig. 12 is the point at which the lowest impeller develops the 3-3 cavity structure. The transitions on the upper impellers occur, as given by eq. 5 above, and these points are joined by the line (U). This latter line is virtually horizontal, reflecting a remarkable constancy in the relative power demand.

The form of the power drop differs in detail from that of a single turbine as a result of involving differing flow regimes for the various impellers comprising the agitator assembly. The power demand is essentially the kinetic energy imparted to the liquid by the agitator. Since the pumping action of a gassed impeller is less efficient than that of an unaerated one, in a multi-turbine assembly the distribution of power draw will depend on the hydrodynamic conditions of the individual turbines.

Without attempting a quantitative description of the composite power demand curve, it appears that in the region between the lines (B) and (U) on Fig. 12 the power falls almost linearly with increasing gas flow. This contrasts with the similar curves for a single impeller which have a point of inflection at the moment of large cavity development and distinct curvature in the power draw relaionship in that region. (Warmoeskerken 1986)

CONCLUSIONS.

The criteria for the establishment of the various hydrodynamic regimes on the upper and bottom impellers of a tall tank (H/T = 3) have been established and related to the known behaviour in a standard tank (H/T = 1). Cavity development behind the blades of the upper impellers is controlled by recirculation in the relevant section of the vessel. The flatter form of the gassed power curve in multiple impeller vessels has been explained in terms of the different loading conditions on the upper impellers.

The hydrodynamic information has been presented in a generalised flow map which is valid for all vessels agitated with multiple Rushton turbines dispersing gases in low viscosity liquids. The flow map is based only on the Froude and Gas Flow Numbers and is expected to apply to vessels of any scale.

Notation.

a	Specific interphase contact area	m^2/m^3
C_i	Distance between impellers	m
D	Impeller diameter	m
g	gravitational acceleration	m/s^2
H	Liquid height	m
$k_l a$	Mass transfer factor	s^{-1}
L	Impeller blade length	m
n	Number of impellers on shaft	-
Po_n	Power demand (n impellers)	W
Po_1	Power demand, single impeller	W
T	Tank diameter	m
Q_g	Gas flow rate	m^3/s
W	Impeller blade (vertical) width	m
ϵ	gas fraction	(-)
ρ	liquid density	kg/m^3
μ	liquid viscosity	Pa s

Dimensionless groups.

Fl	Gas Flow Number,	$= Q_g/ND^3$
Fr	Froude Number,	$= N^2D/g$
Re	Impeller Reynolds Number	$= ND^2\rho/\mu$

References

Bruijn, W., Riet, K. van´t, and Smith, J.M., Trans. Inst. Chem. Engrs. (London), 52, 88 - 104, (1974).

Dierendonck, L.L.van, Fortuin, J.M. and Bos, D. van der, Proc 4th Eur. Conf. on Chem. React. Eng. Brussels, 205 - 217, (1968)

Hicks, R.W. and Gates, L.E., Chem.Eng., 141, July 19, (1976).

Kuboi. R. and Nienow, A.W., Proc. 4th. Eur. Mixing Conf. Noordwijkerhout, Netherlands. (BHRA, Cranfield, U.K.), Paper G2. 247-261, (1982).

Kurpier, P., Steiff, A. and Weinspach, P.M., VDI-GVC Working Party meeting, Progress in Mixing, (1984), (unpubl).

Machon, V., Vlcek, J. and Skrivanek, J., Proc. 5th. Eur. Mixing Conf., Würzburg, W.Germany, (BHRA, Cranfield, U.K.), Paper 16, 155-169, (1985).

Nienow, A.W. and Lilly, M.D., Biotech. and Bioeng. 21 2341, (1979).

Nienow, A.W., Wisdom, D.J. and Middleton, J.C., Proc. 2nd. Eur. Mixing Conf. (BHRA, Cranfield, U.K.), Paper F1, 1-16, (1977)

Nienow, A.W., Warmoeskerken, M.M., Smith, J.M. and Konno, M., Proc. 5th Eur. Mixing Conf. (BHRA, Cranfield, U.K.), 143-155, (1985)

Nishikawa, M., Nishioka, S. and Kayama, T., J.Chem. Eng. Japan, 17, 541-543, (1984).

Roustan, M., Proc. 5th. Eur. Mixing Conf., Würzburg, W.Germany, (BHRA, Cranfield, U.K.), Paper 14. 127-141, (1985).

Rushton, J.H., Gallagher, J.B., and Oldshue, J.Y., Chem.Eng. Prog. 52, 319-323, (1956)

Warmoeskerken, M.M. and Smith, John M., 3rd World Congress of Chemical Engineering, Tokyo. Paper 8k-201, p350 (1986)

Warmoeskerken, M.M., Feijen, J. and Smith, J.M., Inst. Chem. E. Sympos. Ser. 64, J 1-14, (1981).

Warmoeskerken, M.M. and Smith, John M., Chem. Eng. Sci.,40, 2063 - 2071, (1985).

Warmoeskerken, M.M., Gas-Liquid Dispersing Characteristics of Turbine Agitators. Thesis, Delft University NL, (1986).

Aerated and Unaerated Power and Mass Transfer Characteristics of Prochem Agitators

G.J. BALMER
I.P.T. MOORE
A.W. NIENOW
Chemical Engineering Department
University of Birmingham
P.O. Box 363, Edgbaston
Birmingham B15 2TT, United Kingdom

Gas–liquid hydrodynamics and power consumption have been studied using a Prochem hydrofoil agitator in water and sodium carboxymethylcellulose (CMC) solutions in a 0.3 m diameter protofermenter. Some k_La data based on a steady-state technique using respiring yeasts are presented and a brief comparison is made with a Rushton turbine and a pair of Intermig agitators. The work shows that: (*a*) in the turbulent region, Prochem and Intermig both have lower power numbers than Rushton (~1.0 and ~0.7—for the pair—compared to ~5); (*b*) the fall in power on aeration is less with the Prochem and Intermig compared to the Rushton; (*c*) under certain conditions and especially with CMC, the Prochem may operate in regions of torque and flow instability as found with many downward pumping agitators; (*d*) large agitator vibrations are found with Intermig agitators at certain speed ranges confirming what has previously been reported; (*e*) in water, very similar k_La values for all three agitators are found at equivalent power and aeration rates.

Traditionally disc (Rushton) turbines have been used to agitate production scale non-Newtonian fermentation broths. These provide high shear in the impeller region and relatively good oxygen transfer, but have high power requirements.

Increasing energy costs and the desire for higher fermentation yields (higher oxygen demand) has led to the search for impeller designs which improve the oxygen transfer efficiency (k_La for a given specific power input). The Prochem impeller (Fig 1) is of a novel hydrofoil design and has been shown to improve oxygen transfer efficiency by 20% on the production scale (1), and 30-40% on the small scale (1,2). It was assumed that this benefit was due to an improvement in the bulk mixing provided by the Prochem impeller.

This work is a detailed study of the Prochem impeller in a small tank to identify the physical mixing processes which are responsible for the reported improvement in mass transfer efficiency. To date the two-phase bulk flow, impeller hydrodynamics, and gassed and ungassed power consumption have been studied using existing equipment and techniques. The results and comparison with a standard disc turbine and a dual Intermig combination are presented herein.

EXPERIMENTAL

A flat-bottomed cylindrical Perspex vessel of 30cm diameter (T) was surrounded by a square cross-section Perspex water jacket to give distortion-free flow visualisation. The tank was fitted with 4 baffles of width T/10, and the working liquid height was T.

The Prochem impeller of diameter T/2 has 5 hydrofoil blades and a large open-bottomed hub (Fig 1). The impeller was shaft mounted at a clearance from the tank bottom of T/4. A point sparger was centred below the impeller. Impeller speeds and gas rates up to 1000 rpm and 4 vvm (volume of air per minute per volume of liquid in the vessel) respectively were used.

Unfortunately, the de-rotational prism technique which has been successfully used to observe the cavities of a disc turbine (3) cannot be used for a downward pumping impeller since the cavities are not visible from the tank bottom. Hence, cavity observation was achieved by using a stroboscopic light at the side of the tank, and operated at 5 times the impeller speed (13). Power was measured by a shaft mounted strain gauge and telemetry equipment (3).

The fluids used were tap water and solutions of sodium carboxy methyl cellulose (CMC), with the make-up of the latter given by Solomon (4). The CMC is a non-Newtonian fluid, shear-thinning and viscoelastic but only mildly so at the concentrations used here. It has the advantage of transparency for flow visualisation. The shear stress-shear rate flow curves for the CMC were obtained using a Contraves cup and bob rotary viscometer. The rheological data was fitted to the power law model, i.e., $\tau = K \gamma^n$ and is shown in Table 1.

The method of Metzner and Otto (5) was used to account for the spatial variation of shear rate throughout a shear-thinning fluid:

$$\dot{\gamma}_{AV} = k_s N$$

and a value $k_s = 11.5$, that generally used for a disc turbine, has been used for the Prochem impeller too. Hence the Reynolds Number becomes:

$$Re = \frac{\rho N^{2-n} D^2}{K k_s^{n-1}}$$

RESULTS AND DISCUSSION

Two Phase Bulk Flow

Water. The two-phase bulk flow for a disc turbine in water at constant gas rate and increasing impeller speed is well established (6). For a hydrofoil impeller with an axial flow component a very different bulk flow exists. The following stages are observed with the Prochem impeller (Fig. 3) in which N(a)<N(b)<N(c)<N(d);

(a) Little or no dispersion. Large bubbles rise directly through the impeller and to the surface.

(b) Gas dispersion just touches tank bottom. Circulation is from the centre outwards across the tank bottom, with some bubbles rising to the surface without further circulation and others recirculating either into the impeller or just away from the impeller tips. The tank upper region acts as a bubble column, with bubble-free zones close to the wall and surface. Very few large bubbles are present.

(c) More vigorous recirculation with random upward swirls of bubbles. Recirculation is now from above the impeller, with the upper tank still acting as a bubble column though reduced in volume.

(d) At still higher speeds, the upwards bubble swirls become sufficient to induce two-phase mixing (including downflow of gas) in the upper tank, e.g., at 1vvm, N ~ 600rpm.

Between stages (b) and (c), large cavities form (see Fig 4(e)), but the effect on gas dispersion is only an initial slight disruption to circulation. The transitions between the 4 stages are very gradual with no distinct flooding-loading or complete dispersion point as for a disc turbine (as defined by Rushton and Bimbinet (7) and by Nienow et al (8)). The Prochem impeller is obviously flooded at low speeds (Fig 3, stage (a)), but the complete dispersion condition defined as circulation throughout the tank (6) has no equivalent. Stage (b) can be considered as complete dispersion below the impeller, and the speed required is only a weak function of the gas rate, e.g., at 0.5 vvm, 195 rpm; 1.0 vvm, 205 rpm. At constant speed, considerably higher gas rates are required to cause flooding.

Gas from the sparger can reach a downwards pumping impeller by two routes and is important in determining the gas dispersion and the mechanism of cavity formation (see later). For a given speed, at low gas rates the bubbles are swept downwards by the pumping action of the impeller and any gas which enters the impeller does so via recirculation (indirect loading). At higher gas rates, the bubbles rise directly into the impeller (direct loading). Vice-versa applies for constant gas rate and increasing speed. Warmoeskerken et al (9) describe impeller loading for a pitched blade impeller pumping downwards and a ring sparger of smaller than impeller diameter.

The point sparger feeds gas directly into the open-bottomed hub of the Prochem impeller. Thus, the bottom of the hub acts as a ring sparger of the same diameter and clearance from the tank bottom; when directly loaded the bubbles which leave the hub are large, few in number and randomly distributed. The impeller direct-indirect loading transition occurs between stages (a) and (b) of Fig 3.

Nienow et al (6) observed a certain degree of hysteresis in flow patterns for a disc turbine on decreasing speed at constant gas rate. This was not seen with the Prochem impeller, though the gradualness of the stage transitions would make all but a large hysteresis effect unnoticeable.

TABLE 1
Rheological Data for CMC Solutions (Hercules 7H4F)

CMC Concentration % by wt.	K $N\ m^{-2}\ s^{n}$	n
0.1	0.11	0.64
0.4	0.73	0.59
0.8	4.64	0.43

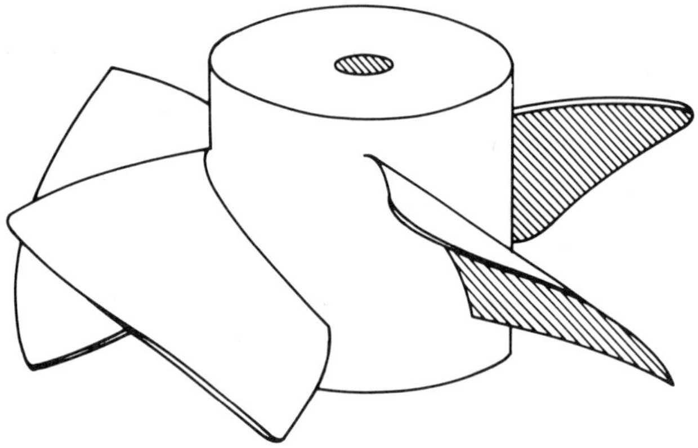

FIGURE 1. Five-blade Prochem impeller.

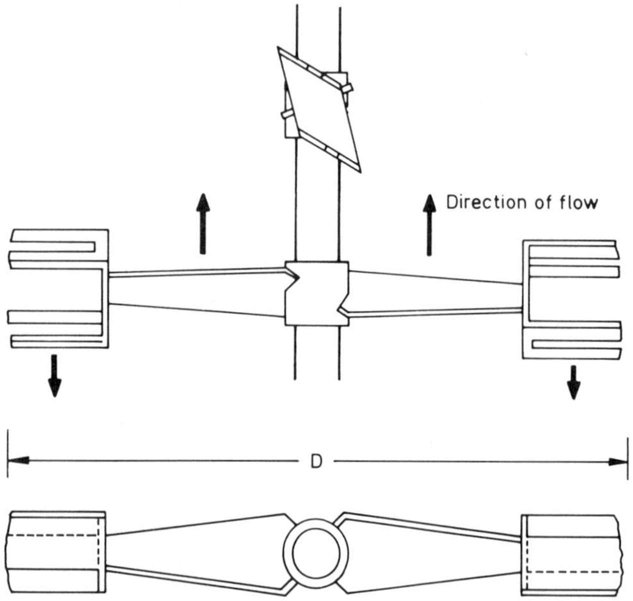

FIGURE 2. Dual Intermig combination (two identical Intermigs shaft mounted at an angle of 90° to each other; impeller separation = $T/2$; clearance from tank bottom of lower impeller = $T/4$; both impellers pump in the same direction, usually inner blades upwards as shown).

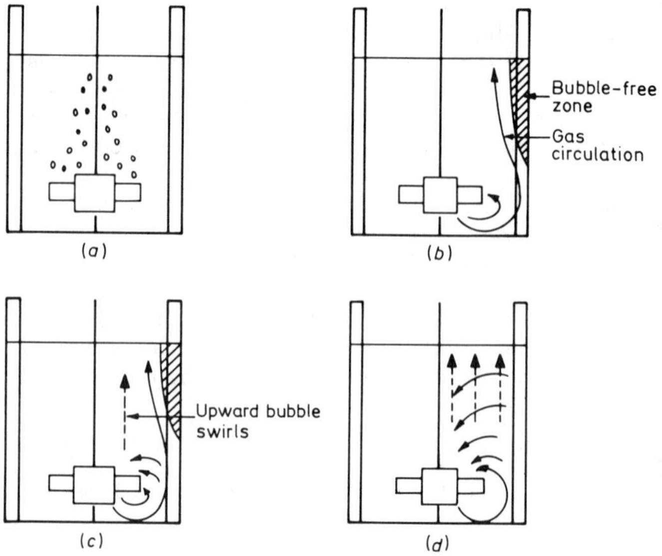

FIGURE 3. Two-phase bulk flow for Prochem impeller in water at constant $Q(N(a) < N(b) < N(c) < N(d))$.

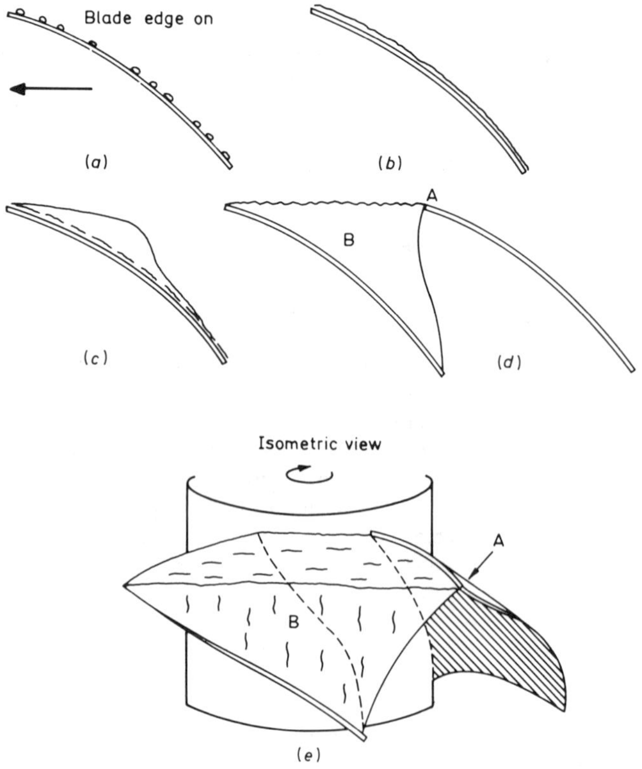

FIGURE 4. Cavity formation for Prochem impeller in water at constant $N(Q(a) < Q(b) < Q(c) < Q(d) < Q(e))$.

CMC. The two-phase bulk flow in 0.4% CMC is characterised by a wide bimodal bubble size distribution. Many tiny bubbles follow the liquid flow with a consequently long residence time, and a few larger bubbles of high buoyancy rise rapidly (10,11).

Gas dispersion by the Prochem impeller is similar to that in water, except that stage (b) of Fig 3 is dependent on the formation of cavities when the first significant gas circulation below the impeller occurs. Between stages (b) and (d), there is a range of speeds when the bulk flow is slow and is periodically interrupted by surging, giving a well-dispersed condition. The start of surging is gradual and it increases in magnitude but decreases in frequency until stage (d) is reached (a distinct point, e.g., at 1.0 vvm, 840 rpm) with a corresponding increase in the gas circulation rate into the impeller region from below.

Hickman (12) used a disc turbine (D=0.5T, T=30cm) in CMC solutions and also observed gas dispersion similar to that in water at low concentrations (≤0.5%) but with the formation of a cloud of bubbles within the cavern or well-mixed region of the tank at higher concentrations (≥0.8%). The Intermig design (Fig 2) achieves a combination of both upward and downward pumping action and hence a complex liquid flow pattern. The two-phase bulk flow has not been reported, but Hickman (12) concluded that the dual Intermig combination (D=0.58T, T=30cm), even at high speeds, does not provide as good gas-liquid mixing as other large diameter dual impeller systems in viscous fluids.

Impeller Hydrodynamics

Water. The formation of gas-filled cavities in the low pressure region behind the impeller blades has been extensively studied since the original work of Nienow and Wisdom (13) and van't Riet (14). Fig 4 shows the development of cavities in water for the Prochem impeller at constant speed (300 rpm) and increasing gas rate such that $Q(a)<Q(b)<Q(c)<Q(d)<Q(e)$;

(a) At a low gas rate, a number of small bubbles become attached to the back face of the impeller blade.

(b) The discrete bubbles become a thin vibrating gaseous layer.

(c) This gas layer billows outwards towards the following blade, but is unstable and often reverts to the layer form.

(d) At a small increase in gas rate, the cavity elongates and connects the upper edges of the blades along the majority of the blade width. The cavities are unstable and vibratory when formed.

(e) The cavity enlarges by the outward movement of region B and the lowering of point A down the blade face. The cavities are still vibratory.

Visual observation showed all 5 cavities to form simultaneously, but could not determine if they were all alike or in a dynamic configuration. Smith and Warmoeskerken (15) concluded that it was the influence of a disc on gas distribution which results in a dynamic configuration for a disc turbine with an odd number of blades, with all cavities alike for a hub mounted impeller. Whilst this seems valid for when cavities are formed from directly sparged gas, the Prochem impeller is indirectly loaded at the first formation of cavities. A considerable hysteresis was found with the Prochem impeller on decreasing the gas rate at constant speed (Fig 4, stages (e) to (a)).

An important difference between the Prochem impeller and the disc turbine is that, for the former, the gas rate required for the formation of cavities decreases with increasing speed (up to ~400 rpm when the gas dispersion makes visual observation of the cavities impossible). The opposite relationship holds for the disc turbine (15). A contributory factor to the different behaviour of the Prochem impeller is that the cavities are formed entirely by recirculating gas and hence the higher the speed, the greater the proportion of sparged but recirculated gas which enters the impeller.

CMC. For water at 300 rpm, the Prochem impeller is indirectly loaded on the formation of cavities. In 0.4% CMC, the impeller is directly loaded at 300 rpm. Thus the development of cavities is unlike Fig 4. At low gas rates, a low frequency pulse of gas rings the hub and partially covers the blade back faces. Eventually full cavities are formed and are the same shape as in water, but less vibratory. As the gas rate is further increased, large bubbles break off the tops of the cavities and become larger and more frequent on increasing gas rate. At higher speeds, the Prochem impeller is indirectly loaded and one may expect a different development of the cavities; but at such speeds the cavities are formed at very low gas rates. When the sparged gas is turned off, the cavities gradually diminish in size until they disappear. This process takes only 20s at 550 rpm, with the cavities immediately collapsing at 350 rpm. On shutting off the gas in a disc turbine agitated non-Newtonian fluid, the cavities remain either indefinitely (10) or slowly disperse depending on the viscoelasticity and yield stress of the fluid (4).

Ungassed Power Number

Fig 5 is a standard power number-Reynolds number plot. Direct comparison with the disc turbine Newtonian curve of Bates et al (16) is difficult because several workers (10,17,18) have shown that, in the transition regime, the power number for this impeller in viscoelastic fluids is considerably less than for inelastic ones. Therefore data from Hickman (12) for a disc turbine (D=0.5T, T=30cm) in CMC is also given in Fig 5. Data from Allsford (19) for the dual Intermig combination (D=0.6T, T=61cm) is provided, based on his experimentally determined value for the Metzner-Otto constant, k_s, of 16 compared to the usual literature value of 11.5 for both the disc turbine (12) and the assumed value of 11.5 for the Prochem impeller. The value of k_s does not significantly alter the shape of the curve but shifts the data to a slightly higher Reynolds number for a higher value of k_s.

At high Re, Nienow et al (17) found the curves for Newtonian and slightly shear-thinning CMC to be in reasonable agreement and this is true for the Prochem impeller in 0.1% CMC and water. The ungassed power number in the turbulent regime for the Prochem impeller is 1.0 (1.1 found by Gbewonyo et al (2)) compared to ~5.0 for the disc turbine (16)).

Since Bujalski et al. (20) reported Po_u for disc turbines to be a function of scale and disc thickness, very precise comparison with other work on the literature is of doubtful value. However, they also reported Po_u increasing to a maximum and then falling again, the change being of the order of 5% confirming earlier reports of this finding. No such variation was observed with the Prochem impeller, although a fall in Po_u was found at the onset of surface aeration (21) at ~500 rpm in water and at higher speeds in progressively more concentrated CMC solutions.

It is more important to study the transition regime of Fig 5 as the apparent viscosities encountered here are closer to those in a fermentation broth (2). The very different shapes of the curves for the Prochem impeller and disc turbine result in the initially large difference in Po_u (at high Re) decreasing rapidly, and at Re < 100, the Po_u for the Prochem impeller is greater. The Intermig combination and the Prochem impeller both show a steady fall in Po_u on increasing Re but the Prochem impeller power consumption is always greater.

Effect of Aeration on Power Demand

Gassed power data can be plotted as Po_g - Fl_g at either constant impeller speed or constant gas rate. Both curves can be related to different aspects of the bulk and impeller hydrodynamics.

Water. Firstly, consider the gassed power curve at constant speed for the Prochem impeller (Fig 6). At low speeds (200 and 255 rpm), there is a gradual fall in the power drawn due to an increasing amount of recirculating gas in the impeller region, although at no gas rate are cavities fully formed (see Fig 4(c)). Nienow et al (8) found a step change in gassed power at the flooding point for a disc turbine in water; for D/T < 0.4, a step increase on increasing gas rate; and for D/T = 0.5, a step decrease. The cavity configuration prior to flooding was different for the two impeller-to-tank diameter ratios. The flooding condition was observed using the Prochem impeller at both 200 and 255 rpm (at ~3.5 vvm) but there is no corresponding step change in the power drawn.

At higher speeds, the steepest portions of the gassed power curves correspond to the visually observed point of large cavity formation (see Fig 4(e)). Although the impeller region is not visible due to the gas dispersion at higher gas rates, the shapes of the curves are attributable to the cavity structure and the effect it has on streamlining the impeller and reducing the pumping capacity. Hence, larger cavities are formed at higher speeds and a maximum cavity size is reached at all but the highest speed (750 rpm).

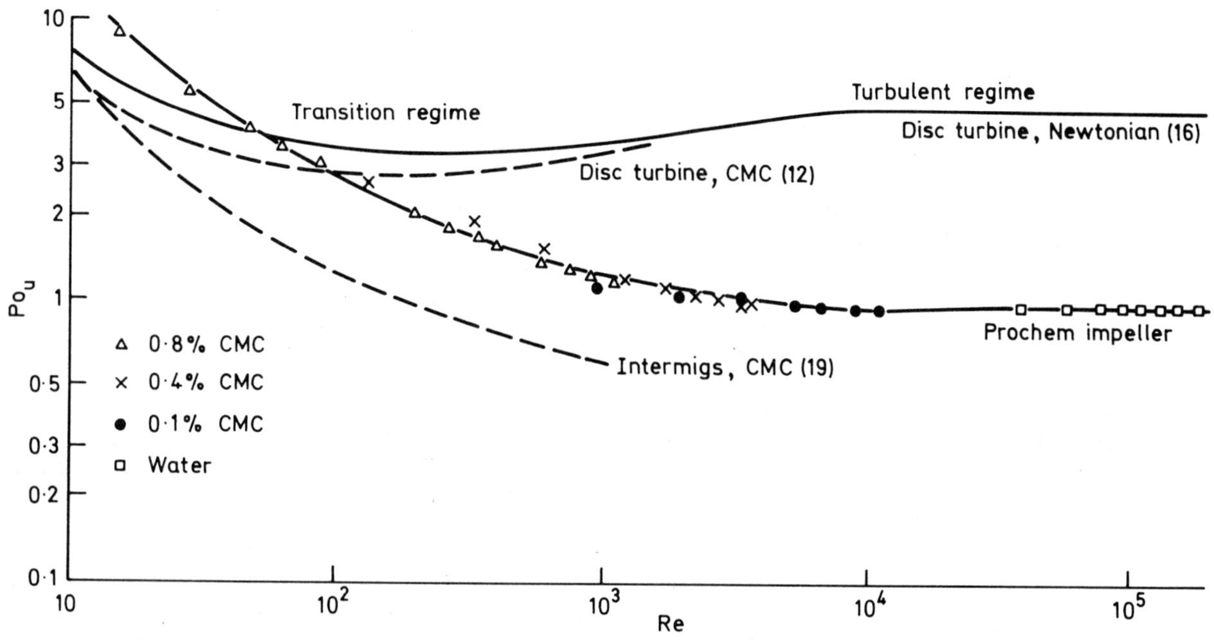

FIGURE 5. Po_u–Re for Prochem impeller, disc turbine, and dual Intermig combination.

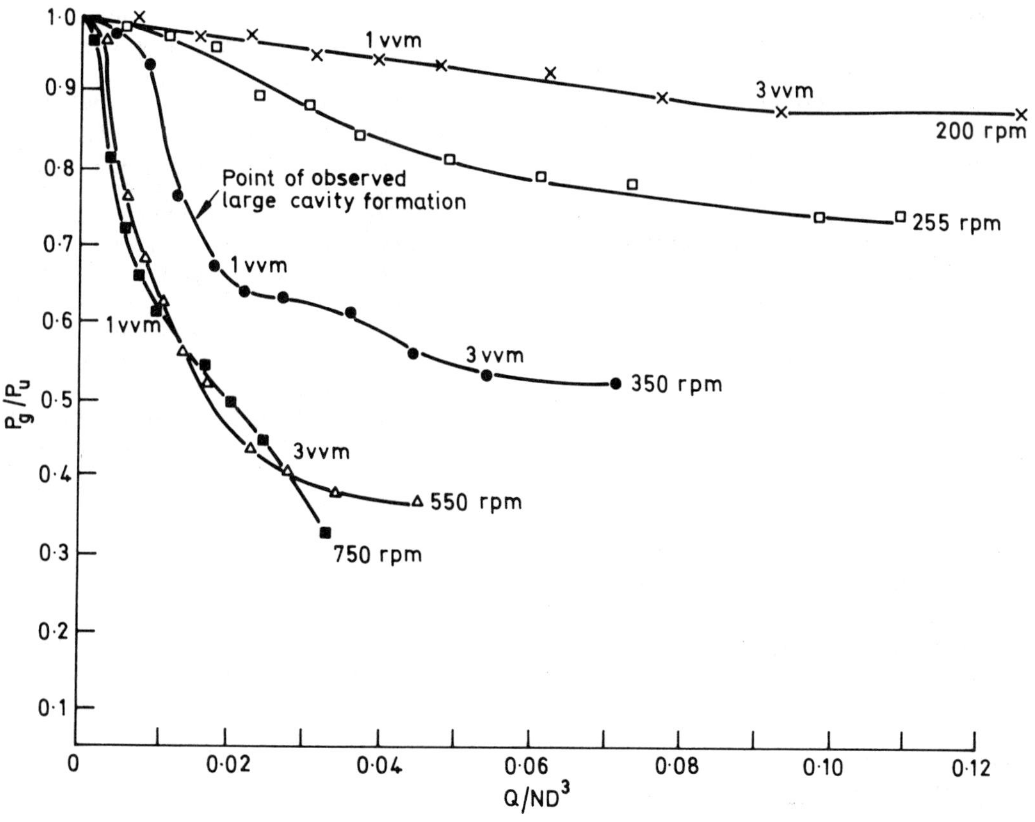

FIGURE 6. P_g/P_u – Fl_g at constant N for Prochem impeller in water.

The formation of large cavities corresponds to torque fluctuations. These are larger on initiation (±20% about mean) and then slowly decrease but are always present thereafter, suggesting that the cavities remain unstable after formation, perhaps due to the downflow of water between the impeller blades as it recirculates.

In Fig 7, the gassed power curves at two speeds are compared with data (22) for a disc turbine (D=0.4T, T=44cm). At 350 rpm and when cavities are fully formed, the power loss on gassing is significantly lower for the Prochem impeller. The reduction in the pumping capacity of the Prochem impeller may thus be less than for the disc turbine. The cavities only bridge the blades at the blade upper edges (see Fig 4(d) and (e)) and the angle of the blades is such that the hindrance of the cavities to the pumping action of the impeller is smaller than for 90° blades.

This confirms the findings of Tyler et al (1) who found a three Prochem impeller system to have a lower power loss on gassing compared to disc turbines on both the small scale (400 l) and in a production scale fermenter (65 m^3).

Fig 8 shows Po_g-Fl_g at constant gas rate (1.0 vvm) for the Prochem impeller. The discontinuity at a low speed corresponds well with the visually observed transition from direct to indirect impeller loading. Again, the steepest portion of the curve is due to the formation of large cavities, and at higher speeds the increase in size is small. Nienow et al (6) give the characteristic gassed power curve at constant gas rate for a disc turbine in water. It indicates a minimum Po_g with the cavities tending towards the vortex structure at higher speeds. Fig 8 for the Prochem impeller shows only a continual increase in cavity size and is consistent with the inverse relationship between speed and gas rate mentioned earlier.

CMC. Fig 9 shows Po_g-Fl_g at constant speed for the Prochem impeller. At 200 rpm, cavities are not formed; and at 350 rpm, they form and then collapse, with the collapse becoming less frequent at higher gas rates (e.g., every 10-20s at < 0.25 vvm, 1-2 min at >1 vvm). Above 350 rpm, cavities are formed at very low gas rates (< 0.25 vvm) and up to 490 rpm, the power drawn is independent of the gas rate. As the speed increases above 490 rpm, successively higher gas rates are required before power consumption is constant. For a disc turbine in shear-thinning fluids, Nienow et al (17) detected a dependency of power on the gas rate at Re > 900, and Ranade and Ulbrecht (10) similarly for Re > 850. For the Prochem impeller, the threshold speed of 490 rpm corresponds to a Reynolds Number of 1620; but even at higher Re, the power consumption is independent of the gas rate above a particular value (e.g., 1.0 vvm at 750 rpm). This is related to a phenomenon described below which was not apparent with water.

From ~600 to ~1000 rpm on increasing gas rate, there is a distinct point at which the gas becomes only periodically well-dispersed and torque fluctuations are present. Fig 10 illustrates this stability transition on increasing gas rate. A significant hysteresis was found on decreasing the gas rate and the reverse transition is more gradual than before. The torque fluctuations on just achieving the unstable condition increase in magnitude with speed (e.g., ±9% about the mean at 700 rpm, ±20% at 900 rpm). Importantly, the combination of gas rate and impeller speed at which a step change occurs in the P_g/P_u versus Fl_g curve of Fig. 9 corresponds to the stability transition of Fig 10, i.e. in the unstable region of Fig. 10, the average (but fluctuating) value of P_g/P_u is constant for a given agitator speed.

Significant torque instabilities have been reported in water using either a small diameter (T/4) pitched-blade impeller pumping downwards or a disc turbine at low clearance. The former instabilities were explained by the opposing upward sparged gas-liquid flow and the downward pumping action of the impeller and the latter to bulk flow pattern transitions (23). Solomon (4) and Hickman (12) both describe the trapping of gas below a downward pumping impeller in CMC solutions. At particular speeds the upward buoyancy of the gas is matched by the downward force of the pumped liquid and gas is trapped below the impeller. The amount of gas increases until the buoyancy overcomes the downflow of liquid and the gas is released and enters the impeller. In this work, the opacity of the aerated CMC solution made it impossible to see the impeller region; but the existence of the open-bottomed hub of the Prochem impeller as a gas entrapment device makes this last possibility the most likely explanation of the torque instabilities.

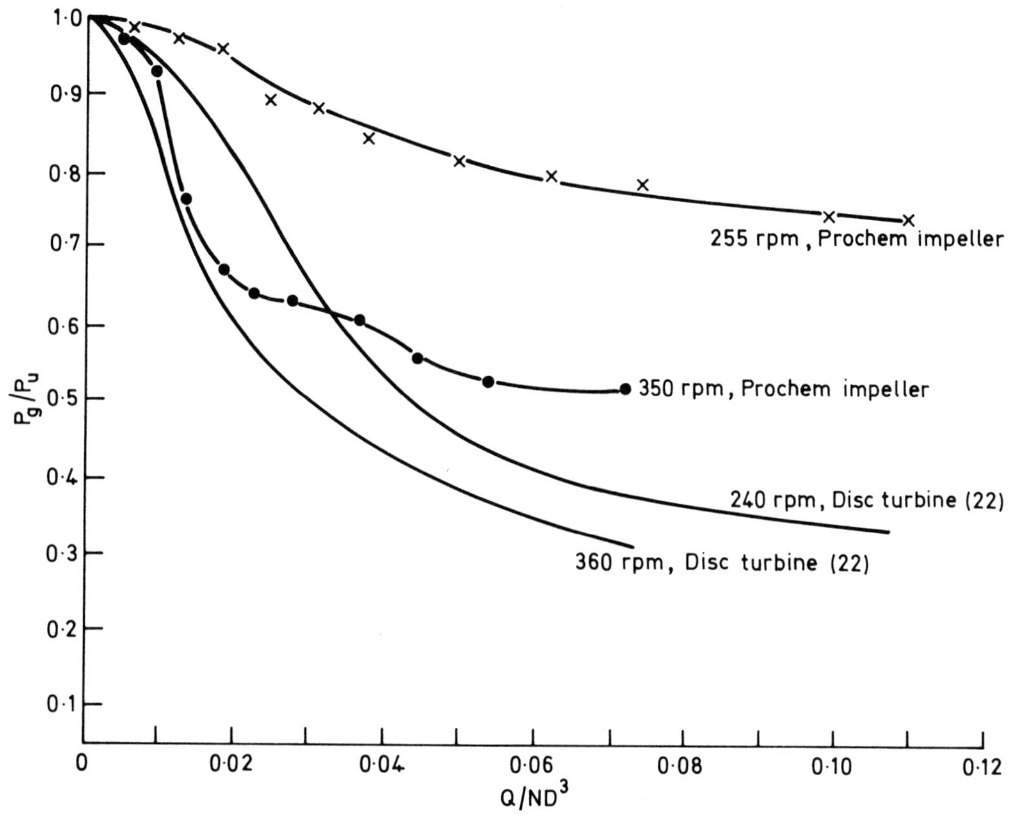

FIGURE 7. $P_g/P_u - Fl_g$ at constant N for Prochem impeller and disc turbine in water.

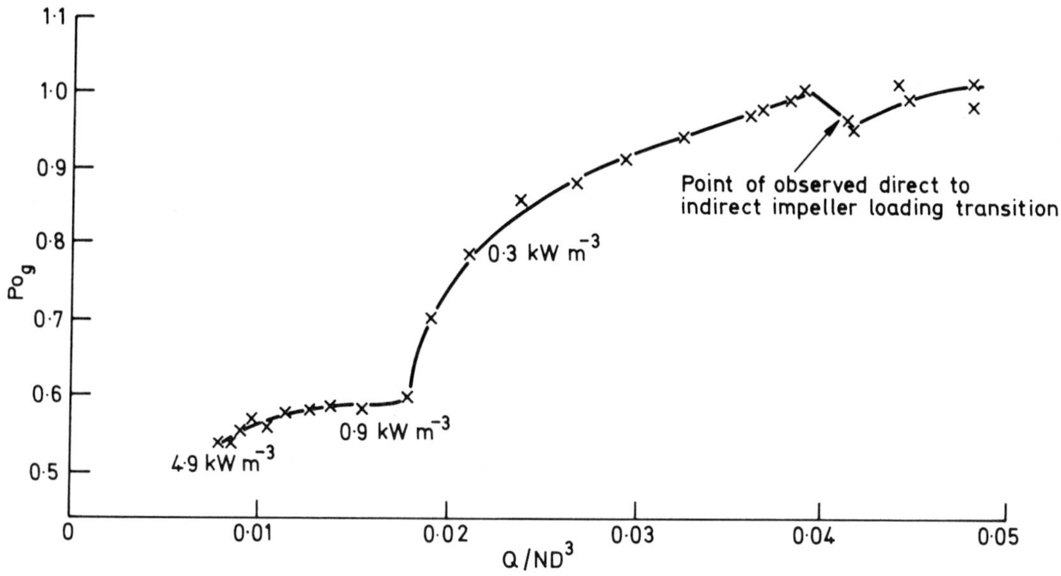

FIGURE 8. $Po_g - Fl_g$ at constant Q (1.0 vvm) for Prochem impeller in water.

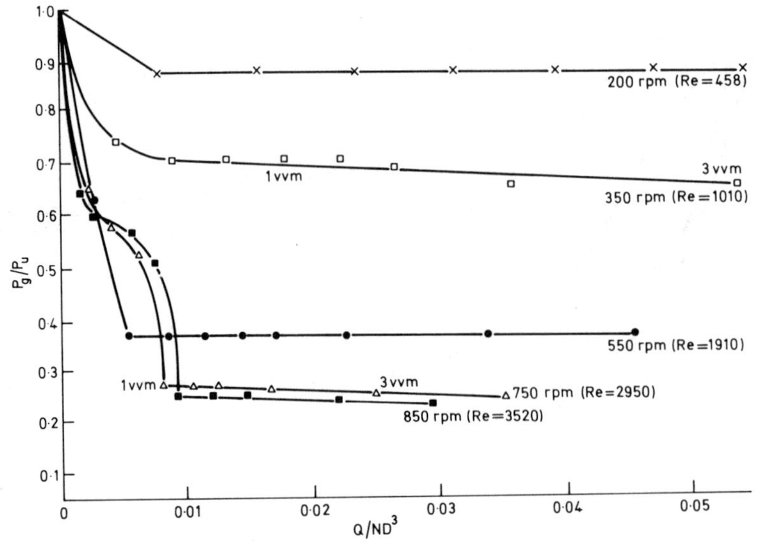

FIGURE 9. $P_g/P_u - Fl_g$ at constant N for Prochem impeller in 0.4% by wt CMC.

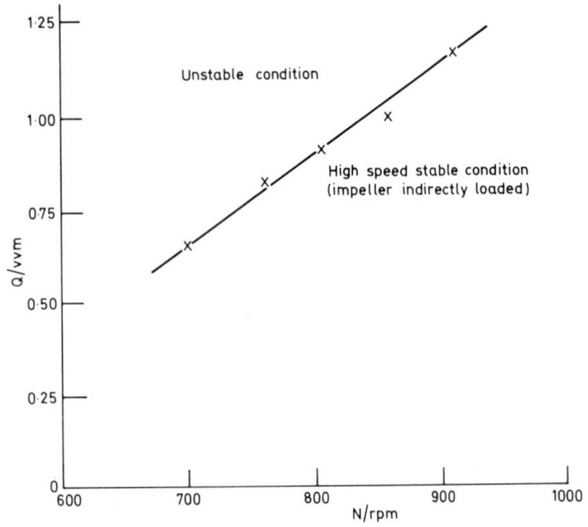

FIGURE 10. Stability transition for Prochem impeller in 0.4% by wt CMC.

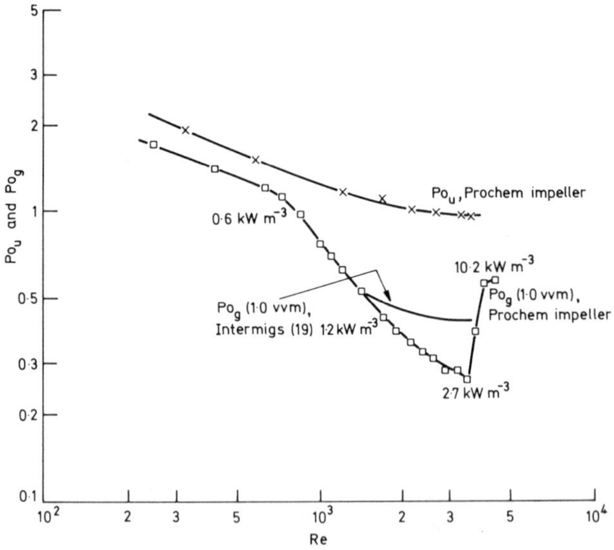

FIGURE 11. *Po-Re* for Prochem impeller and dual Intermig combination in 0.4% by wt CMC.

Hickman (12) observed severe tank vibrations using the Intermigs at high speeds to agitate viscous fluids. Sittig (24) and Kipke (25) noted the same and suggested a resonance effect.

In Fig 10, the lower right region corresponds to the downflow of liquid dominating over the gas flow, resulting in what may be termed a high speed stable condition. In the upper left region of Fig 10, the gas buoyancy and liquid flow forces are evenly matched producing the unstable condition; repeated build up of gas below the impeller (low torque and poor bulk flow) which is then released (higher torque and better gas distribution). At lower speeds than shown in Fig 10, there is likely to be another transition from the unstable condition to a low speed stable condition when the upward gas flow is dominant. Low speed stability corresponds to direct loading of the impeller and high speed stability to indirect loading.

Fig 11 gives Po_u and Po_g (1.0 vvm) in the high Re end of the transition regime for the Prochem impeller. Nienow et al (17) and Ranade and Ulbrecht (10) state that for a disc turbine agitated shear-thinning fluid, at Re <~50 all bubbles are large and do not influence the impeller hydrodynamics; nor do cavities form. Hence $Po_g = Po_u$. Hickman (12) gives Re = 200 as the point when Po_g deviates from Po_u for a disc turbine in 0.4% CMC. Above this Re value, large cavities form immediately on all blades and Po_g falls. For the Prochem impeller there is a small and constant reduction in power number for 250 < Re <800 even though this is prior to large cavity formation and is probably due to the effect of direct impeller loading on reducing the drag exerted by the impeller.

Once large cavities form at Re ~800, the gassed power number falls rapidly with increasing speed to a minimum. The point of sharply increasing Po_g in Fig 11 corresponds with the transition to the high speed stable condition given in Fig 10 at 1.0 vvm. It is interesting to note that this transition was found to be rather dependent on the vessel internal pressure. So too was the value to which Po_g increased, i.e., $Po_g = 0.6$ occurs at N ~900 rpm at 1.0 atm, but at N ~800 rpm at 150 mm Hg gauge.

Data from Hickman (12) for the Intermig combination in Fig 11 shows a gradual fall in Po_g on increasing Re. The ungassed power number for the Intermigs is lower than for the Prochem impeller (see Fig 5), but the opposite is true on gassing (Fig 11). Hence, the power loss on gassing is considerably less for the Intermig combination.

Mass Transfer

Preliminary measurements of the gas-liquid mass transfer have been made using the Prochem impeller in distilled water. The steady-state method was employed using respiring yeast as the oxygen sink (26). The liquid phase oxygen concentration was followed using an oxygen electrode and gas phase oxygen and carbon dioxide concentrations using gas analysers. This information gives the steady state driving force and the mass flux.

k_La values for given specific power inputs were found to be very similar to those obtained using a disc turbine (D/T = 0.5) and Intermigs in the same equipment and using the same technique (26). Many workers have noted that impeller design has little influence on k_La in low viscosity fluids.

CONCLUSION

This work provides a basis for the characterisation of the Prochem impeller in non-Newtonian fluids.

The most obvious aspect of the performance of the Prochem impeller is the presence of gas dispersion and torque instabilities over a range of speeds and specific power inputs in model fermentation fluids similar to those used in industrial fermenters. The effect of this phenomenon on the mass transfer, as well as the dependency on scale and number of impellers used, requires investigation. Mixing time studies will also be important.

ACKNOWLEDGEMENT

One of us (GJB) would like to thank SERC and ICI (New Science) for financial support during this work.

NOTATION

D impeller diameter, m

Fl_g Flow Number (= Q/ND^3)

K consistency index, N $m^{-2}s^n$

k_La mass transfer coefficient, s^{-1}

k_s Metzner-Otto constant

N impeller speed, s^{-1} or min^{-1}

n flow behaviour index in power law model

P_u ungassed power, W

P_g gassed power, W

Po_u ungassed Power Number $(= P_u/\rho N^3D^5)$

Po_g gassed Power Number $(= P_g/\rho N^3D^5)$

Q gas rate, m^3s^{-1} or vvm

Re Reynolds Number $(= \rho ND^2/\mu_a)$

T tank diameter, m

$\dot{\gamma}$ shear rate, s^{-1}

τ shear stress, Pa

Subscript

AV average

LITERATURE CITED

1. Tyler, J., W. Henry, and J. Park, "Characterisation of Hydrofoil Impellers in Large-Scale Fermentation Processes", 10th Eng. Foundation Conf., Henniker, USA (1985).

2. Gbewonyo, K., D. DiMasi, and B.C. Buckland, Proc.Int.Conf. on Bioreactor Fluid Dynamics, Cambridge, England (1986); BHRA, Cranfield, (1986), pp.281-300.

3. Kuboi, R., A.W. Nienow, and K. Allsford, Chem.Eng.Commun., 22, 29 (1983).

4. Solomon, J., "Mixing, Aeration and Rheology of Highly Viscous Fluids", PhD Thesis, University of London (1980).

5. Metzner, A.B. and R.E. Otto, A.I.Ch.E.J., 3, 3 (1957).

6. Nienow, A.W., D.J. Wisdom, and J.C. Middleton, Proc. 2nd Euro.Conf. on Mixing, Cambridge, England (1977); BHRA, Cranfield, (1978)), pp.F1-1 to F1-16 and X53 to X55.

7. Rushton, J.H. and J.J. Bimbinet, Can.J.Chem.Eng., 46, 16 (1968).

8. Nienow, A.W., M.M.C.G. Warmoeskerken, J.M. Smith and M. Konno, Proc. 5th Euro.Conf. on Mixing, Wurzburg, West Germany (1985); BHRA, Cranfield, (1985), pp.143-154.

9. Warmoeskerken, M.M.C.G., J. Speur, and J.M. Smith, Chem.Eng.Commun., 25, 11 (1984).

10. Ranade, V.R. and J.J. Ulbrecht, Proc. 2nd Euro.Conf. on Mixing, Cambridge, England (1977); BHRA, Cranfield, (1977), pp.F6-83 to F6-100.

11. Machon, V., J. Vlcek, A.W. Nienow, and J. Solomon, Chem.Eng.J., 19, 67 (1980).

12. Hickman, A.D., "Agitation, Mixing and Mass Transfer in Simulated High Viscosity Fermentation Broths", PhD Thesis, University of Birmingham (1985).

13. Nienow, A.W. and D.J. Wisdom, Chem.Eng.Sci., 29, 1994 (1974).

14. Riet, K. van't, "Turbine Agitator Hydrodynamics and Dispersion Performance", PhD Thesis, Delft Technical University, Netherlands (1975).

15. Smith, J.M. and M.M.C.G. Warmoeskerken, Proc. 5th Euro.Conf. on Mixing, Wurzburg, West Germany (1985); BHRA, Cranfield, (1985), pp.115-126.

16. Bates, R.L., P.L. Fondy, and R.R. Corpstein, Ind.Eng.Chem., (PDD), 2, 311 (1963).

17. Nienow, A.W., D.J. Wisdom, J. Solomon, V. Machon, and J. Vlcek, Chem.Eng. Commun., 19, 273 (1983).

18. Hocker, H., G. Langer, and U. Werner, Ger.Chem.Eng., 4, 51 (1981).

19. Allsford, K.V., "Gas-Liquid Dispersion and Mixing in Mechanically Agitated Vessels with a Range of Fluids", PhD Thesis, University of Birmingham (1985).

20. Bujalski, W., A.W. Nienow, S. Chatwin, M. Cooke, Chem.Eng.Sci., accepted for publication, 1986.

21. Clark, M.W. and T. Vermeulen, A.I.Ch.E.J., 10, 420 (1964).

22. Warmoeskerken, M.M.C.G., J. Feijen, and J.M. Smith, I.Chem.E.Symp.Ser. No.64, (1981), pp.J1 to J14.

23. Chapman, C.M., A.W. Nienow, M. Cooke, and J.C. Middleton, Chem.Eng.Res.Des., 61, 82 (1983).

24. Sittig, W., J.Chem.Tech. and Biotech., 32, 47 (1982).

25. Kipke, K.D., Proc. 4th Euro.Conf. on Mixing, Leeuwenhorst, The Netherlands (1982); BHRA, Cranfield, (1982), pp.355-370.

26. Hickman, A.D. and A.W. Nienow, Proc.Int. Conf. on Bioreactor Fluid Dynamics, Cambridge, England, 1986; BHRA, Cranfield, (1986), pp.301-316.

Characterization of Oxygen Transfer and Power Absorption of Hydrofoil Impellers in Viscous Mycelial Fermentations

K. GBEWONYO
Merck Sharp & Dohme Research Laboratories
P.O. Box 2000
Rahway, New Jersey 07065

D. DiMASI
University of Virginia
Charlottesville, Virginia 22901

B. C. BUCKLAND
Merck Sharp & Dohme Research Laboratories
P.O. Box 2000
Rahway, New Jersey 07065

Current trends to maximize cell densities impose high oxygen demands on the mass transfer capabilities of existing fermentors. One way to meet this challenge is to employ more efficient agitator designs. In filamentous mycelial fermentations, high non-Newtonian viscosities further reduce gas-liquid mass transfer rates in the broth and also cause poor bulk mixing in the liquid phase especially in large scale fermentors (1,2,3). In some fermentations, mixing becomes even more critical to performance when strategic nutrients like glucose are slowly fed such that a small but finite concentration must be maintained to avoid complete starvation of local cell populations. Similarly during acid/base pH control, effective mixing is necessary to dissipate extremes in local pH levels at the point of shot addition.

In some fermentations, the sensitivity of the cells to mechanical shear damage imposes a further constraint on the choice of the agitator system (4). However, in most aerobic fermentations the overriding consideration still remains the adequate supply of dissolved oxygen to the cells.

Traditionally, radial flow Rushton turbine impellers have been used in fermentation plants. The mechanism of action of this impeller induces highly turbulent shear rates around the impeller zone promoting gas dispersion and bubble break-up. This results in an oxygen-rich envelope around the impeller; however in the regions away from the impeller limited mass transfer occurs and the availability of dissolved oxygen decreases. This segmentation is amplified in viscous mycelial cultures and in large production tanks. Various models have been proposed to describe this phenomenon. Notably Reuss and Bajpai (5) proposed the concept of circulation of mycelial cell aggregates between the dissolved oxygen-rich and oxygen-poor zones such that the circulation time becomes the critical factor which determines the overall effectiveness of oxygen uptake by the cells. In line with this view, a number of investigators have suggested larger diameter Rushton turbines to improve liquid pumping capacity in the fermentors (6). However, this would be accompanied by substantial increases in power draw ($\propto Di^5$) which may not be permissible given the limited horsepower capacity of existing agitator motors.

The application of axial flow impellers which provide good bulk mixing have recently been receiving attention as an alternative to the dominance of the Rushton turbines in fermentation processes. A number of novel designs of axial flow mixers are being tested for fermentation applications such as the Lightining A310 (Mixing Equipment Co., Rochester) (7); Intermig impellers (Ekato GmbH, Germany) (8); and the Maxflo (Prochem Ltd., Canada) (9,10). These designs have been reported to induce more uniform energy dissipation patterns within the entire volume of the fermentor. Although the maximum shear rates at the impeller tip is considerably less than the Rushton turbines, these axial flow mixers show a considerably higher average shear rate throughout the entire volume of the fermentor. This flow regime would drastically affect gas dispersion, bubble break-up and coalesence phenomena. Thus, the effectiveness of these axial flow mixers would be determined primarily by their ability to provide adequate overall oxygen transfer rates compared with the Rushton turbines.

At the Merck Biochemical Engineering Pilot Plant, we have been investigating the Prochem Maxflo hydrofoil impellers in some viscous mycelial fermentations. The data obtained in the avermectin fermentation would be discussed in this paper. This fermentation is used to produce an important class of antiparasitic drugs for veterinary applications as well as agricultural insecticides and for potential use against river blindness in humans (11).

FERMENTOR EQUIPMENT AND INSTRUMENTATION

A schematic diagram of the Prochem hydrofoil impeller is shown in Figure 1. The design consists of five hydrofoil blades set at a critical angle on a central hub. These hydrofoil blades are designed to work on the principles of aircraft wing technology. During rotation, a high hydrodramatic thrust (lift) is created in the upward direction to increase the liquid pumping capacity of the blades. The design also minimizes the drag forces associated with the motion of the blades such that energy losses due to friction are reduced. During operation, it was observed that the hydrofoil blades tend to churn up the liquid in the fermentor in a mild top-to-bottom flow pattern. This is completely different from the mixing patterns created by the Rushton turbines where the liquid is thrown radially outwards from the impellers in a more turbulent agitation pattern.

Figure 1 also shows a schematic diagram of a standard flat blade disc turbine, consisting of six adjustable blades which could be extended outwards to give impeller diameters of 10 to 12 inches.

One of the 800 liter fermentors was retrofitted with two Prochem hydrofoil impellers of 14 inch diameter. The fermentor dimensions are shown in Figure 2. The tank diameter is two and one half feet and the height is five and one half feet providing a working volume of 600 L. The agitator motor is rated at seven and one half horsepower capacity and the motor is controlled by a variable frequency drive which provided agitation speeds ranging from 30-340 rpm to the impellers. The agitator shaft is fitted with two impellers spaced at 20 inches apart, with the lower impellers located at eight inches from the bottom of the tank. An open pipe air sparger one inch diameter is located below the lower impeller. Four baffles two and one half inches width, are located along the walls of the tank. Table 1 summarizes the key tank dimensions.

The mass transfer coefficient K_La is calculated as a ratio of the oxygen transfer rate and the concentration gradient of oxygen between the bulk liquid and the equilibrium concentration at the gas-liquid interface according to the equation:

$$K_La = \frac{OUR}{\left(1 - \frac{D.O.}{100}\right)\left(\frac{P_{O2in} + P_{O2out}}{2}\right)\left(\frac{P_T + P_L}{H}\right)}$$

The equilibrium concentration is determined using Henry's Law, and the average partial pressures (inlet (P_{O2in}) and exhaust (P_{O2out}) air streams), the overhead pressure (P_T) and the mean hydrostatic head pressure (P_L). The constant H, is determined with respect to the operating temperature. The bulk dissolved oxygen concentration is obtained from the % saturation reading of the probe and the equilibrium concentration as determined above. For the size of our tanks, we find that the mean dissolved oxygen concentration gradient as calculated above, is adequate for calculating the K_La, whereas a log mean gradient would be more appropriate for larger scale fermentors.

RESULTS

Power Absorption in Water-Air System

Figure 4 shows a log-log plot of power absorption for the Prochem and Rushton impellers at various agitation speeds in

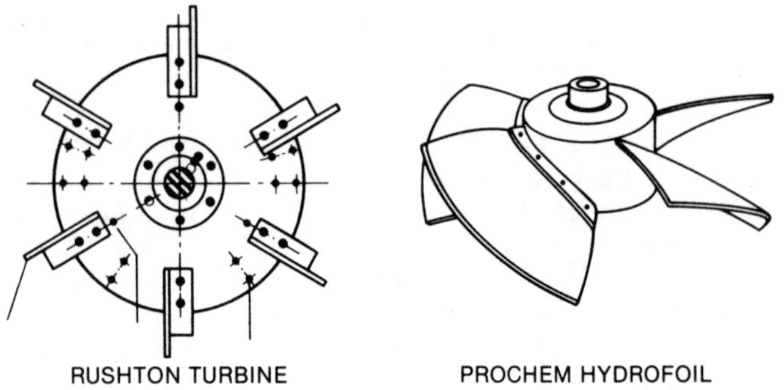

FIGURE 1. Diagram of Prochem Maxflo hydrofoil impeller and flat-blade Rushton turbine.

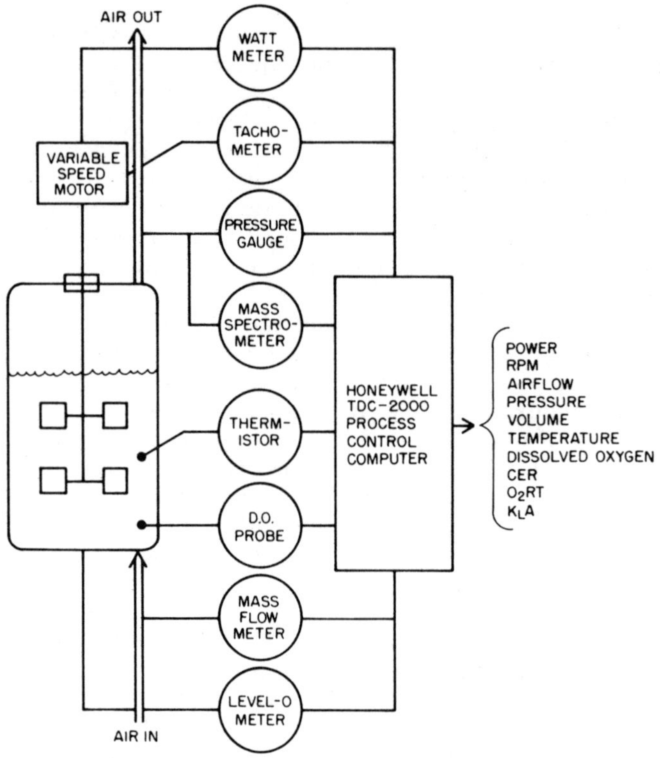

FIGURE 3. Outline of Merck Computer Process Control and Data Acquisition System—showing instrumentation computer inteface and output variables.

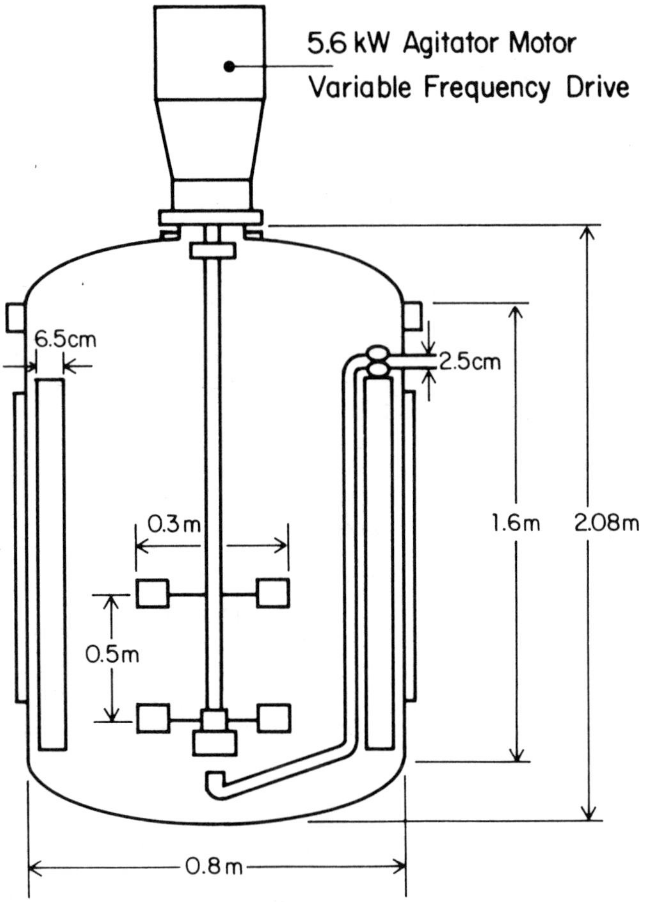

FIGURE 2. Schematic diagram of 800-L pilot plant fermentor showing tank dimensions, baffles, sparger, impeller spacing and motor rating.

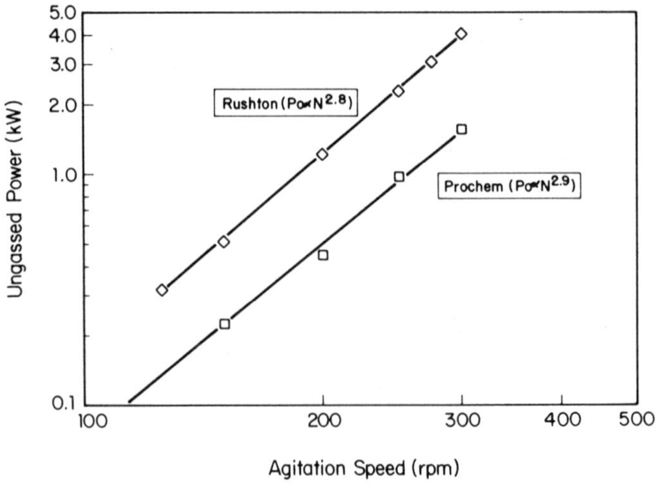

FIGURE 4. Ungassed power absorption by Prochem and Rushton impellers as a function of agitation speed.

unaerated liquid. As to be expected, the power increased as a cubic function with speed for both impellers, however it is interesting to note that the Prochem draws between 40-50% less power than the Rushton at any given speed. Further analysis of the data shows that the power number for the two Prochem impellers was only 2.2 while the Rushton gave a power number of 13 within the same range of Reynolds numbers (>10^5).

During aeration the power drawn by the Prochem impellers decreased down to about 60% of the ungassed power draw (Figure 5). Similar trends have been reported for the Rushton turbine. This factor provides an important criteria for predicting motor load during sterilization, and when aeration is disrupted due to compressor failure.

FERMENTATION STUDIES

The avermectin fermentation is marked by an early period of rapid cell growth during which broth viscosities increase up to about 500 cP. Subsequently, the growth slows down and the cells begin to fragment leading to a decrease in viscosity (Figure 6).

Parallel fermentations were set up with the Prochem and Rushton turbines under identical operating conditions (pressure, airflow, agitation speed). The oxygen uptake rate profiles in Figure 7 show a marked increase in peak oxygen uptake rate with the Prochem impellers. This effect has been observed in a number of fermentations though the magnitude of the difference varies between runs. One possible explanation is that the improved bulk mixing with the Prochem impellers enhances the respiratory activity of the cells. However, variations due to media sterilization and inoculum state may also account for this effect.

An even more interesting outcome of the experiment is demonstrated by the power consumption curves of the two impeller systems in Figure 8. Notably, the Prochem agitator required only about 60% of the power drawn by the Rushton impellers to maintain the fermentation at a comparatively high oxygen uptake rate. The two agitators were run at constant speed of 300 rpm after the first five hours of the fermentation.

The primary objective of oxygen transfer in our fermentation processes is to maintain the dissolved oxygen concentration above a specified minimum level found to be desirable for optimum activity of the cultures. Consequently, we employ a cascade control strategy whereby the agitation speed is automatically varied to keep dissolved oxygen above the desired setpoint value. Figure 9 shows the operation of the Prochem and Rushton impellers under this control strategy at a dissolved oxygen setpoint of 30% saturation. This mode of operation provides an identical basis (equal DO) for assessing the efficiency of the two agitators. The ratio of oxygen uptake rate and the power consumption (corrected for no load values) was used to characterize oxygen transfer efficiency (Figure 10). At peak performance, the Prochem agitator delivered almost twice as much oxygen per unit power consumed than the Rushton turbine.

Detail characterization of the effect of agitator power input on the mass transfer coefficient, K_La, were conducted in further experiments in which agitation speeds were changed independently to determine K_La. The trends plotted in Figure 11 show clearly that at any given power input the Prochem agitator provides considerably better gas-liquid mass transfer capability than the Rushton turbine. Typically the mathematical correlations obtained between 20-50 hours of the fermentation are summarized as follows:

Prochem: $K_La = 129\ (P/V)^{.585}$

Rushton: $K_La = 51\ (P/V)^{.585}$

DISCUSSIONS

The data presented here has far-reaching implications on current thinking as to the rate-determining processes in gas-liquid oxygen transfer in viscous fermentations. It is clear that the local dispersive forces generated by the Rushton turbine become ineffective in promoting adequate mass transfer in viscous fermentation broths. Apparently, outside the immediate range of the impellers, bubble coalesence phenomenon supercedes break-up thus reducing the overall interfacial area available for oxygen transfer.

On the other hand, the bulk flow regimes induced by the Prochem hydrofoil impellers eliminate the coalesence-prone dead zones in the fermentor. Further, the uniform energy distribution of the Prochem impellers enhances bubble break-up throughout the whole liquid volume of the fermentor. This hypothesis would explain the significant increases in K_La obtained with the Prochem

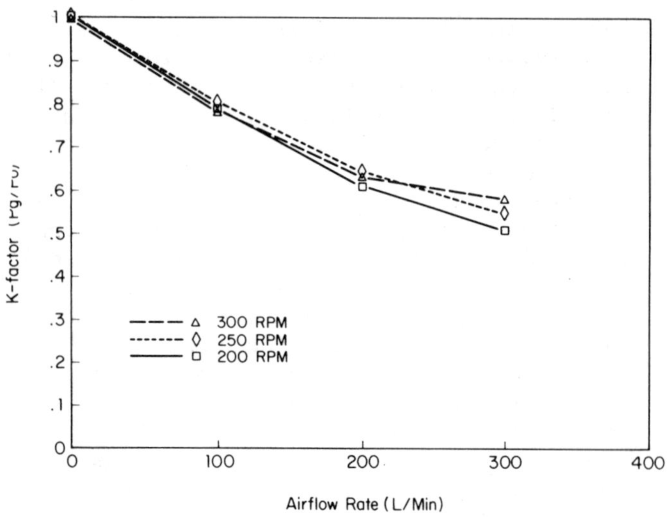

FIGURE 5. Effect of airflow on power absorption by Prochem impeller at different speeds.

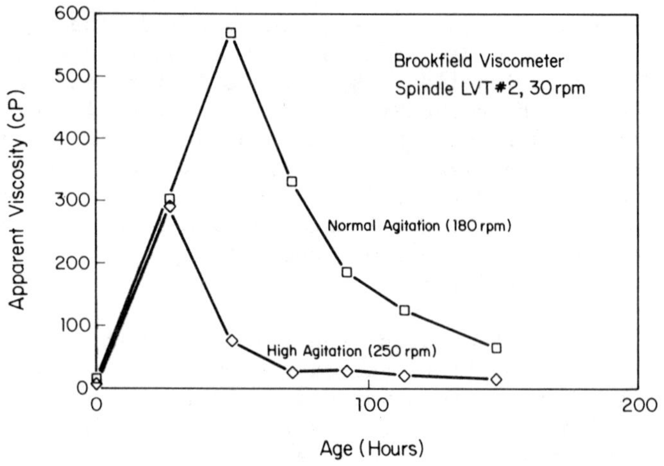

FIGURE 6. Viscosity profiles in the Avermectin fermentation apparent viscosity of samples determined at different ages on the Brookfield viscometer using spindle LVT #2 and rotation speed of 30 rpm. The effect of impeller shear shown by the viscosity profiles at high agitation 250 rpm vs. 180 rpm.

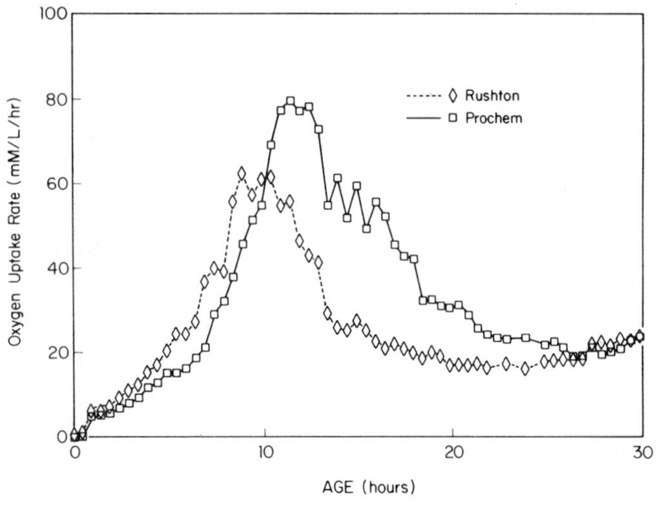

FIGURE 7. Time profile of oxygen uptake rates in Avermectin fermentation batches run with the Prochem and Rushton impellers under identical operating conditions.

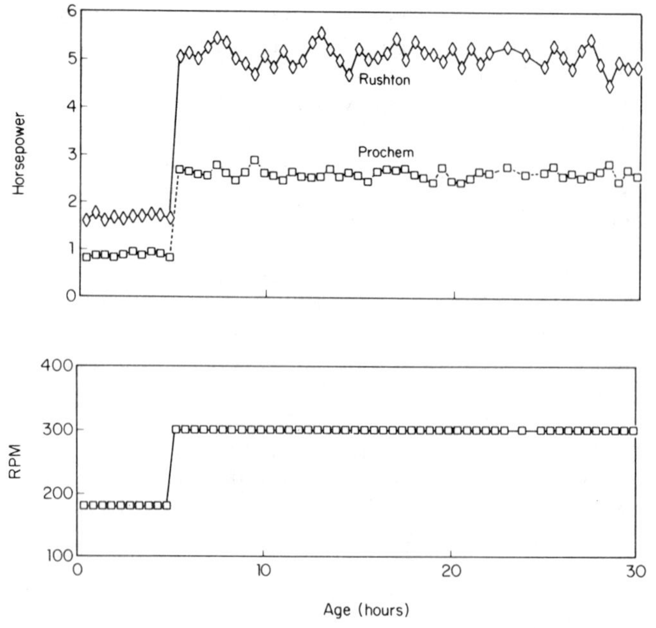

FIGURE 8. Power absorption by the Prochem and Rushton impellers during the Avermectin fermentation. Agitation speeds were maintained at 180 rpm (0–5 hr) and increased to 300 rpm after 5 hr.

TABLE 1

	Hydrofoil	Rushton Turbine
Tank Operating Volume	500L	500L
Tank Diameter D_T	32"	32"
Liquid Height H_L	64"	64"
H_L/D_T	2	2
Impeller Diameter D_i	14"	12"
D_i/D_T	0.44	0.38
Impeller Spacing S	20"	20"
S/D_i	1.43	1.67
No. of Blades	5	6
Motor Capacity	7½ Hp	
Gear Ratio	5:1	
Speed Range	1750 RPM/33 - 338 RPM	

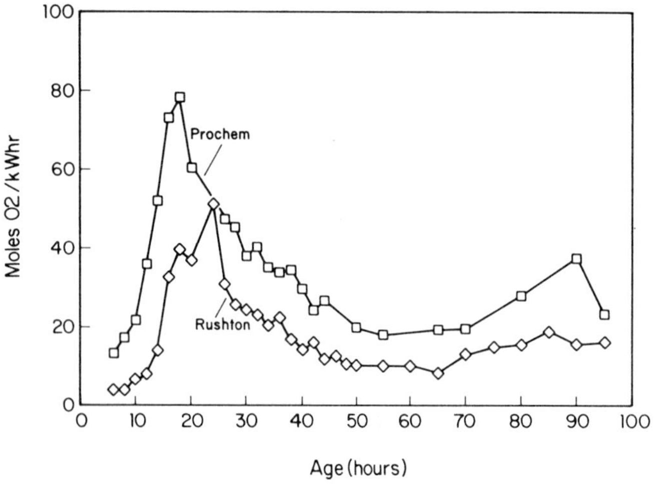

FIGURE 11. Correlation of K_La with power input for the Prochem and Rushton impellers during the Avermectin fermentation. Power was varied by changing agitator speed while airflow rate and pressure are kept constant at 0.4 BBM and 0.7 kg/cm^2, respectively. Data were obtained during the period of 25–45 hr of the fermentations. Log–log plot of power regression analysis of the data is shown.

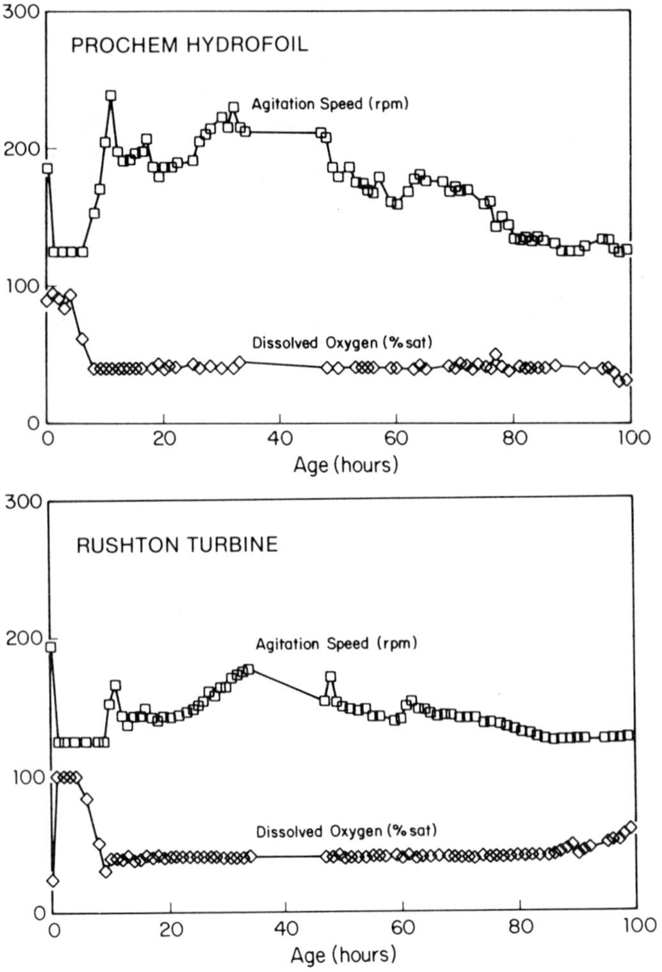

FIGURE 9. Cascade control of dissolved oxygen concentration with agitation speed for the Prochem and Rushton impellers during the Avermectin fermentation. Agitation was varied automatically between 120 and 250 rpm to maintain a minimum dissolved oxygen concentration of 30% saturation.

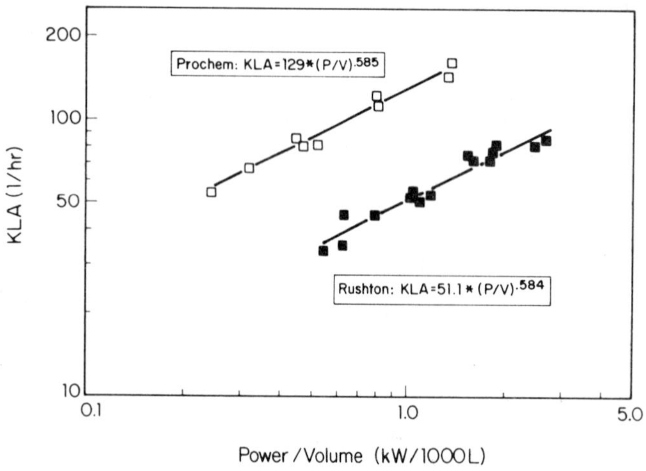

FIGURE 11. Correlation of K_La with power input for the Prochem and Rushton impellers during the Avermectin fermentation. Power was varied by changing agitator speed while airflow and pressure are kept constant at 0.4 WM and 0.7 kg/cm^2, respectively. Data were obtained during the period of 25–45 hr of the fermentations. Log–log plot of power regression analysis of the data is shown.

agitators.

CONCLUSIONS

We have shown that the Prochem agitator enhances oxygen transfer rates in a typical viscous mycelial fermentation. Further, this improved performance was obtained with only about 60% the power consumed by the traditional Rushton turbines. Peak oxygen transfer efficiencies of 80 moles O_2/kWhr and 60 moles O_2/kWhr were noted for the Prochem and the Rushton turbines, respectively. At any given power input K_La values for the Prochem agitator were about double those of the Rushton turbine. The enhanced gas-liquid mass transfer effects were attributed to the more effective bulk mixing provided by the Prochem agitator.

It is clear from these results that retrofitting fermentors with the Prochem agitators would enable a more efficient utilization of the horsepower capacity of these tanks for oxygen transfer. This would create opportunities for developing fermentations with higher oxygen demands using enriched nutrient media and more potent cultures.

ACKNOWLEDGEMENTS

We are grateful to Dr. Alvin Nienow (University of Birmingham, U.K.), Dr. Stephen Drew and Dr. Michael Midler for very useful discussions and advice during this study. Our thanks also to Ms. Terri Sinisi for typing this manuscript.

LITERATURE CITED

1. Deindoeffer, F. H. and E. L. Gaden, Applied Microbiology, 3, 253 (1955).

2. Steel, R. and W. D. Maxon, Biotechnology and Bioengineering, 8, 97 (1966).

3. Wang, D. I. C. and R. C. J. Fewkes, Developments in Industrial Microbiology, 18, 39 (1977).

4. Midler, M. and R. K. Finn, Biotechnology and Bioengineering, 8, 71 (1966).

5. Reuss, M. and R. K. Bajpai, "Oxygen Consumption in Filamentous Broths - An Approach Based Upon Mass and Energy Distributions". Paper presented at the Annual Meeting of the ACS, New York, (1981).

6. Nienow, A. W., "Mixing Studies on High Viscosity Fermentation Processes - Xanthan Gums". The World Biotech. Report, Vol. 1, (On-line publication, U.K.), 293 (1984).

7. Krone, L., Stieber, R. W. and A. Kirpekar, "Comparison of Mixer Types in Large Scale Fermentations". Engineering Foundation Conference, Henniker, N.H. (1985).

8. Kipke, K. K., "Improvement of Fermentation Parameters by a Judicious Choice of Impeller Type". Publication of Ekato, Rinhr-und Mischtechnik GmbH, D-7860, Schopfheim.

9. Tyler, J., Henry, W. and J. Park, "Characterization of Hydrofoil Impellers in Large-Scale Fermentation Processes". Engineering Foundation Conference, Henniker, N.H. (1985).

10. Gbewonyo, K., Di Masi, K. and B. C. Buckland, "The Use of Hydrofoil Impellers to Improve Oxygen Transfer Efficiency in Viscous Mycelial Fermentations". Proceedings of the International Conference on Bioreactor Fluid Dynamics, Cambridge, England (1986).

11. Campbell, W. C., Fisher, M. H., Stapley, E. O., Albers-Schonberg and T. A. Jacob, Science, 221, 823 (1983).

Transfer Oxygen Potential of an Air-Pulsed Continuous Fermentor

M. DONDÉ CASTRO*
G. GOMA
G. DURAND
Institut National des Sciences Appliqueés
Toulouse, France

A new type of air-pulsed continuous reactor is presented. Mathematical models were constructed for main performance features including gas hold-up and oxygen transfer coefficient. The results show that oxygen transfer potential and energetic yield are superior to those by other agitation techniques.

Gas-liquid contact operation has a great industrial importance. It is a principal factor in cost and productivity of fermentation. This is especially true in single cell protein production where oxygen must be supplied at the necessary rate to mantain a certain concentration, higher than a critical value in the medium.

Several technologies have been proposed to carry out gas-liquid contact operation, among those mechanical agitation, bubble columns, etc. Although the parameters employed for their evaluation are in common use, the comparisons between those technologies are difficult when they are applied to fermentations. This is because the media of fermentation have features quite different from those of chemical systems from which determinations are usually made. On the other hand, the media of fermentation properties change during the experiment, and generalizations are difficult to make.

The parameters normally used for evaluating the aeration-agitation technologies are basically the transfer capacity of oxygen and the energetic cost of the transfer. The determination of the transfer potential of oxygen represented by the coefficient k_La is carried out by several procedures. One of these procedures is the Cooper sulfite oxidation method (1), which is the oldest and most used. It is based on the catalytical oxidation of sodium sulfite. The method proposed by Taguchi and Humprey in 1960 (2) was the first one which was possible to use in a microbial culture. It is based on the critical observation of the variation of dissolved oxygen in the culture. Such variation is produced primarily by the subsequent readministration of oxygen.

The Cooper method tries to measure the rate of the purely physical process of oxygen absorption. In order to carry out this measurement it is necessary to satisfy certain conditions. The fulfillment of those conditions assures that the rate of reaction of sulfite is small enough to be negligible in comparison to the rate of physical absorption. In this case the enhancement coefficient E=1, and

$$v = k_La \; (c^* - c) \tag{1}$$

The fulfillment of those conditions also assures that the oxygen concentration c in the above equation is nul. This then allows us to make the followuing equation:

$$k_La = \frac{v}{c^*} \tag{2}$$

The conditions referred to above are presented by Danckwerts (3) in the form of equations in which diverse properties of the system are involved, such as oxygen diffusivity, oxygen interface concentration and kinetical constants. These properties determine the limits of temperature, pH and catalyzer concentration.

---

* M. Dondé Castro is now with Instituto Tecnologico de Merida. Yucatan, Mexico.

For example, the condition by which the chemical reaction in the film is negligible is expressed by the equation:

$$\frac{k D c^*}{k_L^2} << 1 \qquad (3)$$

This condition is difficult to satisfy mainly if temperature rises during the experimentation with the consequent increase of the kinetical constant. If this condition is not satisfied, higher values of k_La than those due to physical diffusion will be obtained.

Scientific literature rarely explains the precautions that are taken in order to satisfy these conditions, so that Equation (2) is valid. Because of this, the correlation of results is difficult to make.

Various papers demonstrate that, at least in certain conditions, enhancement factor $E > 1$ and its resulting value must be considered for calculating k_La by Equation (2). Apparently, the single presence of sodium sulfite and sulfate would have an influence on coefficient k_L and would make $E > 1$ (4), (5). Linek and Benes (6) conclude that the sulfite method is not reliable in measuring the oxygen transfer potential because greater values than those corresponding to purely physical absorption are obtained.

If we try now to apply the values of transfer capacity of oxygen calculated with physico-chemical systems to biological systems, we have to determine if the presence of cells and macromolecules would not produce, in turn, enhancement of oxygen absorption in reference to purely physical phenomenon. Many papers have been published about this subject. The opinions are very diverse. So, while some investigators conclude that $E_f > 1$ (7, 8, 9, 10, 11, 12) others affirm that $E_f = 1$ (13, 14, 15).

Merchuk (16) established a mathematical model for oxygen transfer trough a cell film. Such a model leads to the conclusion that $E_f \geqslant 1$. E_f is the fermentation enhancement coefficient defined by $E_f = k_{L_f}/k_L$

On the other hand, given that the interfacial area a is unknown in agitated fermentors, the importance of E' is rather theoretical.

In accordance with the arguments presented above, we have to define an enhancement coefficient E' in order to evaluate our fermentor. This can be done using the equation:

$$E' = \frac{(k_La)f}{(k_La)p} \qquad (4)$$

where $(k_La)p$ is the value of transfer coefficient measured by physicochemical methods. $(k_La)f$ is measured during the fermentation.

E' defined by Equation (4) is of great importance because the interfacial area a is not known.

Although the sulfite method es commonly used to evaluate the transfer capacity of fermentors, few results are given in the literature which permit a comparison of sulfite values to fermentation results. Schultz and Gaden made a reference to this (17). Values of E' will be usually variable during a batch fermentation because of the variation of diverse properties of the medium. For example, Jaral et al (18) found values of E' between 0.25 and and 0.75 while Yagui and Yoshida (13) determined E'= 1.

MATERIALS AND METHODS.

Fermentor Description.

The fermentor consists of a perforated-plate column with a 15 cm. interior diameter and a height of 1.65 m. (Figure 1). It contains 60 stainless-steel plates with 2.5 mm. diameter perforations in such a quantity that the area of the perforations represents 22% of the total surface area of the plates.

The agitation is produced by pulsing the liquid. At one side of the perforated-plate column, there is a secondary column with an interior diameter of 4 cm. to which air is sequentially supplied. Two check valves prevent the liquid inside the reactor from oscillating. The return of liquid from the main column to the side column is done by spilling.

The instrumentation used allows the monitoring of dissolved oxygen concentration in the liquid, the oxygen pressure, and CO_2 concentration in the gas output. It also allows for the control of pH and supplied air flow.

Oxygen Transfer Potential.

The determination of k_La was accomplished by the sulfite oxidation method with a 30 g/liter sodium sulfite concetration and a 10^{-5} catalyzer concentration ($Co\ Cl_2$). An oxygen probe placed in the upper chamber and calibrated at the same temperature and pressure used during the experiment, permitted the determination of oxygen in the gas output.

In order to determine k_La during the fermentation, a mass balance of the gas was made. A second oxygen probe was installed in the middle of the main column. In this case, the calibration of both probes was made before the inoculation under identical conditions of temperature and pressure as during the fermentation.

The concentration of carbon dioxide in the gas output of the fermentor was also measured.

The general equation obtained from mass balances of oxygen, nitrogen, and carbon dioxide is:

$$v = k_La\ (c^* - c) = 0.21A - \frac{Y_{O2}\ (F_{N2} + F_{CO2})}{Y_{O2}} \qquad (5)$$

where F_i is the flow rate; Y_{O2} is the oxygen mol fraction in the gas output, and A is the air flow rate supplied to the reactor (mmol/liter h). Constant gas flow rate through the reactor has been assumed. If we call R_2 the reading of the oxygen probe placed in the gas output, and R_1 the reading of the probe submerged in the fermentor liquid, we can obtain the following equations:

$$k_La\ (c^* - c) = 0.21\ A\left[1 - \frac{R_1\,(0.79 + Y_{CO2})}{1 - 0.21\,R_1}\right] \qquad (6)$$

$$c^* - c = 0.21\,P\,H\,(R_1 - R_2) \qquad (7)$$

The Henry constant H and probe performance features were assumed to remain constant, during the entire experiment.

According to studies published by Serizawa et al (19), the effect of direct contact of bubbles with the submerged probe must be taken into account to calculate k_La with equation (6). This can be done introducing the factor $(1-\varepsilon)$ in the above equations. The equation for k_La becomes:

$$k_La = \frac{A(1-\varepsilon)}{P\,H\,(R_1 - R_2)}\left[1 - \frac{R_1\,(0.79 + Y_{CO2})}{1 - 0.21\,R_1}\right] \qquad (8)$$

In order to determinate k_La with the sulfite oxidation method, the same Equation (8) was used. In this case, $Y_{CO2} = 0$ and $R_2 = 0$.

Dissipated Power in the Reactor.

The lost power in the reactor is due to both pulsation and aeration operations. Dissipated power in pulsation (Wp) was calculated from the pressure drop in the reactor ($-\Delta P$). Since the pressure drop evolution during the air supplying period of the pulsation (to) is unknown, the uppermost limit of lost power was calculated using the maximum value of ΔP :

$$W_p = \frac{-\Delta P.\ S.\ l.}{(t_o + t_c)V} \qquad (9)$$

the power dissipated by the air dispersion in the reactor was calculated from the polytropic compression work formula, as follows:

$$W_a = A\int_1^2 PdV \qquad (10)$$

Energetic Yield in Oxygen Transfer (OER).

The energetic yield of oxygen transfer can be easily calculated with Equation (11) which was derived from Equations (6), (9), (10):

$$OER = \frac{k_La\ \ (c^* - c)}{W_p + W_a} \qquad (11)$$

Fermentation Experiments.

K. fragilis was grown in a glucose solution in a batch culture. The biomass was determined by filtration in a membrane with pores 0.45 mm. diameter. Glucose concentration was determined by the DNS method.

RESULTS.

Gas Hold-Up.

The adjusted equation to experimental results is:

$$\%\,\varepsilon = 19.92(VVM)^{0.428} - 7.426(VVM)^{0.018}\,(af)^{0.2} \qquad (12)$$

This equation is represented in the Figures 2, 3, in which a decrease of ε is noted as the intensity of agitation increases from 0 to 1.4 cm/sec. However ε increases as the intensity of aeration increases.

Oxygen Transfer Potential.

In the Figure 4 the experimental results obtained by the sulfite oxidation method are shown. When the aeration rate (VVM) is equal to 0.4 min^{-1} , the effect of agitation on k_La is negligible. However under the same condition, an increase in aeration intensity causes an increase in k_La .

The flooding point occurs when VVM = 2.5 min^{-1} (Figure 5). The value of

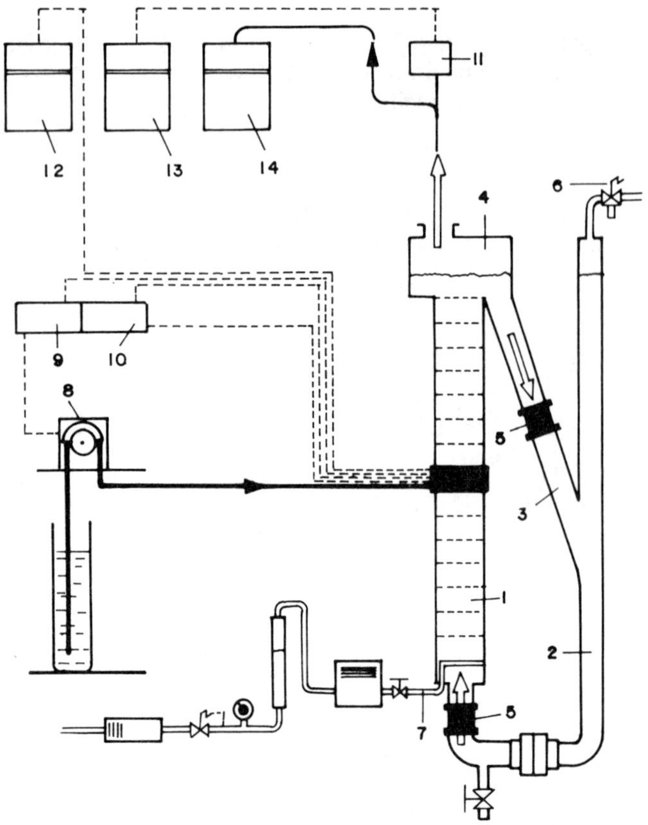

FIGURE 1. Diagram of pulsed reactor; 1—main column; 2—side column; 3—liquid return; 4—expansion chamber; 5—check valve; 6—air supply for pulsation; 7—air supply for aeration; 8—ammonia supplier for controlling pH; 9—pH controller; 10—temperature controller; 11—probe to measure oxygen tension in gas output (R1); 12—dissolved oxygen concentration recorder; 13—oxygen tension recorder; 14—carbon dioxide recorder.

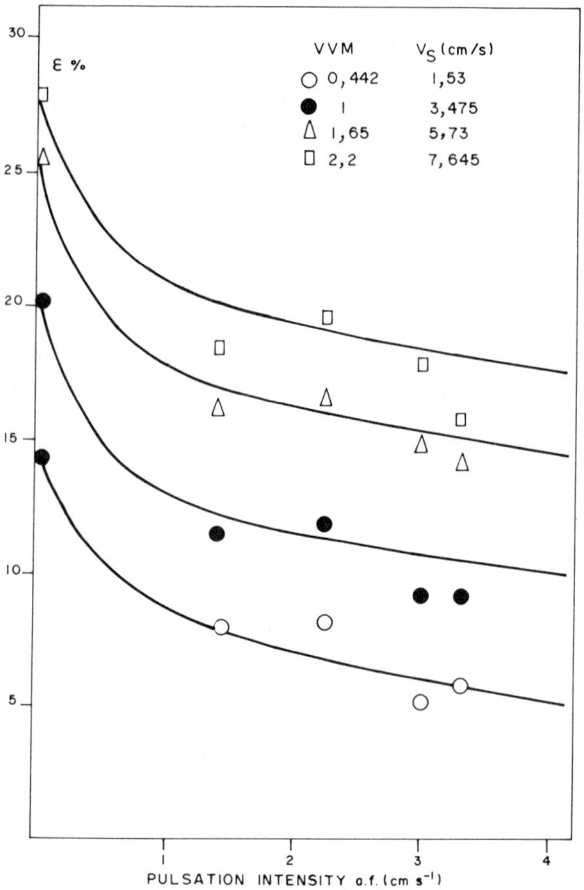

FIGURE 2. Gas hold-up (ε) vs. pulsation intensity and aeration rate.

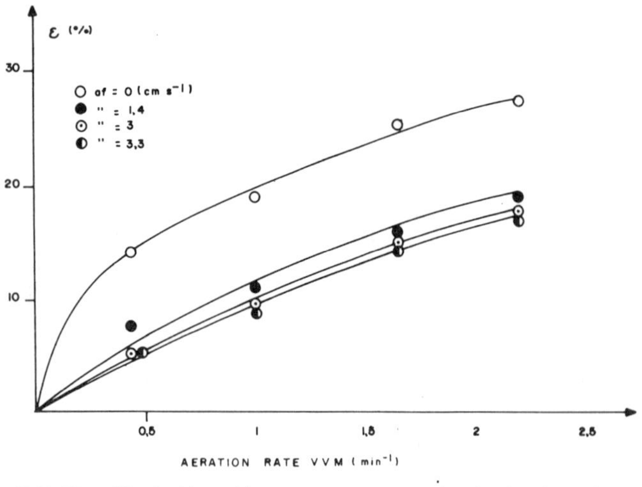

FIGURE 3. Gas hold-up (ε) vs. aeration rate and pulsation intensity.

k_La during the fermentation was virtually constant. The adjusted equation to experimental results for $af = 1$ is:

$$k_La = 1089\,(VVM)^{0.486} + 222\,(VVM)^{1.337}\ h^{-1} \qquad (14)$$

The average values in diverse conditions can be seen in Table 1. In this same table we can see the values of k_La obtained under the same conditions with the sulfite method as well as the enhancement coefficient $E' = (k_La)\ ferm/(k_La)\ sulf$.

In the Figure 6 we can see the results for a typical fermentation.

Dissipated Power and Energetic Yield in Oxygen Transfer.

Using as the maximun value of ΔP the difference between the internal and external air pressure of the air tank, and Equation (9), a series of values of the lost power in agitation (Wp) was calculated. It was necessary to measure the liquid displacement in the side column and the periods to, tc.

Equation (15) was then adjusted to this series of values (the value of correlation coefficient was 0.98).

$$Wp = 95\ (af)\quad watt/m^3 \qquad (15)$$

With respect to lost power in aeration, Equation (16) was obtained, using the maximum observed value of entrance pressure (1.3 bar) and a temperature equal to 293º K.

$$Wa = 610\ (VVM)\quad watt/m^3 \qquad (16)$$

The energetic yield can be calculated with Equation (17) which was derived from Equations (14), (15) and (16). It was also necessary to use a unit conversion factor and c* = 0.16 mmol/lt.

$$OER = \frac{5.575(VVM)^{0.486} + 1.137(VVM)^{1.337}}{0.095\ (\ f\ ) + 0.061\ (VVM)} \quad \frac{Kg\ O}{kwh} \qquad (17)$$

These results can be used for comparisons with other reactors.

The mass transfer coefficient k_La measures the potential of the reactor to transfer during the fermentation must be less than those obtained by the sulfite oxidation method, since the dissolved oxygen tensions in the fermentation medium will never be nul. Consequently, the energetic yield will be less than the one calculated from the sulfite method experiments.

DISCUSSION OF RESULTS AND CONCLUSIONS.

In the Figures 2, 3, 4 and 5, it is apparent that the pulsation decreases the value of gas hold-up. This effect is greater when $af \leq 1.4$ cm/sec. In the same way, the value of k_La increases if the values of the aeration rate (VVM) are not less than 0.4 min^{-1}. The energetic consuption of the pulsation is very little especially if we consider the improvement in the oxygen transfer of the reactor. For example, if the aeration intensity is $af = 1$ cm/sec and VVM= 0.16 min^{-1} the power dissipated by the pulsation will be less than that dissipated by aeration.

The energetic yield of oxygen transfer decreases if the intensity of agitation and/or aeration increases. For example if af=1 cm/sec, the values shown in Table 2 are obtained.

Performance features of the reactor are better than other performance features that have been published in the literature. The gas hold-up is less than those in conventional plate columns.

The mass transfer coefficient (k_La) determined with the sulfite method is greater than others values found in the literature.

The oxygen transfer rate during fermentation as well as the energetic yield are greater than those reported for other pilot-scale fermentors (Table 3).

Table 3 was prepared from some of the results presented in reference (20).

TABLE 1
E' During Fermentation

FERMENTACION No.	VVM min^{-1}	$\overline{k_La}$ h^{-1}	k_La sulf h^{-1}	$\frac{k_La\ ferm}{k_La\ sulf} = E'$
1	1.37	989	1473	0.670
2	1.37	954	1473	0.647
3	1.0	609	1227	0.496
4	0.82	554	1142	0.485
5	0.82	425	1142	0.372
6	0.40	149	762	0.195

TABLE 2

VVM min^{-1}	OER Kg 0/KW (sulfite method)	k_La h^{-1}
0.4	11.34	
0.8	9.34	1044
1.2	9.07	1782
1.5	8.61	2394

TABLE 3

TYPE OF FERMENTOR	v mmol 02/ ℓ h	OER Kg 02/ KWh (VVM = 0.8 h^{-1} af = 1 cm/sec)
Stirred Tank (50 ℓ)	200	0.4
Bubbling Column (200 ℓ)	50 - 100	2.54 - 0.4
Air Lift (200 ℓ)	50 - 100	7.125 - 5.36
Pulsed Reactor (37 ℓ)	150 - 200	4.47

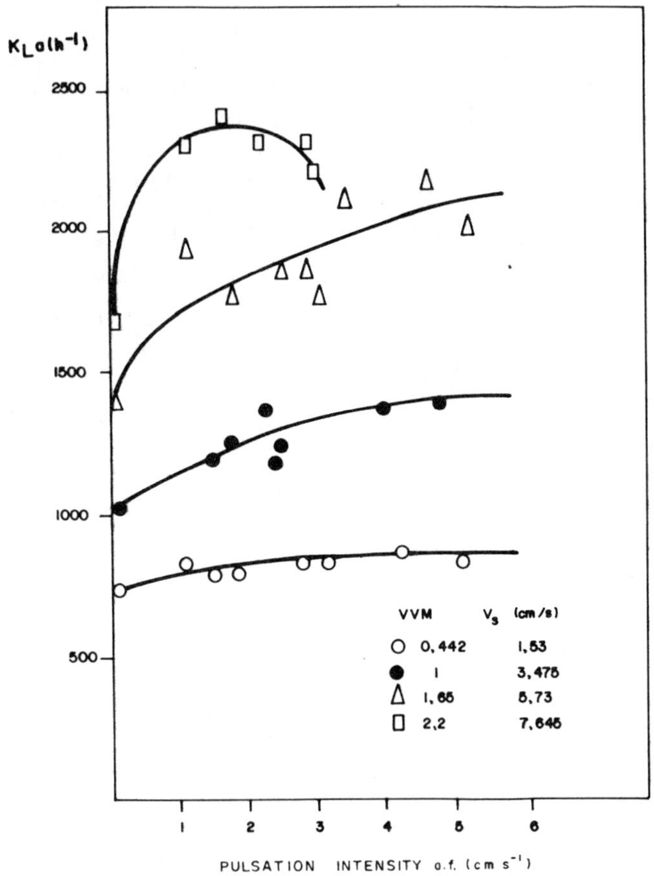

FIGURE 4. Mass transfer coefficient (k_La) vs. pulsation intensity and aeration rate.

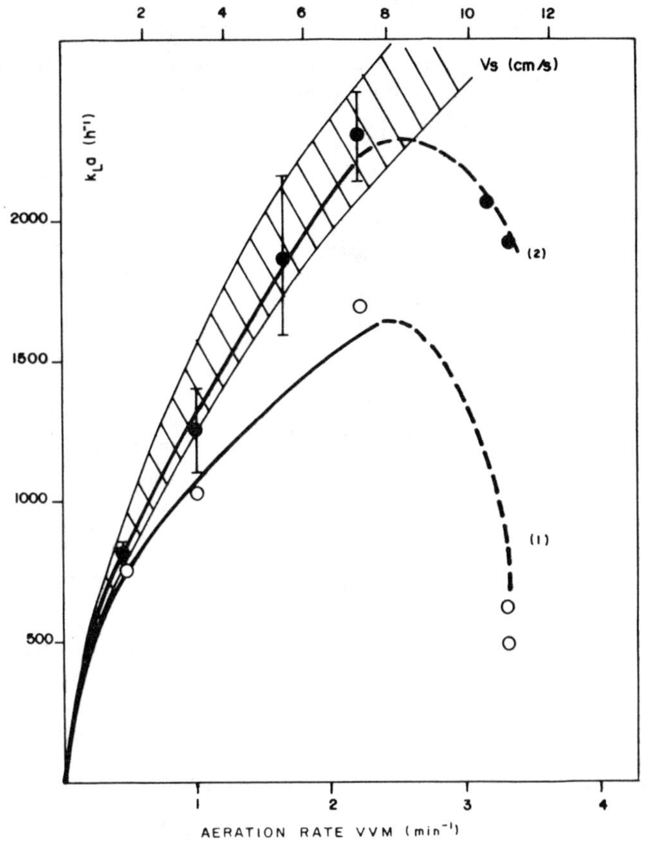

FIGURE 5. Mass transfer coefficient (k_La) vs. aeration rate without pulsation (1) and with pulsation (2) (af = 1 cm/s).

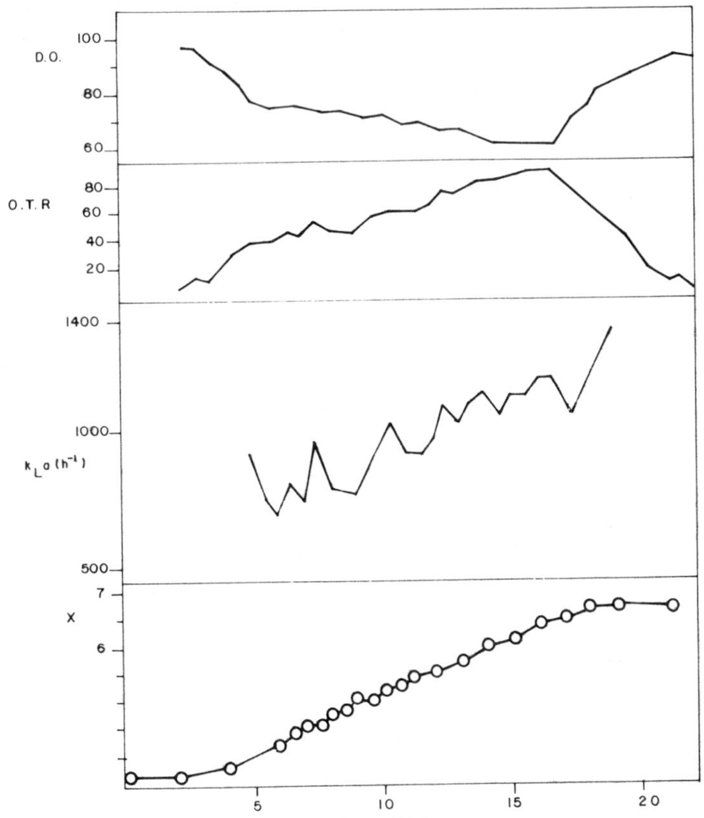

FIGURE 6. Results in fermentation. Units: Cells concentration (X), g/L; mass transfer coefficient (k_La), H^{-1}; oxygen transfer rate (O.T.R.) mmol/L h; dissolved oxygen concentration (D.O.), % saturation at 20°C.

ACKNOWLEDGMENT.

The author would like to thank the Consejo Nacional de Ciencia y Tecnología (México) wich supported in part this research and the Centro de Investigación Científica de Yucatán for the material support when preparing this paper.

NOTATION.

a	Gas-liquid area per unit liquid volume, cm^{-1}
A	Air flow rate, m mol/liter h
af	Pulsation intensity, cm/sec
C	Oxygen concentration in the liquid bulk, m mol/liter
D	Diffussivity coefficient, cm^2/sec.
E'	Modified enhancement coefficient defined by equation (4)
F	Flow rate, m mol/liter h
H	Henry's constant, mmol/liter bar
k	Reaction kinetical constant, vol/mass time
k_L	Mass transfer coefficient, cm/h
k_La	Oxygen transfer potential, h^{-1}
k_{Lf}	Mass transfer coefficient in fermentation, cm/h
l	Displacement of the liquid in the side column, meter
OTR	Oxygen transfer rate, mmol/liter h
P	Total pressure, bar
R_1,R_2	Oxygen probe readings
OER	Oxygenation efficiency rate
S	Area of the side column section m^2
tc	Period where the pulsation air is stopped, sec
to	Period where the pulsation air is supplied, sec
V	Oxygen transfer rate, mmol/liter h
v	Volume of liquid in the reactor, m^3
VVM	Aeration rate, min^{-1}
Wa	Dissipated power in aeration, w/m^3
Wp	Dissipated power in pulsation, w/m^3
x	Cell concentration, gram/liter
Y	Mol fraction
ε	Gas hold-up

LITERATURE CITED.

1. Cooper, C., Fernstrom, G., Miller, S., Ind. Eng. Chem., 36, 504 (1944)

2. Taguchi, H., Humprey, A.E., J. Ferment Technol., 44, 881 (1966)

3. Danckwerts, P.V., Gas-liquid reactions. Mc Graw Hill (1970)

4. Marrucci, G., Nicodemo, L., Chem. Eng. Sci., 22, 1257 (1967)

5. Zieminsky, S., Whittemore, R., Chem. Eng. Sci., 26, 509 (1971)

6. Linek, V., Benes, P., Biotechnol. Bioeng. 20, 697 (1978)

7. Tsao, G., Biotechnol. Bioeng., 10, 765 (1968)

8. Tsao, G., Biotechnol. Bioeng., 20, 159 (1978)

9. Wise, D., Wang, D., Mateles, R., Biotechnol. Bioeng., 11, 647 (1969)

10. Lee, Y., Tsao, G., Ind. Eng. Chem., 27, 1953 (1972)

11. Lee, Y., Tsao, G., Chem. Eng. Sci., 27, 1601 (1972)

12. Tsao, G., Lee, D., AIChE, J., 21, 979 (1975)

13. Yagui, S., Yoshida, F., J. Ferment. Technol. (Japan) 5, 905 (1974)

14. Yagui, S., Yoshida, F., Biotechnol. Bioeng. 27, 1083 (1975)

15. Linek, V., Benes, P., Biotechnol. Bioeng., 19, 565 (1977)

16. Merchuk, J., Biotechnol. Bioeng., 19, 1885 (1977)

17. Schultz, E., Gaden, E., Ind. Eng. Chem., 48, 12,2209 (1956)

18. Jaral, M., Gyory, E., Tombor, J., Biotechnol. Bioeng. 11, 605 (1969)

19. Serizawa, A., Kataoka, I., Michyoshi, I., Int. J., Multiphase Flow 2, 221 (1975)

20. Quintero, R., Ingeniería Bioquímica, First Ed., P. 99, Alhambra, México (1981).

Modeling of Interspecies Hydrogen Transfer in Microbial Flocs

SADETTIN S. ÖZTÜRK
BERNHARD Ø. PALSSON
Department of Chemical Engineering
University of Michigan
Ann Arbor, Michigan 48109

JURGEN THIELE
J. GREGORY ZEIKUS
Michigan Biotechnology Institute
East Lansing, Michigan 48824

Interspecies hydrogen transfer in microbial flocs is believed to play an important role in regulating the syntrophic interactions in anaerobic digestion. The efficiency of this process is reported to be better in larger flocs, which along with thermodynamic calculations, suggest that intra-floc diffusional limitations are important. In this work, a mathematical analysis is carried out to assess the existence of diffusional resistances in microbial flocs, that are composed to acetogenic and methanogenic bateria. When applied to a model digestion system where ethanol is used as a substrate for methane formation, the model predicts the dynamics of interspecies hydrogen transfer accurately and explains the higher biomethanation efficiency of larger flocs. A further thermodynamic analysis of the energetics of this system led to the important conclusion that the acidogenic and methanogenic reactions are coupled by another electron-carrier than molecular hydrogen.

The anaerobic digestion is receiving increased attention as it provides a convenient way of eliminating wastes by energy production through methane formation. The digestion is a complex process involving a number of strongly interacting microorganisms. A dynamic balance in these syntrophic interactions is essential for determining the overall performance stability and efficiency of the anaerobic digester (1).

The overall structure of the digestion process has been established and the process is believed to proceed in three stages (2) as seen in Fig. 1. The first stage involves hydrolysis of biomass into smaller units, amino acids, sugars and fatty acids. The second stage involves acidogenesis where these units are converted into acetic acid and hydrogen (H_2) as precursors for methane (CH_4) formation. Intermediates such as butyric, lactic and propionic acids and ethanol are also produced and these are converted into acetic acid and H_2 by acetogenic bacteria. In the third and last stage methanogenic bacteria utilize the H_2 and acetate to form methane.

The interaction between acetogenic and methanogenic bacteria has been subjected to considerable number of studies and it has been shown to play an essential role in the digestion. The term interspecies hydrogen transfer (IHT) has been used to describe the interaction. It has also been shown that the growth and metabolism of acetogenic bacteria is inhibited by H_2 and unless H_2 is removed the acetogens do not function which leads to an accumulation of the in-

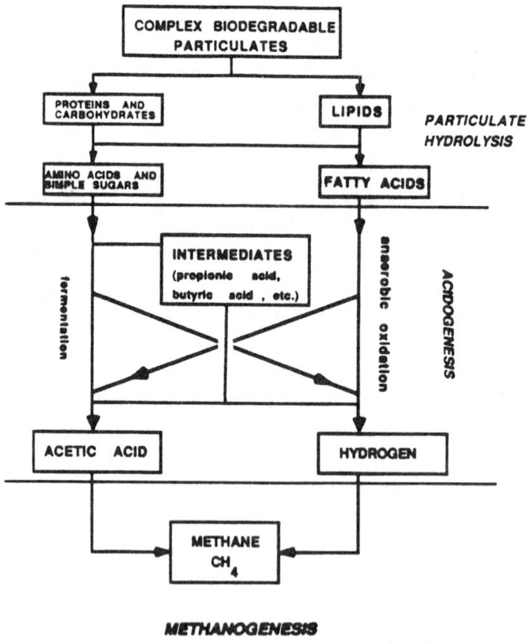

FIGURE 1. Overall structure of anaerobic digestion (after Bryers, 2).

termediate acids, a drop in pH which ultimately results in digester failure. The success of the digestion therefore becomes dependent on the efficiency of IHT.

The acetogenic and methanogenic bacteria are not only metabolically coupled but also they are mostly found attached each other in the form of microbial aggregates or flocs. The bacteria held together by an extracellular matrix. One can disrupt the flocs to obtain smaller sizes but after a certain period of time the bacteria re-aggregate. Electron microscopic pictures of the flocs shows that they contain a variety of bacteria. Heterogenity and "juxtapositioning" of the species within the floc is clear and there is no special organization of the bacteria.

For a reliable design and control of anaerobic digestion reactors, a complete understanding of the structure and dynamic behaviour of the syntrophic interactions is essential. These interactions must be studied both biologically and kinetically, characterized and quantified. This work presents a quantitave investigation of the interactions between acetogenic and methanogenic bacteria. A model system where the ethanol is converted to methane is considered to avoid the complexity of the complete process and to gain more insight into the dynamics of the IHT.

DYNAMICS and ENERGETICS of the MODEL SYSTEM

In the work by Thiele et al (3) flocs from a whey digester were seperated, purified and examined. In this model system, ethanol is converted into acetate and H_2 by the acetogenic bacteria and the hydrogen is believed to be transfered to methanogenic bacteria by IHT and converted into methane. *Desulfovibrio vulgaris* and *Methanobacterium formicicum* were the predominant acidogen and methanogens respectively. Here two different preperations for the flocs were used; intact and disrupted flocs. The intact case corresponded to the original purified flocs from the digester while disrupted case used fragments obtained from the original, intact flocs by gentle disruption.

Fig. 2 shows the time courses for the system obtained for intact and disrupted floc preperations. It is clear from the figure that ethanol is converted into acetate and hydrogen and the hydrogen is converted into methane. It was found

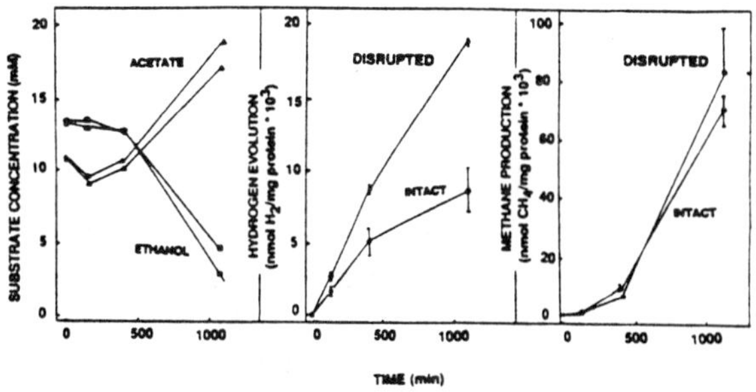

FIGURE 2. Kinetic behavior of ethanol methanogenesis with microbial flocs.

that more than 95% of the total H_2-equivalents produced are not released from the flocs, but converted to methane. The IHT efficiency is defined as the ratio of H_2 used for methane formation to the total H_2 formed which calculated from ethanol consumption. In this respect, the more H_2 levels in the medium for the same amount of ethanol converted means less efficient IHT. An interesting result from this data is that the efficiency of process is dependent on the floc size; reducing the size from $100\mu m$ in the intact case to $10\mu m$ in the disrupted case increases the H_2 evolution approximately twofold (Fig. 2). Since all other conditions were identical this data suggests that diffusional resistances existed inside the floc.

The energy metabolism in the acetogenic bacteria during ethanol oxidation is believed to involve a substrate level phosphorylation (SLP). The conversion of ethanol to acetate drives ATP production as illustrated by Fig. 3 and for each molecule of ethanol consumed one molecule of ATP is formed. The vital dependency of acetogenic bacteria on a methanogen or any other electron accepter (e.g. sulfate reducer) is also clear from this scheme; ferredoxin must be continually oxidized. The production of H_2 by this mechanism is reversible and the produced H_2 must be removed for the process to continue. The ATP production continues as long as enough free energy is provided by the ethanol oxidation. When H_2 accumulates in the media, the free energy extracted is lowered and the acetogenic bacteria are inhibited. This energy metabolism via substrate level phosphorylation requires a free energy input of about $-40kJ/mole$ from the ethanol oxidation reaction.

The free energies during the time course of the experiments were calculated in Thiele et al. (3), for both intact and disrupted cases based

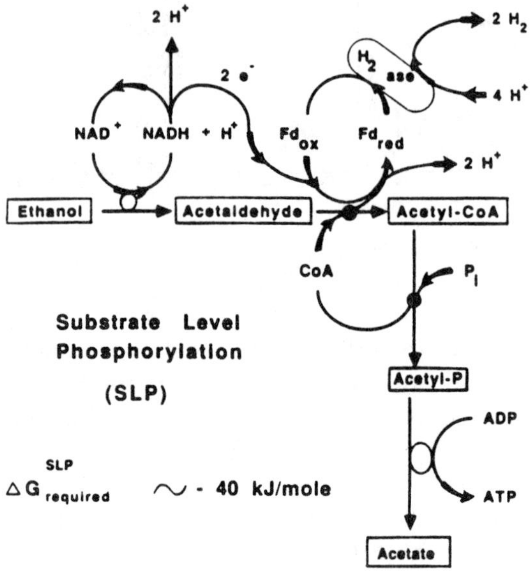

FIGURE 3. Hypothetical energy metabolism during acetogenic ethanol oxidation.

on the known bulk concentrations (Table 1). It can be seen that the free energy provided from ethanol oxidation is only enough for SLP at the beginning of the experiments. However the ethanol oxidation continues at the same rate under these unfavorable conditions for SLP.

Hence there is an appearent contradiction in the proposed energy metabolism scheme which assumes that H_2 formation is used in the regeration of ferridoxin in SLP and this H_2 is the transfering species between the acetogenic and methanogenic bacteria. The acetogens metabolize even under the conditions that the free energy is less than $-20kJ/mole$.

Based on this contradiction it was postulated, that either the concentrations in the floc were very different from the bulk values, which would be in accord with the difusional limitation hypothesis or, that the proposed energy metabolism is erroneous.

It is known that diffusional resistances in active microbial aggregates or flocs create differ-

TABLE 1
Comparison of the Free Energy Changes during Ethanol Oxidation by Anaerobic Digester Flocs (3). The Calculations Are Based on the Measured Bulk Concentrations and H_2 Partial Pressures

condition	time (min)	ΔG (kJ/ reaction)	H_2 partial pressure (Pa)	(ethanol) (mM)	(acetate) (mM)
intact	0	-51.2	0.85	13.5	11.0
disrupted	0	-45.3	2.73	13.5	10.7
intact	136	-28.1	78.8	13.0	9.5
disrupted	136	-27.3	89.9	13.5	9.0
intact	405	-22.2	232	12.7	10.5
disrupted	405	-21.5	276	12.6	10.0
intact	1100	-14.4	384	3.0	19.0
disrupted	1100	-13.7	589	4.7	17.0

ences in concentrations (see for example ref. 4). Hence the bulk concentrations do not represent the concentrations at the reaction site. The differences become more important for the cases with low diffusivities and/or high reaction rates (4). Therefore the free energies calculated from the bulk values may not correspond the actual free energies at the metabolic site. The concentrations at the metabolic site, or inside the flocs, are however not assesible for measurement at the present time.

In this paper, we present a simple reaction-diffusion model for dynamics of interspecies hydrogen transfer. The aim is to identify the diffusional resistances during IHT and make a critical analysis of the data published in the literature (e.g. ref. 3). The model equations and the assumptions made in their derivation are presented in detail. Measurements of diffusion coefficients in the floc phase for hydrogen, ethanol and acetate and the evaluation of reaction rate constants are included.

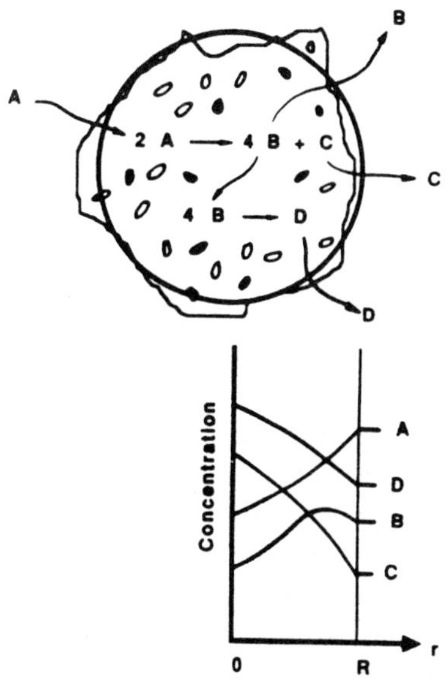

FIGURE 4. Schematic representation of the reaction diffusion processes and the concentration profiles in the floc.

DYNAMIC DESCRIPTION of MODEL ANAEROBIC DIGESTION SYSTEM

Microbial Flocs. The overall reactions taking place in the floc for the model system where ethanol was used as a substrate are assumed to be described by the following reaction mechanism:

Acetogens:

$$2C_2H_5OH + 2H_2O \rightarrow 4H_2 + 2CH_3COO^- + 2H^+ \quad [1]$$

Methanogens:

$$4H_2 + HCO_3^- + H^+ \rightarrow CH_4 + 3H_2O \quad [2]$$

For convenience we will use A to denote ethanol (C_2H_5OH), B to denote hydrogen (H_2), C to denote acetate (CH_3COO^-) and D to denote methane (CH_4).

We will treat the microbial flocs as spheres[1] in which acetogen and methanogens are uniformly distributed[2]. A schematic of an idealized spherical floc is shown in Fig. 4 where the reactant A diffuses into the floc and reacts according to reaction [1]. The concentration of A is lowered in the floc due to the simultaneous diffusion and reaction processes. The product of reaction [1], species B, Either undergoes reaction [2] or diffuses out of the floc. The products from reaction [1] and [2], C and D diffuse out of floc into the surrounding liquid medium. The concentration of of these species must be higher in the floc in order to have diffusion out of the floc. A relatively complex concentration profile for B, as illustrated in Fig. 4, can be obtained depending on the relative magnitudes of the reaction rates for reaction [1] and [2] and mass transfer by diffusion. We will now set out to evaluate the parameter combination that give concentration profiles that are consistent with the experimental data by Thiele et al. (3).

The equations describing the concentration profiles in the flocs are obtained by taking a microsopic mass balance in spherical coordinates on A, B, C and D yielding:

[1]Extensive results of reaction-diffusion modeling work has shown that in general the results obtained are relatively insensitive to geometry (e.g. ref. 5)

[2]Thin sectioning electron microscopy of purified flocs reveals a quasi homogeneous mixture of bacterial cells and the absence of any microcolony formation (3).

$$\frac{D_A}{r^2}\frac{d}{dr}(r^2\frac{dC_A}{dr}) = r_1 \quad (1)$$

$$\frac{D_B}{r^2}\frac{d}{dr}(r^2\frac{dC_B}{dr}) = r_2 - 2r_1 \quad (2)$$

$$\frac{D_C}{r^2}\frac{d}{dr}(r^2\frac{dC_C}{dr}) = -r_1 \quad (3)$$

$$\frac{D_D}{r^2}\frac{d}{dr}(r^2\frac{dC_D}{dr}) = -\frac{1}{4}r_2 \quad (4)$$

where r is the radial coordinate, D_A, D_B and D_C diffusion coefficients, r_1 and r_2 are the reaction rates describing the acetogenesis and methanogenesis respectively. Taking into consideration that the system is buffered at $pH = 7$, H^+ concentration is assumed to be uniform everywhere and therefore no equation is needed H^+. Also changes in the total number of the water molecules is neglected.

The boundary conditions for the floc phase differential equations are:

- At the center, $(r = 0)$:

$$\frac{dC_A}{dr} = \frac{dC_B}{dr} = \frac{dC_C}{dr} = \frac{dC_D}{dr} = 0 \quad (5)$$

This condition implies that the concentration profiles are symmetrical around the center.

- At the surface, $(r = R)$:

$$C_A = C_A^0 \quad C_B = C_B^0 \quad C_C = C_C^0 \quad C_D = C_D^0 \quad (6)$$

This condition means that the concentration of the floc at the surface is the same as in the well mixed bulk phase, denoted by the zero superscript.

For the mass balances given above, only two of the four are independent and the concentrations of C and D can be determined via overall mass balances. For the concentration of C we can justify that:

$$C_C = C_C^0 + \delta(C_A^0 - C_A) \quad (7)$$

where δ is a diffusivity ratio, $(\delta = D_A/D_C)$ and superscript 0 denotes bulk concentrations.

The reaction rates, in general, can be expressed in Michealis-Menten form. As mentioned before, the acetogenic reaction (ethanol consumption) is inhibited by its product, H_2 and reversible. Although the kinetic inhibition has not been identified, the reversibility of the reaction can be considered in the following form of the rate expression:

$$r_1 = \frac{k_1(C_A - C_B^2 C_C/K'_{eq})}{K_{m,A} + C_A} \quad (8)$$

On the other hand, the kinetics of methanogenic reaction can simply be represented by:

$$r_2 = \frac{k_2 C_B}{K_{m,B} + C_B} \quad (9)$$

When the reaction rate expressions are inserted into differential equations (Eq.1 and 2) the final form of the model equations is obtained:

$$\frac{D_A}{r^2}\frac{d}{dr}(r^2\frac{dC_A}{dr}) = \frac{k_1(C_A - C_B^2(C_C^0 + \delta(C_A^0 - C_A))/K'_{eq})}{K_{m,A} + C_A} \quad (10)$$

$$\frac{D_B}{r^2}\frac{d}{dr}(r^2\frac{dC_B}{dr}) = k_2\frac{C_B}{K_{m,B} + C_B} - 2\frac{k_1(C_A - C_B^2(C_C^0 + \delta_1(C_A^0 - C_A))/K'_{eq})}{K_{m,A} + C_A} \quad (11)$$

Bulk liquid. In the batch digestion of ethanol considered here, liquid phase containing flocs is in contact with a gas phase. The total volume is constant and both phases can be considered well mixed. The mass transfer rate between the gas and liquid phases and between the flocs and the liquid are rapid. A simple time constant comparison shows that the transients inside the floc are much faster than the concentration changes in the bulk. Hence, we can use the solution of steady state floc phase equations and write the following balance equations to describe the variations of the bulk concentrations with time by using C_A^0 and C_B^0 for the bulk concentrations of ethanol and dissolved hydrogen respectively:

for A

$$V\frac{dC_A^0}{dt} = -a_v D_A(\frac{dC_A}{dr})_{r=R} \quad (12)$$

for B

$$V(1+\alpha)\frac{dC_B^0}{dt} = -a_v D_B(\frac{dC_B}{dr})_{r=R} \quad (13)$$

where V is the liquid volume, a_v is the specific interfacial area of the flocs, and α is the distribution coefficient for H_2 between gas and liquid phases which is defined as:

$$\alpha = \frac{\text{mole in gas}}{\text{mole in liquid}} = (\frac{V_G}{V})(\frac{P}{RTHe}) \quad (14)$$

where He is the Henry's constant for species i, P is H_2 partial pressure, T is temperature, R is gas constant and V_G is gas volume.

Solution Technique: The system of two differential equations which comprise the mathematical model is nonlinear and subject to the indicated boundary conditions. An analytical solution to this system does not exist and numerical solutions are hard to obtain. The equations were solved numerically using the method of orthogonal collocation (6). On the other hand the concentration-time profiles are evaluated by the solution of first-order differential equations after the derivatives are calculated. In this work we used Runge-Kutta-Gill method to carry out these calculations (7). At each point of the calculations the required derivatives for these equations are obtained from the solution of differential equations 10 and 11.

Results obtained from the solution of floc phase mass balance equations showed that different cases are obtainable depending on the magnitudes of the model parameters. Fig. 5 shows the concentration profiles of H_2 inside the floc phase as a function of a combined parameter, Thiele modulus ($\phi = R\sqrt{(k_1/K_{m,A})/D_A}$). This figure is prepared for a special case for which $K_{m,A} \gg C_A^0$, $K_{m,B} \gg C_B^0$ and for $C_A^0 \gg C_B^0$ or first order kinetics for both the reactions. From the figure we see that the floc concentration can be equal to (low ϕ), greater (intermediate ϕ) or smaller than (high ϕ) the bulk concentration.

Another parameter, ($\psi = R\sqrt{(k_2/K_{m,B})/D_B}$) was found to be equally important being another Thiele modulus. These type of parameters (Thiele moduli) are consist of rate constants and diffusivities and are commonly encountered in such diffusion-reaction problems. Accurate experimental determination of these parameters are essential as the models are very sensitive to these values (Fig. 5).

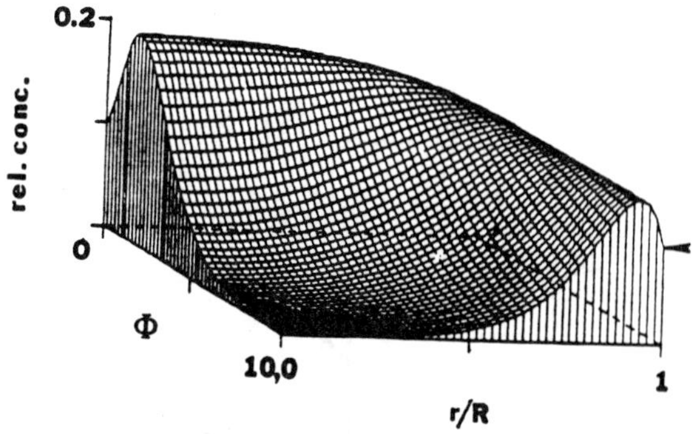

FIGURE 5. Model predictions for H_2 concentration in the floc phase (the arrow shows the bulk concentration).

MEASUREMENT of DIFFUSION COEFFICIENTS

Little is known about the diffusion coefficients in microbial flocs. Experimental data are scarce and the values reported vary largely, because accurate measurements are difficult and tedious. Onuma et al. (8) measured the diffusivities for glucose and oxygen in microbial aggregate and found that they are not different from those for diffusion in free aqueous solutions. There was a slight dependency on the C/N ratio. On the other hand Bailey and Ollis (4) report $50 - 100$ times lower coefficients for O_2. For precise determination of the parameters appear in our model, independent measurements were carried out for H_2, ethanol and acetate at a temperature of 37 $°C$.

Floc material from a whey digester was collected and seperated by centrifugation. It was heated at 80 $°C$ for 10 minutes to stop the biological activity. The flocs were resuspended in a phosphate buffer (0.0816 M, pH=7.0) containing 500 μM $HgCl_2$. A part of the flocs were examined for biological activity and none was found. After centrifugation at 100,000 g for 30 minutes a paste that mimics the physical structure of the flocs in the digester was obtained and used for the diffusion experiments.

Diffusivity of ethanol in the floc phase was determined in a diffusion cell of 50 ml (Fig. 6).

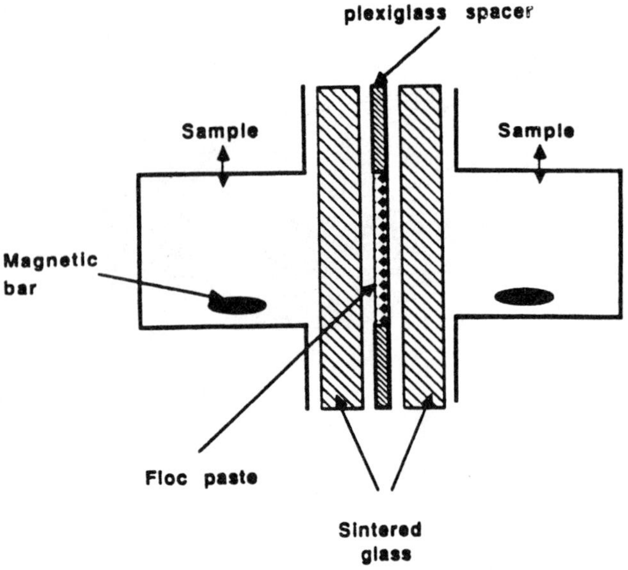

FIGURE 6. The diffusion cell used for the determination of ethanol and acetate diffusivities.

The floc paste was placed into a circular plexiglass spacer between two filter papers (Milipore, Cat. no = HAWP 04700, filter type HA, pore-size = $0.45\mu m$). The spacer was 2 mm thick with a diameter of 3 cm. Two sintered glass disks (4 mm thick), each being one side of the spacer was used to hold it between the cells. Magnetic stirrers were used to mix the content of the cells.

The chambers were filled with a phosphate buffer ($0.0816\ M, pH = 7$) containing $500\mu M$ $HgCl_2$. A known amount of ethanol was injected to one side and samples from the injection and receiver sides were taken. During sampling 0.5 ml cell content was replaced with the original buffer solution to keep the volume constant and this dilution was accounted for in the analysis. The samples are acidified with H_3PO_4 (1 N final) and subsequently analyzed on a model 419 Packard gas chromotograph. Experiments were run for 5-7 days and sampling time was varied between 12-24 hours.

For the determination of the diffusion cell constant, the spacer was removed and measurements were done with filter papers and sintered glasses only. Same composition of the liquid and the procedure were applied. Diffusion coefficients were then calculated from the standard procedure (e.g. ref. 9).

The determination of diffusion coefficients using a O_2 electrode is simple, fast and reliable (10, 11). However, such a measurement for H_2 has not been been well-established. A Clark electrode (Diamond Electrotech) was modified to measure H_2 amperometrically. A teflon attachment with an inner diameter of $3mm$ and length of $5mm$ was manufactured and placed on the top of the electrode to hold the paste. The electrode was enclosed in a temperature controlled jacket. The gas phase was kept in contact with the electrode as the jacket was closed by two stopers. The total volume was 40 ml. A small amount of water (1 ml) was added at the bottom and an anular wetted tissue was placed on the top of the attachment to provent the paste from drying.

Two readings, one with pure water and one with floc paste were carried out. After the temperature equilibration to $37\ ^\circ C$, H_2 was injected to the gas phase initially containing 100% N_2 by a syringe simultaneously using a second syringe to avoid the pressure change. The gases were mixed, final mixture was 40% H_2. The gas composition was measured by gas chromotography using a Hewlett Packard 5890A gas chromotograph equipped with a thermal conductivity cell. The overpressure, if present, was eliminated by punching the top stopper by a needle. Electrode signals were amplified and recorded. After 2-2.5 hours a steady state was reached. The consumption of H_2 during the experiments was found to be negligible. As the electode characteristics was observed to be changing from run to run calibrations were carried out immediately after the experiments.

The diffusion coefficients measured here were found not to be very different than those for aqueous solutions (Table 2). The values were about two (for hydrogen) and four (for ethanol and acetate) times lower than the values in water. This can be explained with the appearent diffusivity of the floc phase. The decrease in diffusion coefficients higher in the latter species as their diffusional path is restricted by the porosity and tortousity of the floc material.

RATE CONSTANTS

The Michealis-Menten constants for acetogenic and methanogenic bacteria are determined by Chartrain (12) in a similar system. The rate constants, k_1 and k_2 are determined from the data of Thiele et al. (3). The data for disrupted case (Fig. 2) (appearently with less or negligible

TABLE 2
Summary of the Model Parameters

CONSTANTS		
k_1	$= 6.7 \cdot 10^{-5}$ M Ethanol/sec	1
k_2	$= 1.83 \cdot 10^{-4}$ M H_2/sec	1
K'_{eq}	$= 2.4 \cdot 10^{-2}$ $(atm)^2 = 2.19 \cdot 10^{-10}$ M^2	2
$K_{m,A}$	$= 2.5 \cdot 10^{-4}$ M	3
$K_{m,B}$	$= 6.5 \cdot 10^{-6}$ M	3
He	$= 1.62 \cdot 10^3$ atm/M	4
D_A	$= 1.2 \cdot 10^{-6}$ cm^2/sec	1
D_B	$= 1.39 \cdot 10^{-5}$ cm^2/sec	1
D_C	$= 1.71 \cdot 10^{-6}$ cm^2/sec	1

[1] determined in this study. The volumetric rate constants are calculated with a measured apparent protein density of 0.051 g/cm^3 and the diffusivities are averaged over at least three measurements.

[2] based on a standard free energy of $\Delta G^{o\prime} = 9.62 kJ/reaction$

[3] from Chartrain (12)

[4] from Thiele et al (3)

diffusional resistances) are used for the fitting of rate equations and for the extraction of rate constants. With the information on the volume fraction of the flocs in the liquid it was possible to determine the rate constants per mg of floc protein. A measured floc protein density allowed to calculate the volumetric reaction rate constants. The results are included in Table 2.

RESULTS and DISCUSSION

After having all the relevant coefficients in model equations the data by Thiele et al. (3) was simulated. The predicted H_2 concentration profiles inside the floc is given in Fig. 7. We see that the floc phase always has a higher concentration than the bulk. Although the bulk concentration changes with time the center sees a constant H_2 concentration. The diffusional resistances inside the floc is considerable and are more important initially when the bulk concentration is low.

Fig. 8 shows that the model represents the experimentally obtained concentration profiles well. The concentrations for ethanol and acetate are predicted within 5% error in both intact and disrupted cases. It must be mentioned that this calculations no fitting procedure has been used for the intact case. The rate constants obtained from the disrupted case and the diffusion coefficients measured are used directly.

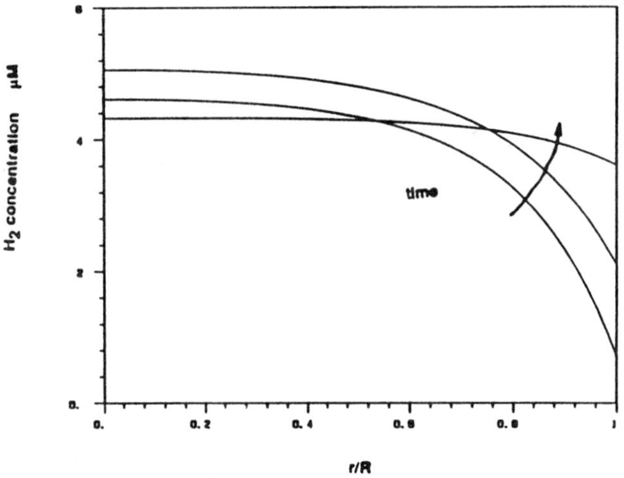

FIGURE 7. Model predictions of H_2 concentrations in the floc under experimental conditions.

The hydrogen concentrations in the bulk are also predicted successfully by the model (Fig. 9). The difference in the concentrations arises

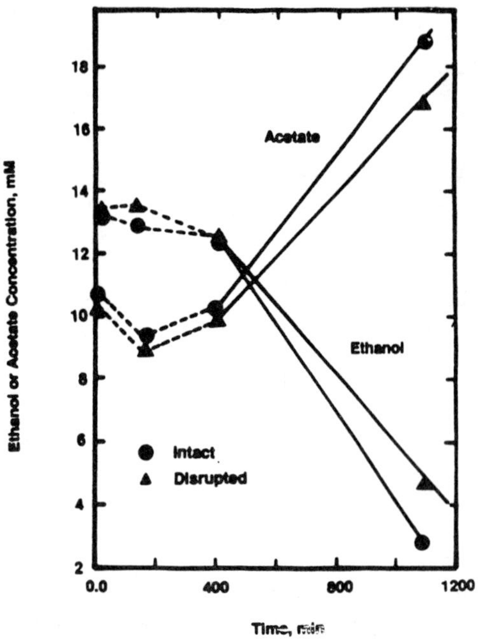

FIGURE 8. Comparison of the model predictions with the experimental data by Thiele *et al.* (3): Ethanol and acetate concentrations.

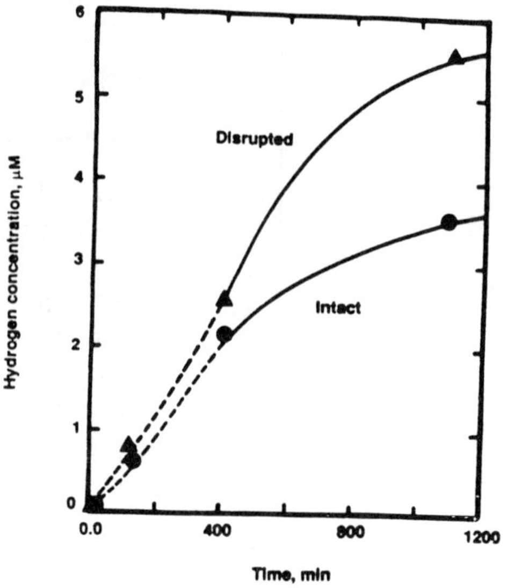

FIGURE 9. Comparison of the model predictions with the experimental data by Thiele *et al.* (3): Hydrogen concentrations.

TABLE 3
Predicted Concentrations and Free Energies at the Center of the Microbial Flocs

condition	time (min)	ΔG (kJ/ reaction)	H_2 partial pressure (Pa)	(ethanol) (mM)	(acetate) (mM)
intact	0	-15.7	820	13.3	11.0
disrupted	0	-36.3	15.15	13.5	10.7
intact	136	-16.5	755	12.7	9.5
disrupted	136	-23.1	203.2	13.5	9.0
intact	405	-15.7	828	12.7	10.7
disrupted	405	-20.7	356	12.6	10.0
intact	1100	-11.3	705	2.8	19.2
disrupted	1100	-13.7	589	4.7	17.0

mainly from the change in the surface areas and the fluxes into the bulk liquid. The system is mildly diffusion controlled as ten-fold increase in surface area gives only twofold increase in the release of H_2 from the flocs.

It was found that the concentrations inside the floc are not significantly different from the bulk values. The concentrations at the center of the floc are presented in Table 3. When the concentrations at the center are compared with the bulk values (Table 1) it is clear that only H_2 concentrations differ significantly. The calculated free energies at the center are also presented in this table. Here we see that even at the beginning of the experiments the free energy is not enough to provide the acetogenic bacteria enough free energy for ATP production via the postulated energy metabolism (SLP).

It is therefore concluded that although the kinetics of the model system can be explained successfully the energetics connot be resolved with the present IHT mechanism. There are two possible explainations:

- the acetogenic bacteria do not produce ATP by SLP, but by an ATPase coupled to transmembrane gradients. The authors are unaware of any biochemical data that support this hypothesis.
- the electron carrier is not H_2, but some other chemical species. Recent unpublished data obtained at MBI (Michigan Biotechnology Institute) strongly supports that the electrons are carried by formate which is produced by reducing bicarbonate

NOTATION

a_v	:	specific interfacial area of the flocs in the liquid
α	:	distrubition coefficient
C	:	concentration
D	:	diffusivity
δ	:	diffusivity ratio
He	:	Henry's constant for H_2
k_1	:	volumetric reaction rate for acetogenesis
K_2	:	volumetric reaction rate for methanogenesis
K'_{eq}	:	equilibrium constant
K_m	:	Michealis constant
r	:	radial coordinate
r	:	reaction rate
R	:	radius
ρ_f	:	floc density
V	:	liquid volume
V_G	:	gas volume

Subscript and superscripts

A	:	ethanol
B	:	hydrogen
C	:	acetate
D	:	methane
o	:	bulk
0	:	bulk

LITERATURE CITED

1. Harper, S. R., and F. G. Pohland, Biotech. Bioeng.,XXVIII, pp. 582-602 (1986).
2. Bryers, J. D., Biotech. Bioeng.,XXVII, pp. 638-649 (1985).
3. Thiele, J., M. Chartrain, and J. G. Zeikus, "Studies on Interspecies Hydrogen Transfer in Anaerobic Ecosystems", (submitted) (1986).
4. Bailey, J. E. and D. F. Ollis, "Biochemical Engineering Fundamentals", 2nd Ed., McGraw Hill Co., N.Y. (1986)
5. Aris. R., 'The Mathematical Theory of Diffusion and Reaction in Permeable Catalysts', 2 vols., University Press, Oxford (1975)
6. Ascher, U., J. Christiansen, and R. D. Russel, ACM Trans. Math. Softw., 7, 223 (1981)
7. Carnahan, B., H. A. Luther, and J. O. Wilkes, "Applied Numerical Methods", Wiley, New York (1969).

8. Onuma, M., T., T. Omura, T. Umita, and J. Aizawa, Biotech. Bioeng., Vol XXVII, pp. 1533 (1985)

9. Sherwood, T.K., R. L. Pigford, and C. R. Wilke, "Mass Transfer", McGraw Hill Co., New York (1975).

10. Akita, K., Ind. Eng. Chem. Fundam., 20, 89 (1981)

11. Ju, Lu-Kwang and C. S. Ho (1985), Biotech. Bioeng., XXVII, pp. 1495-1499.

12. Chartrain, M., Ph.D Thesis, University of Wisconsin (1985).

BIOPROCESS SCALE-UP

Scale-up of Fluid Mixing Equipment

VINCENT W. UHL
University of Virginia
Charlottesville, Virginia 22901

JOHN A. VON ESSEN
Philadelphia Mixers
Palmyra, Pennsylvania 17078

The various approaches available for scale-up in mixing operations are described and evaluated. It is shown that application of the principles of similarity may be conceptually useful, but it has little direct application value. An example is given that the four "translation equations" yield the same scale-up result. The place of performance indices or scale numbers in scale-up is discussed.

The profusion of papers (1) dealing with the scale-up of mixing operations suggests both the importance of this problem, and the difficulties confronting the process engineer. Factors which contribute to the complexity of scale-up are: the great number of different process operations conducted in agitated equipment; the range of charge characteristics - single to multiphase systems - high viscosity to mobile fluids; process conditions from batch to continuous, and semicontinuous; and yet one must get a handle on, develop an approach to this plethora of situations, an approach which can possibly lead to a ready answer, or a minimum program for a sound, reasonable solution. This calls for exploiting these techniques, and generally in this order:

1. Direct comparison to an existing successful installation of similar size.
2. The extrapolation or interpolation of test data generally secured for at least two scales.
3. The extrapolation of the result from a single lab or plant scale, preferably with the aid of an appropriate "translation equation."
4. Use of an available, appropriate equation.

The scale-up equations for different process results are generally in terms of performance indices such as power per unit volume (P/V), torque per unit volume (T_Q/V), or speed ratio (N_2/N_1). These are termed "translation equations." Often rate equations in terms of dimensionless groups are used. The existence of these many forms of relations, often for the same phenomenon has been the source of needless confusion, one which this paper endeavors to allay.

At the outset one must be aware of both the power and limitations of the various techniques. For instance, mixing operations have gone directly from the bench scale to commercial size, e.g., the Shell hotacid alkylation process (2). Conversely, processes which worked poorly or not at all on a small scale have been industrial successes, e.g., Hall's electrolytic process for the manufacture of aluminum (3). It must be recognized that scaling either up or down (termed modeling) is not always successful. Explanations for unexpected behavior include:

1. Indicated piloting not undertaken because of high cost or lack of time.
2. Inconsistency of available data or equations.
3. The desired process result is

inadequately defined.
4. A "mixed regime"[a] is involved.
5. In scaling a change occurs in the fluid regime.

The lessons here are that the process result must be understood and the capability of available methods to predict performance on a much different scale must be appreciated.

Over the past several decades, the advances have been both theoretical, leading mainly to a better understanding of what the process results entail, and experimental, generating better techniques, sophisticated instrumentation, and the statistical design of experiments. The goal to reduce the amount of expensive piloting has been achieved in some quarters (1,4); but piloting is still required for the complex operations.

PRINCIPLE OF SIMILARITY

It has been de rigueur for authors to consider the principle of similarity when introducing scale-up. The similarity states generally recognized as applicable for agitation are geometric, kinematic, and dynamic. Geometric similitude is illustrated in Figure 1. Reviews of the formal material on the principles of similarity and the related theory of models for process operations and particularly on their application to agitation can be found in both Johnstone and Thring (1) and Rushton (5).

While a useful purpose is served by this approach - it makes one aware of the significant characteristics of systems and how they vary with scaling - seldom is it possible to apply it directly. Geometric similarity is often maintained in going from bench to pilot plant tests; however, in the commercial size the configurations are often changed. For example, the ratio of the liquid height, Z, to the tank diameter, T, (Z/T), may go from 0.5 to 2.0 or more; the ratio of the impeller diameter, D, to the tank diameter, T, (D/T), may decrease from say 0.5 to 0.3, for process and for economic reasons. The maintenance of kinematic similarity is not adhered to because the increase of liquid velocities with scale would result in an increase in power intensity, or P/V, which would be excessive. Except in a few cases, it is impossible to apply dynamic

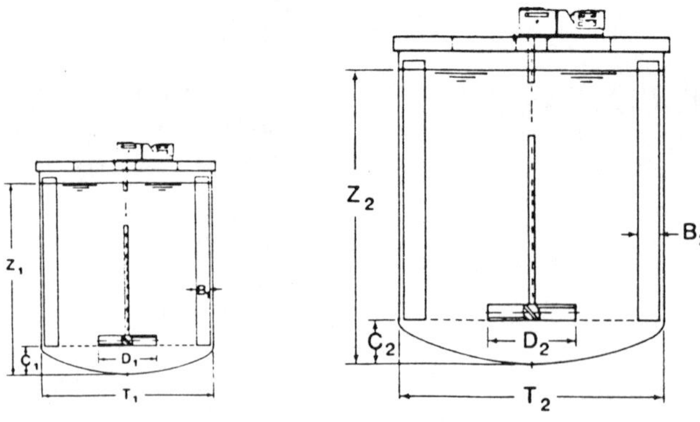

$$\frac{D_1}{T_1} = \frac{D_2}{T_2} \quad \frac{Z_1}{T_1} = \frac{Z_2}{T_2} \qquad \frac{C_1}{T_1} = \frac{C_2}{T_2} \quad \frac{B_1}{T_1} = \frac{B_2}{T_2}$$

FIGURE 1. An example of geometric similarity.

similarity, i.e., to hold more than one significant force ratio constant: Reynolds number, N_{Re}; Power number, N_P; Froude number, N_{Fr}; etc. An exception is the case of the power correlation for the turbulent regime in baffled tanks, where there is only one ratio, the Power number, because it is essentially constant (see Figure 2). Equations which predict other process results usually include more than one force or other dimensionless ratio.

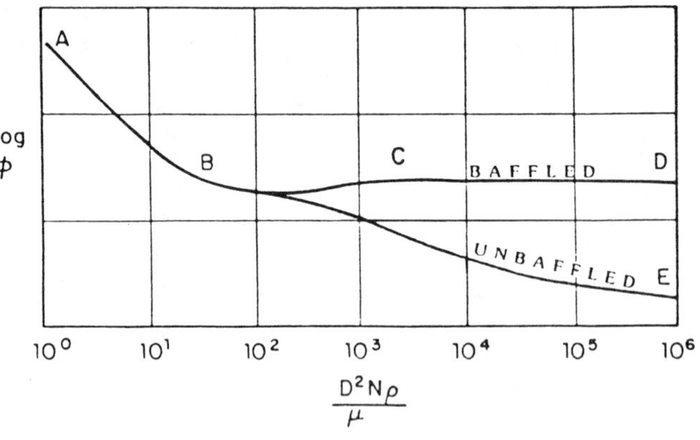

FIGURE 2. Power characteristics of a mixing impeller. After Rushton (*16*).

This is a propitious place to introduce a related consideration frequently overlooked in scaling; it is that the surface area of the vessel increases with volume approximately as $(V)^{2/3}$ which makes it impossible to maintain dynamic similitude with the same process fluid. This complicates scale-up because modified fluid properties called for to maintain dynamic similarity are virtually impossible to realize. The actual process fluid is normally used in a scaling test, and the balance between circulation rate and intensity of turbulence cannot be maintained. This complicates scale-up.

---

[a] A "mixed regime" is a condition for which the overall rate of change is significantly influenced by two processes having incompatible similarity criteria.

major heading "special considerations"). As indicated in Figure 4, this dimensional ratio can have a considerable effect on power invested to attain a particular process result.

Correlations for other common operations have been attempted with varying degrees of success; these are gas-liquid (absorption) and liquid-liquid (extraction). Besides referring to Nagata and Yamamoto (18) and Uhl et al. (19) regarding this situation, also see the subheading "dimensional equations."

Translation equations. Translation equations, also called scale-up rules, have become popular in recent years. Tatterson (20) delineates three approaches which are reproduced in Tables I and II. The earliest by Penney (21), which uses ε, (P/V), embraces seven different operations for turbulent mixing and five for viscous mixing. The relations displayed in Table II use these three scale-up parameters; ε, P/V, by Penney (21); τ, (T_Q/V), by Von Essen (22); and agitator speed, N, by Rautzen, et al. (23). They are compared only for the five operations in the turbulent regime given by Penney as shown from Tatterson (20) in Table I.

It appears that the first and most common scale-up rule was constant power input per unit volume, ε. It is often expressed as Hp/1000 gal. Over the years it was advanced for a number of operations by Büche (24),

TABLE I

Scale-up Recommendations by Penney (*21*) per Tatterson (*20*). Adapted from Uhl and Von Essen (*35*).

Laminar or viscous mixing, impeller Reynolds number <300		Turbulent mixing, impeller Reynolds number >300	
Criteria	Scale-up procedures	Criteria	Scale-up procedures
1) Equal heat transfer per unit volume	$\frac{(P/V)_2}{(P/V)_1} \sim \left(\frac{V_2}{V_1}\right)^1 \sim \left(\frac{D_2}{D_1}\right)^3$	1a) Equal mass and heat transfer co-efficients to particles, bubbles, or drops	$\frac{(P/V)_2}{(P/V)_1} \sim 1$
		1b) Equal bubble or drop diameter	
2) Equal heat transfer coefficients	$\frac{(P/V)_2}{(P/V)_1} \sim 1$	2) Equal heat transfer coefficients on stationary surfaces	$\frac{(P/V)_2}{(P/V)_1} \sim \left(\frac{V_2}{V_1}\right)^{0.15} \sim \left(\frac{D_2}{D_1}\right)^{0.45}$
3) Equal blend time	$\frac{(P/V)_2}{(P/V)_1} \sim 1$	3) Equal blend time	$\frac{(P/V)_2}{(P/V)_1} \sim \left(\frac{V_2}{V_1}\right)^{0.66} \sim \left(\frac{D_2}{D_1}\right)^2$
4) Equal tip speed	$\frac{(P/V)_2}{(P/V)_1} \sim \left(\frac{V_2}{V_1}\right)^{-0.66} \sim \left(\frac{D_2}{D_1}\right)^{-2}$	4) Equal tip speed	$\frac{(P/V)_2}{(P/V)_1} \sim \left(\frac{V_2}{V_1}\right)^{-0.33} \sim \left(\frac{D_2}{D_1}\right)^{-1}$
5) Equal impeller Reynolds number	$\frac{(P/V)_2}{(P/V)_1} \sim \left(\frac{V_2}{V_1}\right)^{-1.32} \sim \left(\frac{D_2}{D_1}\right)^{-4}$	5) Equal impeller Reynolds number	$\frac{(P/V)_2}{(P/V)_1} \sim \left(\frac{V_2}{V_1}\right)^{-1.32} \sim \left(\frac{D_2}{D_1}\right)^{-4}$
		6) Equal Froude number	$\frac{(P/V)_2}{(P/V)_1} \sim \left(\frac{V_2}{V_1}\right)^{0.15} \sim \left(\frac{D_2}{D_1}\right)^{0.45}$
		7) Equal solids sus-pension	$\frac{(P/V)_2}{(P/V)_1} \sim \left(\frac{V_2}{V_1}\right)^{-0.19} \sim \left(\frac{D_2}{D_1}\right)^{-0.55}$

Hirsekorn and Miller (25), and Miller and Mann (26), among others. Hyman (27) noted that the best documented use of constant power per unit volume as a scaling basis has been in gas-liquid systems carrying out fermentations and sulfite oxidations. These are shear sensitive operations. However,

TABLE 11

Equivalent Scale-up Procedures for Impeller Reynolds number > 300. Adapted from Uhl and Von Essen (35) and Tatterson (20)

Target operations	Penney(21) Eq. (1)		Von Essen(22) Eq. (2)		Rautzen et al.(23) Eq. (3)	
	Criteria	Scale-up procedures	Criteria	Scale-up procedures	Criteria	Scale-up procedures
1	Equal blend time	$\frac{(P/V)_2}{(P/V)_1} \sim \left(\frac{D_2}{D_1}\right)^2$	Equal blend time	$\frac{(T_Q/V)_2}{(T_Q/V)_1} \sim \left(\frac{D_2}{D_1}\right)^2$	Equal blend time	$\frac{N_2}{N_1} \sim 1$
2	Equal Froude number	$\frac{(P/V)_2}{(P/V)_1} \sim \left(\frac{D_2}{D_1}\right)^{0.45}$	Equal surface effects	$\frac{(T_Q/V)_2}{(T_Q/V)_1} \sim \left(\frac{D_2}{D_1}\right)^1$	Equal surface behavior	$\frac{N_2}{N_1} \sim \left(\frac{D_2}{D_1}\right)^{-0.5}$
3	Equal mass transfer coefficients to particles, drops, or bubbles; equal bubble or drop diameter	$\frac{(P/V)_2}{(P/V)_1} \sim 1$	Equal dispersion	$\frac{(T_Q/V)_2}{(T_Q/V)_1} \sim \left(\frac{D_2}{D_1}\right)^{0.66}$	Equal mass transfer (rate)	$\frac{N_2}{N_1} \sim \left(\frac{D_2}{D_1}\right)^{-0.66}$
4	Equal solids suspension	$\frac{(P/V)_2}{(P/V)_1} \sim \left(\frac{D_2}{D_1}\right)^{-0.55}$	Equal dilute solids suspension	$\frac{(T_Q/V)_2}{(T_Q/V)_1} \sim \left(\frac{D_2}{D_1}\right)^{0.5}$	Equal solids suspension	$\frac{N_2}{N_1} \sim \left(\frac{D_2}{D_1}\right)^{-0.75}$
5	Equal tip speed	$\frac{(P/V)_2}{(P/V)_1} \sim \left(\frac{D_2}{D_1}\right)^{-1}$	Equal fluid motion (velocity)	$\frac{(T_Q/V)_2}{(T_Q/V)_1} \sim 1$	Equal liquid motion (velocity)	$\frac{N_2}{N_1} \sim \left(\frac{D_2}{D_1}\right)^{-1}$

[a] From Tatterson (20)

RELATIONS FOR SCALE-UP

The relations for scale-up can be classified by their source:

- Theory, generally for microscale phenomena
- Empirical correlations of test data based on:
  - Dimensionless groups
  - Scale-up parameters (These result in so-called translation equations)
  - Dimensional variables

In addition to the above, there are also qualitative indices, which are characterized by words, such as mild, moderate, vigorous, or by arbitrary scale numbers, which correspond to such characterizations.

Theoretical Equations

These are often rigorous relations based on theory and they describe microscale, or local phenomena. Examples are for:

- Mass transfer to or from a liquid drop or a gas bubble
- Mass or heat transfer to a surface (Boundary layer theory can be applied here)

This approach is seldom applied, not only because of the paucity of useful relations, but also because of the effect of other concurrent phenomena such as the bulk circulation of the fluid.

Empirical Correlations

Scale-up calculations can be made with empirical equations. In some the variables are dimensionless groups and in others, i.e., the translation equations, use is made of mixing criteria, such as torque per unit volume. There are also empirical equations in dimensional form based on correlations of test variables. Unfortunately most of the empirical equations are based on tests run on bench or pilot scale equipment, with idealized geometric ratios and fluid properties so that "extrapolation" to commercial scale often is limited and can be unreliable.

Dimensionless groups. Dimensionless groups as the variables are common for correlations of mixer performance particularly to predict rates. Examples are: power invested (6,7), heat transfer (8,9), and mass transfer (to fixed surfaces) (10,11), blending times (12,13), and the suspension of solids within certain limitations (14,15). For a representation of the relations for power invested see Figure 2. A generalized

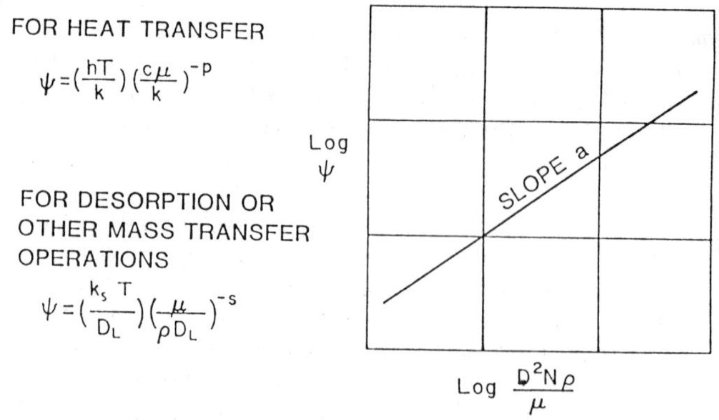

FIGURE 3. Correlation of rate coefficient, fluid properties, and fluid motion for turbulent regime. Adapted from Rushton (*16*).

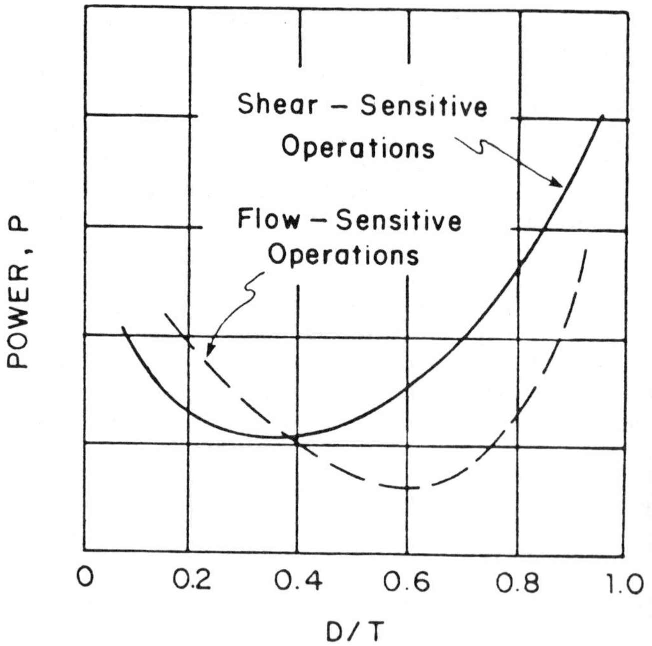

FIGURE 4. Examples of power vs *D/T* ratio for particular constant process results. Shear sensitive curve from Holland and Chapman (*17*). From Uhl and Von Essen (*35*).

correlation for heat and mass transfer is shown in Figure 3. Note that the common correlating group for these examples is the "mixing" Reynolds number, $D^2N\rho/\mu$. In most of the previous references cited, the data are from different size vessels for which geometric similitude was maintained. However, in some correlations of test data, e.g., Fenic and Fondy (13) for blend times, and Nienow (14) for solids suspension, the effect of varying D/T is accommodated (see under note for the translation equations for different process results, which are listed in Table I (Penney) and repeated under Penney in Table II, that the scale-up parameter, ε, is a power function of scale ratio D_2/D_1 in Equation (1) or V_2/V_1 in Equation (1a).

$$\frac{(P/V)_2}{(P/V)_1} = \frac{\varepsilon_2}{\varepsilon_1} = f\left(\frac{D_2}{D_1}\right) = \left(\frac{D_2}{D_1}\right)^{y} \quad (1)$$

$$\frac{(P/V)_2}{(P/V)_1} = f\left(\frac{V_2}{V_1}\right) = \left(\frac{V_2}{V_1}\right)^{y/3} \quad (1a)$$

These results assume that geometric similitude is maintained.

On the other hand, for flow sensitive operations such as blending and solids suspension, another rule, constant torque per unit volume, τ, (T_Q/V), has been invoked, for instance by Connolly and Winter (28), of Philadelphia Mixers, Palmyra, Pennsylvania. They describe a constant torque per unit volume technique that was successfully employed to scale up a coal slurry mixing operation from 50 gallons to 8 million gallons. Also, τ can be expressed as a function of scale for other operations. Then the scale-up relation is:

$$\frac{(T_Q/V)_2}{(T_Q/V)_1} = \frac{\tau_2}{\tau_1} = f\ (D_1/D_2) = (D_2/D_1)^{3x} \quad (2)$$

$$\frac{(T_Q/V)_2}{(T_Q/V)_1} = f\ (V_2/V_1) = (V_2/V_1)^{x} \quad (2a)$$

Relations such as described by Equations (1), (1a), (2), and (2a) can also be presented graphically. An example for Equation (1a) is presented in Figure 5.

Conventional wisdom has suggested constant peripheral speed of the impeller as a criterion for certain operations such as emulsification. Incidentally, it corresponds exactly to constant torque intensity, if geometric ratios are constant. This is demonstrated in detail by Connolly and Winter (28).

Another relation, which we shall term the "speed ratio correlation," appears to have been first introduced by Rushton (16). More recently Rautzen et al. (23) have popularized this approach, which is:

$$N_2/N_1 = (D_2/D_1)^n \quad (3)$$

It requires that geometric similitude be strictly maintained.

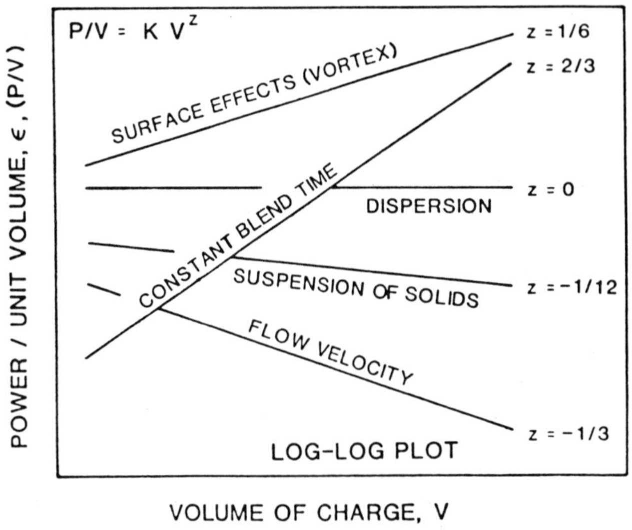

FIGURE 5. General representation of Equation (1a) where the relations are considered to be power functions. Note that the value of *K* will differ with the specific equation for each particular process result. Also the exponent *z* is equal to *y*/3 in Equation (1a). From Uhl and Von Essen (35).

The relations, such as depicted on Figure 2 for power-drawn, and Figure 3 for transport processes, can also be considered in the same manner as the translation equations. For instance, the "process results," "ψ," are expressed as a power function of the Reynolds number:

$$\psi = \psi' D = K\ (N_{Re})^a \quad (4)$$

The exponents a, n, x and y for the four "translation" equations or rules are displayed in Table III for the five operations noted by Tatterson in Table II. In the appendix it is demonstrated that all four of these rules give the same numerical answer for scaling up a given operation. However, some have advantages over others; this is discussed in the following outline. Some useful information on each of the five "target" operations listed in Table II (turbulent regime) is also tabulated.

Equal Blend Time (Constant speed) -

- Difficult to maintain on scale-up because fluid velocities increase

with D; this means the power requirement can become prohibitive on scale-up
- Sometimes necessary for fast chemical reactions
- Usually not necessary for blending applications because time scale of other process equipment (e.g., fill and pump-out times) tends to increase with vessel volume

Surface Effects -
- Vortex formation or elimination
- Dry solids drawdown

Dispersion (Constant Power/Unit Volume)-
- A shear sensitive operation
- Usually use a high-shear radial turbine
- Gas dispersion in liquid
- Two-phase liquid contacting

Solids Suspension-
- Scale-up rules can vary (see under main heading "special considerations")
- Power per unit volume always decreases with increasing tank size
- Torque per unit volume can be constant or increase with tank volume

Equal Fluid Motion (Constant Torque/Unit Volume) -
- Flow (velocity) sensitive operations
- Most common mixing requirement
- Results in same relative motion throughout tank
- Insures same absolute velocity upon scale-up
- Many solids suspension applications, especially dense slurries
- Blend time increase with tank diameter (usually acceptable)
- High viscosity mixing

TABLE III
Exponents for Translation Equations. Adapted from Uhl and Von Essen (*35*).

	Translation equation criteria and corresponding exponents			
Target operations	ϵ, P/V Eq. (1) y	τ, T_Q/V Eq. (2a) x	N_2/N_1 Eq. (3) n	N_{Re} Eq. (4) a
Equal blend time	2	$\frac{2}{3}$	0	$\frac{1}{2}$
Surface effects	$\frac{1}{2}$	$\frac{1}{3}$	$-\frac{1}{2}$	$\frac{2}{3}$
Dispersion	0	$\frac{2}{9}$	$-\frac{2}{3}$	$\frac{3}{4}$
Dilute solids suspension	$-\frac{1}{4}$[a]	$\frac{1}{6}$[a]	$-\frac{3}{4}$[a]	$\frac{4}{5}$[a]
Equal fluid motion (velocity) and dense solids suspension	−1.0	0	−1.0	1.0
Geometric similitude	Some liberty	Insensitive to (22)	Strict (23)	Approximate

[a] Actually a range with actual value depending on process conditions.

Dimensional equations. Dimensional equations are common for gas dispersion (see Uhl et al. [19]) and for suspension of solids, e.g., the classic Zwietering equation (29). Dimensionless relations are often preferred because they can be theoretically sounder and are less susceptible to numerical errors in use. Dimensional relations are mentioned here for the sake of comprehensiveness.

Indices or Scale Numbers

In practice, arbitrary criteria are used to express the level or intensity of process results. Often one of three levels of agitation is specified such as: mild, moderate, or vigorous. However, a given designation such as mild, for a given process result, for solid suspension for example, can require much more power or torque than mild agitation for blending. In other words, such verbal specifications are relative. Scale numbers, in fact from 1 to 10 have been used by The Pfaudler Co., Rochester, NY (30) and Chemineer, Inc., Dayton, OH (31) to characterize specific process results. A sample tabulation of such scale numbers, used by Chemineer, Inc. (31) for solids suspension, is shown in Table IV.

TABLE IV
Scale Numbers for Degree of Suspension of Solids in Agitated Vessels. (Scale numbers are termed Chemscale by Chemineer, Inc., Ohio.) Copied from Gates *et al.* (*31*).

Scale of Agitation	Description
1-2	Agitation levels 1-2 characterize applications requiring minimal solids-suspension levels to achieve the process result. Agitators capable of scale levels of 1 will: - Produce motion of all of the solids of the design-settling velocity in the vessel. - Permit moving fillets of solids on the tank bottom, which are periodically suspended.
3-5	Agitation levels 3-5 characterize most chemical-process-industries solids-suspension applications. This scale range is typically used for dissolving solids. Agitators capable of scale levels of 3 will: - Suspend all of the solids of design-settling velocity completely off the vessel bottom. - Provide slurry uniformity to at least one-third of fluid-batch height.

	- Be suitable for slurry drawoff at low exit-nozzel elevations.
6-8	Agitations levels 6-8 characterize applications where the solids-suspension level approaches uniformity. Agitators capable of scale level 6 will: - Provide concentration uniformity of solids to 95% of the fluid-batch height. - Be suitable for slurry drawoff up to 80% of fluid-batch height.
9-10	Agitation levels 9-10 characterize applications where the solids-suspension uniformity is the maximum practical. Agitators capable of scale level 9 will: - Provide slurry uniformity of solids to 98% of the fluid-batch height. - Be suitable for slurry drawoff by means of overflow.

In other papers, the scale numbers corresponding to process conditions for fluid blending and motion (32) and dispersion of gases (33) are listed. Also, for the particular process conditions described in References (31, 32, 33), the recommended shaft speed and the standard prime-mover horsepower are given for vessels of various capacities corresponding to a desired scale number. The shaft speeds are those for prime-mover speeds commonly available (1750 rpm, 1150 rpm) linked with the gear ratios standard with the American Gear Manufactures Association (AGMA). An abridgement of such a power-speed tabulation is displayed in Table V for suspending solids with a settling velocity of 10 ft/min.

For the three operations noted an analysis of the data, e.g. Table V, indicates that the applicable scale-up rules are:

Operation	Ref.	Criterion	Relation	Exponent
Blending and motion	(32)	Equal fluid motion	Eq. (2)	x = 0
Suspension of solids	(31)	Dispersion	Eq. (1)	y = 0[b]
Dispersion of gases	(33)	Dispersion	Eq. (1)	y = 0

[b] Actually the criteria should be somewhere intermediate for dispersion and equal fluid motion. Note that an exponent for y of -1/4 is given as a range value in Table III. Also, see the footnote [a] for Table III.

TABLE V
Prime Mover Power and Shaft Speed for Solids Suspension. (For particles with settling velocities of 10 ft/min.)[a]

Scale of agitation[c]	Equivalent volume (gal.) 500	5,000	15,000	100,000
1	1/350[b]	5/125	10/84	60/84
		3/84	7.5/68	50/68
		3/68	5/45	40/56
		2/45	3/37	30/37
4	1/155	7.5/84	30/100	200/68
		5/56	26/84	150/56
			15/45	125/45
				100/30
5	1/125	15/156	40/100	300/100
		10/100		250/84
		7.5/68		150/45
		5/45		125/37
6	1/100	10/84	40/84	300/68
			30/68	250/56
			25/56	200/45
			20/37	150/37
10	5/125	50/100	150/84	
		40/84	125/68	
		30/68	100/56	
		25/56	75/45	

[a] Abstracted from Gates et al. (31).

[b] 1/350 denotes nominal horsepower of drive and shaft speed in rpm.

[c] See Table IV.

DESIGN BY DIRECT COMPARISON

Full size installation experience provides a highly prized form of mixing data. If a successful fluid mixing installation is known with a volume similar to a new proposed application, the new mixer can be designed by direct comparison. If volume or geometry are not identical, it may be necessary to extrapolate for a short distance or apply geometric correction factor, but if changes are small, the reliability of the design will be high.

This is a very popular industrial design technique. It is not usually necessary to have any knowledge of the mixing operation.

Preferably the fluid to be mixed in the new application will be identical to the known installation. If not, a correction may be necessary. The simplest such correction would be a difference in fluid density. If no other fluid properties are different, power requirement would normally be directly proportional to a change in fluid density. Changes in viscosity or solids phase characteristics can be much more complex, and can limit the usefulness of data from a known installation.

EXTRAPOLATION FROM TEST DATA ON MULTIPLE SCALES

When test data are available for several scales, one of the several translation equations can be used to establish a relation to extrapolate to commercial size. The procedure historically described follows either Equation (1) or Equation (1a), where, P/V, is a function of either D, or V, often expressed in thousands of gallons. (See Reference [34]). Another, and a preferred relation is $\tau = f(V)$ which corresponds to Equation (2a). An example of the method is illustrated in Figure 6. Note that the graph-scales are log-log because translation equations are normally linear power functions. The exponent is found from the slope of the straight line through the test points. The units for the ordinate in Figure 6 in inch-pounds/1000 gallons which is common, while the abscissa is volume (gallons) instead of a length scale; this corresponds to Equation (2a).

Data from more than two points are preferred and caution should be exercised when the extrapolation is extensive. The determination of scale-up exponent can be difficult if there are changes in geometry or fluid properties.

SCALE-UP BY EXTRAPOLATION FROM SINGLE RESULT

This technique can also use a graph such as Figure 6 on which the single datum for a test result is plotted. The slope to use must be selected from experience or insight. For Figure 6, this would correspond to the exponent x in Equation (2a). However, if the operation is heat or mass transfer to a surface, a plot as shown in Figure 3 is preferred where the abscissa is the "mixing" Reynolds number.

If the scale-up exponent for the process in question is well-known, a single datum even in small scale can be useful to eliminate questions on geometric ratios and fluid properties. The single test would ideally have identical geometric similitude to the

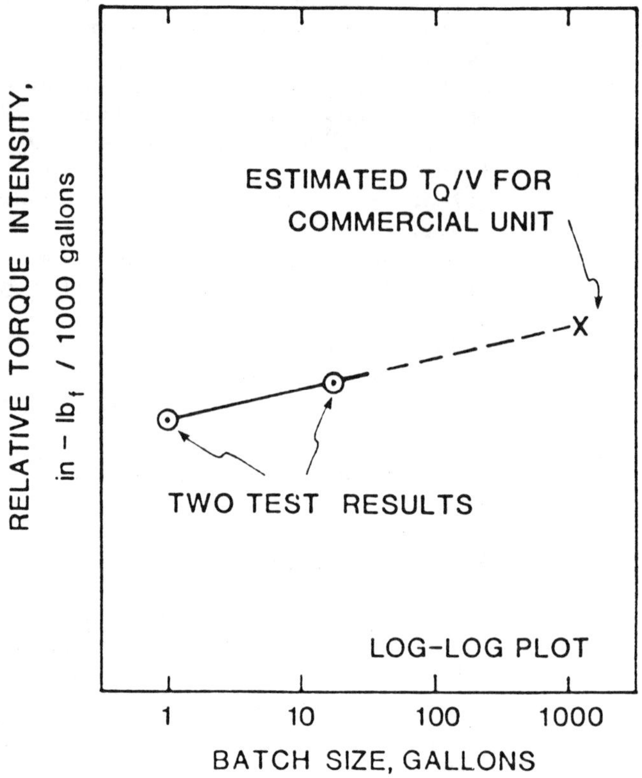

FIGURE 6. Plot developed from test data to extrapolate torque intensity, τ, to attain equal process result on commercial scale. From Uhl and Von Essen (35).

proposed commerical design, and the test fluid should be the proposed commercial fluid. If solids are present, they should be the actual commercial scale solids, including sieve size fractions. The minimum size for such a single datum should be 5 gallons, as smaller sizes may be in a laminar flow regime for the test but turbulent in commercial scale. Also, solid phase particles or gas phase bubbles can be significant in size to the small mixing impellers if tests are run on a very small scale.

DESIGN BY EQUATION

It is important to employ a geometry compatible with design equation if no test data are available. This is the least desirable design technique, but it can be successful if the process is not complex.

SPECIAL CONSIDERATIONS

All scale-up exponents are modified by geometric ratios such as D/T (see Figure 4), impeller location, Z/T, degree of baffling, and impeller type. Volume related indices, such as P/V or T_Q/V are less affected by the vessel aspect ratio, Z/T, than a speed ratio correlation would be. When a Z/T change is made in a scale-up calculation, test data are desired to assess the effect on the scale-up exponent being used. An increase in Z/T (beyond 1.0) may call for two or more impellers to obviate stagnant zones.

In scale-up, geometric similitude is often not maintained in the interest of an optimum design, i.e., one which provides the minimum combination of capital cost and outlay for power. Usually the D/T ratio is reduced and often the Z/T ratio is increased. Their impact can be correlated from tests by modifying the appropriate "translation" relation by a function of the specific dimension ratio found either experimentally or estimated. In this connection Von Essen (22) has demonstrated that torque intensity correlations tend to be influenced less by the D/T ratio than the other dimensional correlations noted.

One process result for which the scale-up exponents vary considerably (e.g., from 1/6 to 0 for the torque correlation) is the suspension of solids. The settling velocity of the solids, and the volume fraction of the solids in the slurry are the principal variables.

In scale-up one must check to be sure that all scales considered are for the same regime, i.e., laminar or turbulent. If one or both sizes are in the transition regime as indicated, for instance by the power correlation, Figure 2, the translation relations given in Tables I and II, and the exponents in Table III are invalid.

When the process result is flow (velocity) sensitive, which corresponds to Equal Fluid Motion in Table III, the criterion could be constant torque intensity, i.e., x = 0 in Equation (2). On the other hand if the prime consideration is constant shear or turbulence intensity, constant power per unit volume is indicated.

CONCLUSION

If one needs to scale-up a mixing operation, an appreciation of the variety of techniques is helpful in selecting the basic approach and if necessary an alternate or check method. In any case, the process result should be clearly declared and the mechanisms by which it is attained should be grasped as completely as possible. This provides a basis for the selection or check of exponents in the translation equations, or the extrapolation of data as in Figure 6.

NOTATION

a	Exponent in Reynolds number correlation, see Equation (4)
B	Width of baffle
C	Distance of impeller from bottom of tank
Cp	Heat capacity
D	Impeller diameter
D_L	Liquid phase diffusivity
d	Diameter of solid particle, liquid droplet, or gas bubble
g	Acceleration due to gravity
h	Heat transfer coefficient
k	Thermal conductivity
k_L	Liquid phase film mass transfer coefficient
K,K' K"	Constants
n	Exponent of diameter ratio in speed ratio correlation, see Equation (3)
N	Impeller rotational speed
P	Power applied to a mixer shaft
Q	Pumping rate of an impeller
Sg	Specific gravity
T	Diameter of vertical cylindrical vessel
T_Q	Torque applied to a mixer shaft
V	Volume of fluid in a mixing vessel
x	Exponent of vessel volume ratio in torque per unit volume correlation, see Equation (2)
X	Mass fraction of solids in a slurry
y	Exponent of vessel diameter ratio in power per unit volume correlation, see Equation (1)
Z	Vertical height of fluid in a mixing vessel from bottom of vessel to fluid level

Dimensionless groups

N_{Fr}	Froude number, $D N^2/g$
N_P	Newton (power) number, $P/\rho N^3 D^5$
N_{Re}	Mixing Reynolds number, $\frac{D^2 N\rho}{\mu}$

Greek letters

ε	Power intensity, P/V
μ	Viscosity
ρ	Density
τ	torque intensity, T_Q/V
ψ	A function (see Figure 3)
ψ'	ψ/D (used in Equation [4])

Subscripts

1 Small size
2 Large size

LITERATURE CITED

1. Johnstone, R. E., and M. W. Thring, Pilot Plants, Models and Scale-up Methods", McGraw-Hill, New York (1957).

2. McAllister, S. H., Oil Gas J., 139 (Nov. 12, 1937).

3. Hall, C. M., Metallurg. Chem. Engrg., 9, 71 (1911).

4. Anon., "Scale-up; new challenges. . . more powerful techniques," Chemical Week, 34-37 (July 13, 1983).

5. Rushton, J. H., "The Mechanics of Similitude Applied to Pilot-Plant Experimentation." Proceedings of the (Second) Midwestern Conference on Fluid Dynamics, pp. 156-174 (1951).

6. Rushton, J. H., E. W. Costich, and H. J. Everett, "Power Characteristics of Mixing Impellers." Chem. Eng. Progress, 46 Pt. I, 395, Pt. II, 467 (1950).

7. Bates, R. L., P L. Fondy, and J. G. Fenic, Chap. 3 in "Mixing: Theory and Practice," Vol. I, Uhl, V. W. and J. B. Gray (Eds.) Academic Press, New York (1966).

8. Uhl, V. W., Chap. 5 in "Mixing: Theory and Practice," Vol. I, Uhl, V. W. and J. B. Gray (Eds.), Academic Press, New York (1966).

9. Bondy, F., and S. Lippa, Chem. Eng., 90 (7), 62-71 (1983).

10. Mack, D. E., and R. E. Marriner, Chem. Eng. Progress, 45, 545 (1949).

11. Hixson, A. W., and M. I. Smith, Ind. Eng. Chem., 41, 973 (1979).

12. Fox, E. A., and V. E. Gex, AIChE Journal, 2, 539 (1956).

13. Fenic, J. G., and P. L. Fondy, "Application of Similarity Analysis to Blending of Miscible Liquids." Paper presented at Annual Meeting of AIChE, Atlantic City (1966).

14. Nienow, A. W., Chem. Eng. Sci., 23, 1453-1459, (1968).

15. Uhl, V. W., "Suspension of Solids in Agitated Tanks." Paper #8, AIChE 72nd Annual Meeting, San Francisco (Nov. 25-29, 1979).

16. Rushton, J. H., "The Use of Pilot Plant Mixing Data," Chem. Eng. Progress, 47 (9), 485-488 (1951).

17. Holland, F. A.., and F. S. Chapman, "Liquid Mixing and Processing in Stirred Tanks." p. 59, Reinhold Publishing, New York (1966).

18. Nagata, S., and K. Yamamoto, "Criteria for Scaling Up Mixing Vessels." Memoirs of the Fac. of Eng. Kyoto Univ., 29, 75 (1967).

19. Uhl, V. W., R. L. Winter, and E. L. Heimark, "Mass Transfer in Large Secondary Treatment Aerators". Chem. Eng. Progress Symp. Ser. 73, No. 167, pp. 33-41 (1977).

20. Tatterson, G. B., "Scale-up Procedures and Power Consumption in Agitated Vessels." Food Technology, pp. 65-70 (May 1971).

21. Penney, W. R., Chem. Eng., 78 (7), 86 (1971).

22. Von Essen, J. A., "Liquid Mixing: Scale-up Procedures," Lecture Notes, Liquid Mixing Short Course, The Center for Professional Advancement, East Brunswick (1980).

23. Rautzen, R. R., R. R. Corpstein, and D. S. Dickey, Chem. Eng., 119, (Oct. 16, 1976).

24. Büche, W., VDI-Zedschrift, 37, 1065-1069 (1937).

25. Hirsekorn, F. S., and S. A. Miller, Chem. Eng. Progress, 49, 459 (1953).

26. Miller, S. A., and C. A. Mann, Trans. A.I.Ch.E., 40, 709 (1944).

27. Hyman, D. in "Advances in Chemical Engineering", Vol. 3, pp. 154, 188, Academic Press, New York (1962).

28. Connolly, J. R. and R. L. Winter, Chem. Eng. Progress, 65 (8), 70-78 (1969).

29. Zwietering, T. N., Chem. Eng. Sci., 8, 244 (1958).

30. The Pfaudler Co., Bulletin 1018, "Agitation Speed-Power Calculator," Rochester (1961).

31. Gates, L. E., J. R. Morton, and P. L. Fondy, "Selecting Agitator System to Suspend Solids in Liquids," Chem. Eng., 144-150 (May 24, 1976).

32. Hicks, R. W., J. R. Morton, and J. G. Fenic, "How to Design Agitators for Desired Process Reponse," Chem. Eng. (April 26, 1976).

33. Hicks, R. W. and L. E. Gates, "How to Select Turbine Agitators for Dispersing Gas into Liquids," Chem. Eng. (July 19, 1976).

34. Bates, R. L., D. L. Fondy, and J. G. Fenic, Ch. 3 in "Mixing Theory and Practice," Uhl, V. W., and J. B. Gray (Eds.), p. 167, Academic Press, New York (1966).

35. Uhl, V. W. and J. A. Von Essen, Chap. 15 in "Mixing: Theory and Practice," Vol. III, ed. by V. W. Uhl and J. B. Gray, Academic Press, New York (1986).

APPENDIX

This exercise demonstrates that the four translation equations predict the same scale-up result.

The ABC Company has obtained laboratory- and pilot-scale data on a complex mixing application and now wants to evaluate a mixer offered for a large vessel. A chemical conversion efficiency of 65% is desired. Both laboratory-and pilot-scale mixers were run at variable speeds and set just fast enough to obtain 65% conversion. Mix density is 1.2 specific gravity Sg. Viscosity μ is 100 centipoises.

Lab: Vessel = 12" diameter x 12" liquid level

(L) Impeller: 4-45° pitched blade turbine (PBT), 5" diameter

Speed = 643 RPM

Bench: Vessel = 36" diameter x 36" liquid level

(B) Impeller: 4-45° PBT, 15" diameter

Speed = 258 RPM

Plant: Vessel = 8 feet diameter x 8 feet liquid level

(P) Impeller: 4-45° PBT, 40" diameter

Solution

1. Try a dimensionless number correlation:

$$\psi' D = K (N_{Re})^a \qquad (1A)$$

$\psi'_L = 0.65$

$D_L = 5/12 = 0.417$ feet

$N_L = 643$ RPM

$Sg = 1.2$

$\mu = 100$ cp

To facilitate the calculation of the mixing Reynolds number this dimensional form of the Reynolds number will be used:

$$N_{Re} = 1{,}480 \frac{D^2 N Sg}{\mu} \qquad (2A)$$

where D is in feet and N is rpm. For ρ, the density, specific gravity Sg is used, and μ is in centipoises.

Then

$$(N_{Re})_L = 1{,}480 \times 1.2 \times 643 \times (0.417)^2/100 = 1{,}986$$

Similarly,

$$(N_{Re})_B = 1{,}480 \times 1.2 \times 258 \times (1.25)^2/100 = 7{,}160$$

Then

$$\left[\frac{(N_{Re})_L}{(N_{Re})_B}\right]^a = \frac{\psi'_L D_L}{\psi'_B D_B}$$

$$a = \frac{\ln\left[\frac{\psi'_L D_L}{\psi'_B D_B}\right]}{\ln\left[\frac{(N_{Re})_L}{(N_{Re})_B}\right]} = \frac{\ln\left[\frac{0.65 \times 0.417}{0.65 \times 1.25}\right]}{\ln\left[\frac{1986}{7160}\right]} = 0.86$$

$$K = \frac{\psi'_B D_B}{(N_{Re})_B^{0.86}} = \frac{0.65 \times 1.25}{(7160)^{0.86}}$$

$$= 3.93 \times 10^{-4}$$

Thus, for $D_P = 40/12 = 3.33$, in Equation (1A), where $\psi' = 0.65$:

$$0.65 \times 3.33 - 3.93 \times 10^{-4} \quad (N_{Re})_P^{0.86}$$

$$(N_{Re})_P = 2.24 \times 10^4$$

and

$$1480 \times 1.2 \times N \times (3.33)^2/100$$
$$= 2.24 \times 10^4$$

N = 114 RPM

2. With a Power per Unit Volume correlation:

The power requirement is calculated from

$$P = N_P \, Sg \, N^3 \, D^5/6.12 \times 10^7 \qquad (3A)$$

for which the values of Sg, N and D are in the same units as for the dimensional Reynolds number equation (A2) above. Then since $N_P = 1.62$ from an appropriate N_P versus N_{Re} correlation

$$P_L = 1.62 \times 1.2 \times (643)^3 \ (0.417)^5/6.12 \times 10^7$$

$$= 0.106 \text{ HP}$$

$$V_L = 7.48 \, \pi \, T_L^2 \, Z_L/ 4$$

$$= 7.48 \times 3.14 \times (1.0)^2 \times 1.0/ 4$$
$$= 5.87 \text{ gallons}$$

$$(P/V)_L = 0.106 \text{ Hp}/5.87 \text{ gal.} \times 10^{-3}$$

$$= 18.1 \text{ Hp}/1{,}000 \text{ gallons}$$

Similarly, $(P/V)_B = 10.5$ HP/1,000 gallons

$$(P/B)_B = (P/V)_L(T_B/T_L)^y$$

$$10.5 = 18.1(3/1)^y$$

and $y = -0.50$

Then $(P/V)_P = (P/V)_B(T_P/T_B)^{-0.50}$

$$(P/V)_P = 10.5(8/3)^{-0.50}$$
$$= 6.43 \text{ HP}/1{,}000 \text{ gallons}$$

$$V_P = 3{,}008 \text{ gallons}$$

$$P_P = 6.43 \times 3{,}008/1{,}000 = 19.3 \text{ HP}$$

Substituting these values in Equation (3A)

$$N_P = \left[\frac{6.12 \times 10^7 \times 19.3}{1.62 \times 1.2 \times (40/12)^5}\right]^{1/3}$$

and

N = 114 RPM

3. Try a Torque per Unit Volume Cor-relation:

$T_Q = 63{,}000$ P/N where P is in Hp and N in rpm

Then $(T_Q)_L = 63{,}000 \times 0.106/643$
$= 10.4$ in. -lb.
$(T_Q/V)_L = 10.4/5.87$
$= 1.77$ in.-lb./gallon

Similarly, $(T_Q/V)_B = 2.56$ in.-lb./gallon

$$(T_Q/V)_B = (T_Q/V)_L(T_B/T_L)^x$$

$$\text{or } 2.56 = 1.77(3/1)^x$$

and $x - 0.34$

From this

$$(T_Q/V)_P = (T_Q/V)_B(T_P/T_B)^{0.34}$$

$$(T_Q/V)_P = 2.56(8.3)^{0.34}$$
$$= 3.57 \text{ in.-lb./gallon}$$

$$(T_Q)_P = 3.57 \times 3{,}008 = 10{,}739 \text{ in.-lbs.}$$

$$N_P = \left[\frac{971\ T_Q}{N_p\ S_g\ D^5}\right]^{0.5} = \left[\frac{971 \times 10{,}739}{1.62 \times 1.2\ (40/12)^5}\right]^{0.5}$$

and

N = 114 RPM

4. Try a Speed Ratio Correlation:

$$N_B = N_L(T_B/T_L)^{-n}$$

$$258 = 643(3/1)^{-n}$$

$$n = 0.83$$

$$N_P = N_B(T_P/T_B)^{-0.83}$$

$$N_P = 258(8/3)^{-0.83}$$

$$\boxed{N = 114 \text{ RPM}}$$

Conclusion

All four methods predict the same plant scale impeller speed. If D/T or Z/T ratios were not identical, the different methods would have predicted different plant scale impeller speeds, and in that case, as has been demonstrated, the torque per unit volume scale-up would have been the most reliable.

Scale-up Strategies for Bioreactors

DAVIS W. HUBBARD
Department of Chemical Engineering
Michigan Technological University
Houghton, Michigan 49931

Fermentation system scale-up methods for stirred tanks are often based on gas–liquid mass transfer correlations, but the non-Newtonian properties of the broths are frequently ignored when making scale-up calculations. The non-Newtonian properties of various types of broths are described using rheological models. The model parameters can be included in gas–liquid mass transfer correlations.

Data for heat and mass transfer to non-Newtonian liquids in stirred tanks are available in the literature, and correlations applicable to fermentation systems can be developed. For Newtonian fluids, dimensional correlations written in terms of power per unit volume and superficial gas velocity are common and have been verified for systems having volumes up to 4.4 m^3. For non-Newtonian fluids, dimensionless correlations are used, but these dimensionless correlations have been verified only for systems having small volumes—0.005 m^3.

The dimensional and dimensionless transfer coefficient correlations are equivalent if limitations typical for bioreactors are imposed. Scale-up strategies which include non-Newtonian properties are developed and compared with other commonly used scale-up strategies in order to reach a fundamental understanding about what makes a particular strategy effective in biochemical processing situations. The processing consequences and the power input for a particular fermentation depend upon which of several different scale-up stategies is chosen.

Scale-up means reproducing in production size equipment the results from a successful fermentation carried out in laboratory-size , 1 liter, or pilot plant size, 100 liter, equipment. The results may be specified in terms of the production rate per unit volume of cells or some extracellular product. The conditions required for growth in a successful biochemical process will have been established in a laboratory fermenter. These conditions may include--

- *Temperature
- *Pressure
- *Shear rate(possibly an upper limit)
- *Nutrient concentration
- *Oxygen concentration
- *pH

Scale-up involves maintaining these conditions no matter what the fermenter volume is. If conditions are the same, and if no mutations occur which might cause the growth kinetics or the metabolic products to change, then the production rate per unit volume should be the same in the large and small systems.

Producing the same conditions no matter what the volume of the system is means paying attention to mixing, aeration, and temperature control. These aspects of fermentation are directly influenced by fluid mechanics. It is the objective of this work to review some of the fluid mechanical aspects of stirred tank bioreactors and show how these are related to scale-up.

In a fermentation system, the growth of micro-organisms is frequently controlled by the availability of oxygen--the other nutrients being supplied in excess. Oxygen is often supplied by bubbling air into the broth. The fluid mechanics aspects of gas-liquid mass transfer are therefore important in scaling-up fermentation processes. The growth rate of micro-organisms also depends on temperature. Temperature control involves heat transfer which is influenced by fluid mechanics. The temperature and the oxygen concentration are usually the critical conditions for scale-up. The pH can be controlled separately, and the nutrients except for oxygen can be added in excess. Scaling-up temperature control systems is difficult, because heat is produced or disappears volumetrically but must be transferred out from(or into) the fermenter at the wetted surfaces. As Anderson et al.(1) point out, in systems which are geometrically similar, where the ratios of corresponding lengths all equal the same constant, the system volume increases as the cube of some characteristic length while the heat transfer area increases as the square of the characteristic length. At least in difficult cases, it will not be possible to scale-up a fermenter in such a way that the heat transfer requirements are met in the same way in the laboratory-size system and in the large-size system. Other measures such as using a refrigerant rather than cooling water or using an external heat exchanger may be required. Thus, scale-up is usually based upon oxygen as the limiting substance and upon maintaining the the oxygen concentration in the broth constant no matter what the volume of the system is. Scale-up by this method has been discussed in monographs published over the past twenty years--Aiba et al.(2,3), Oldshue (4,5,6), Blanch and Bhavaraju(7), Bailey and Ollis (8), Moo-Young and Blanch (9,10), Anderson et al.(1), Prokop(11), and Charles(12) among others.

The objective of the scale-up strategy is to keep the dissolved oxygen concentration, c_L, constant above some critical value as described by Richards(13) or Finn(14). This is done by matching the oxygen uptake rate(OUR) and the oxygen transfer rate(OTR). Maintaining c_L constant by this method cannot be done directly, because the OUR and the OTR must be controlled by manipulating the operating variables. This indirect procedure for controlling c_L is fraught with uncertainty. This is especially true in a batch fermentation where the cell mass and therefore the oxygen demand may be increasing at an exponential rate. This oxygen demand or OUR is of primary importance for cell growth. Oxygen is transferred from a bubble to the liquid and then to the cell wall and across it. The limiting part of this process is taken to be the transfer from the bubble gas-liquid interface to the bulk of the liquid. Evidence that this is true, is that a specific dissolved oxygen concentration can be maintained in the bulk of the liquid. This limits the OTR by establishing a limit for the concentration difference which drives the oxygen transfer.

$$OTR = k_L a \, V \, (c^* - c_L) \qquad (1)$$

If $(c^* - c_L)$ is limited, then OTR must be matched to OUR by altering $k_L a$. This can be done by altering the processing conditions--N, Q, and c^*(by using oxygen enriched air). In spite of the difficulties mentioned, scale-up of fermentation processes has been carried out successfully using this method. One objective of the present work is to relate this scale-up method to the fundamental fluid mechanics of bioreactors.

Much of the discussion of bioreactor scale-up is based on $k_L a$ data obtained for water. Fermentation broths are mostly water, but they contain electrolytes, suspended solids (particles, flocs, and filamentous materials), and dissolved macromolecules. These materials may make the rheological properties and the surface tension of the broth different from water. These differences may affect the heat and mass transfer characteristics in the bioreactor, because the fluid dynamical situation may be quite different from the situation if water was present.

BROTH RHEOLOGY

The rheological properties of a fluid indicate something fundamental about the fluid mechanical behavior to be expected. Fluids respond to stress or applied forces by flowing. The relationship between stress and flow for fluids which have simple behavior is given by Newton's law of viscosity as described by Bird et al.(15).

$$\tau_{xz} = - \mu (dv_z/dx) \qquad (2)$$

μ, the absolute viscosity, is taken to be a property of the fluid when the fluid is considered to be a continuum, and this means that μ is a function only of the state variables--pressure and temperature. For liquids which follow equation (2), the temperature dependence of viscosity is given by

$$\mu = A \exp(B/T) \quad (3)$$

The viscosity can be calculated from molecular collision parameters if the fluid is considered to be a collection of molecules. To do this, a kinetic theory of molecular motion is needed. Such a theory is available for gases, but the molecular theory for liquids is still being developed.

Water and electrolyte solutions behave according to equation (2). Newton's law of viscosity can be generalized to include fluids which contain suspended material or dissolved macromolecules, but this is strictly empirical because of the lack of a good kinetic theory of liquids. The generalized Newton's law can be written

$$\tau_{xz} = - \mu_{app} (dv_z/dx) \quad (4)$$

where μ_{app} is the apparent viscosity which is not a property of the fluid, because it depends on the velocity gradient as well as on temperature and pressure. The main factor of concern as far as scale-up goes is the dependence on velocity gradient. The implications of this will be discussed below.

Empirical mathematical models have been used by many authors to describe the dependence of μ_{app} on velocity gradient. Roels et al. (16) list some which have been applied to fermentation broths.

* Ostwald-deWaele(or "power law")

$$\mu_{app} = K \, |dv_z/dx|^{n-1} \quad (5)$$

* Bingham

$$\mu_{app} = [\, \mu_0 \pm \tau_0/|dv_z/dx| \,] \quad \text{for } \tau_{xz} > \tau_0 \quad (6a)$$

and

$$\mu_{app} \rightarrow \infty \text{ or } dv_z/dx = 0 \quad \text{for } \tau_{xz} < \tau_0 \quad (6b)$$

* Herschel-Bulkley(power law with yield stress)

$$\mu_{app} = K_0 |dv_z/dx|^{n-1} - \tau_0'/|dv_z/dx| \quad (7)$$

* Casson

$$\sqrt{\tau_{xz}} = \sqrt{\tau_C} + K_C \sqrt{dv_z/dx}$$

from which

$$\mu_{app} = \tau_C / |dv_z/dx| + K_C \sqrt{\tau_C} / |dv_z/dx| + K_C^2 \quad (8)$$

The parameters which appear in these mathematical models must be evaluated by fitting experimental data.

The mathematical models described above relate shear stress to the velocity gradient or rate of deformation in the fluid. Some fluids also show some of the characteristics of elastic solids when the shear stress to which they are subjected is changing. Such fluids are called viscoelastic. If the deformations and the rates of change of shear stress are small, then a simple linear model can be used to describe the fluid.

$$\tau_{xz} + \lambda [\partial/\partial t(\tau_{xz})] = -\mu(\partial v_z/\partial x) \quad (9)$$

This equation, called the Maxwell equation, is a superposition of Newton's law of viscosity and Hooke's law of elasticity. λ is the relaxation time of the fluid. Fermentation broths, especially those containing extracellular macromolecules, can exhibit viscoelastic behavior as well as non-Newtonian behavior. Equation (9) is applicable to fermentation broths only in a general sense, because if the the relaxation time is small, the fluid follows Newton's law instead of some other relationship between shear stress and velocity gradient. When a shear stress is applied to a viscoelastic fluid, a stress perpendicular to the direction of the shear stress is observed. An example of this is the flow between parallel flat plates induced by moving one plate parallel to the other at a steady velocity. A shear stress is required to keep the velocity constant, and a normal stress perpendicular to the fluid motion is required to keep the plate spacing constant. This normal stress can be measured, and its magnitude is a measure of the degree to which the fluid is elastic. Fluids which are not viscoelastic do not require a normal stress in this kind of flow situation. Carreau(17) and Bird et al.(18) discuss some rheological models which include normal stresses and which have been applied to polymer solutions. These models are quite complicated with several constants to be evaluated from experimental data. The simplest form is the Carreau equation for viscosity which is a generalized Newtonian model similar to the empirical models described by equations (5), (6), (7), and (8). The Carreau equation contains a relaxation time which is related to the elasticity of the fluid. This model has not been applied in the analysis of impeller agitated fermenters. Ulbrecht(19) and Ranade and Ulbrecht(20) describe viscoelastic behavior in a simple way and discuss how to introduce some viscoelastic fluid properties into the analysis of mixing systems. These properties can be measured using a cone-plate viscometer equipped for axial thrust measurements as well as torque measurements. The viscoelastic property which can be calculated from thrust measurements is the primary normal stress difference. This in turn is related to the characteristic relaxation time of the solution.

$$\lambda = N_1 / [\mu_{app}(dv/dx)^2] \qquad (10)$$

If the shear stress is given by the power law model--equations (4) and (5)--and if a similar form is used to express the primary normal stress difference

$$N_1 = A\ (dv/dx)^b \qquad (11)$$

then

$$\lambda = (A/K)\ (dv/dx)^{b-n-1} \qquad (12)$$

The concept of relaxation time is useful for data correlation purposes as will be discussed later. If λ is large compared to the characteristic time for the process, viscoelastic effects will be important. For processes taking place in agitated tanks, the characteristic process time is usually taken to be (1/N).

Charles(21) gives a review of papers in which the rheological properties of fermentation broths are discussed and presents some of his own unpublished data. He groups the broths into three categories--mycelial cultures, cultures containing extracellular polysaccharides, and bacterial and yeast cultures. Comparing the work of various authors and deducing general information about broth rheology is difficult. The various authors report experimental results obtained using different types of viscometers, and this influences the results obtained even for the same kind of broth. Usually, the range of shear rates used was small--0 to 0.2 s^{-1} or 0 to 3 s^{-1} in most cases. In a few cases, a wider range of shear rates was covered--up to 10, 100, or 1000 s^{-1}.

For mycelial broths, behavior which corresponds to the power law model, to the Bingham model, or to the Casson model has been observed. The magnitudes of the model parameters depend on the cell concentration and on the cell morphology--whether filamentous or pellet-like. High molecular weight substrates such as dissolved cornstarch or a high concentration of undissolved solids can also affect broth rheology, but these effects decrease in importance as the cell mass increases. Some cultures have been reported to be Newtonian, but there is some uncertainty about the viscosity measurement techniques used. If only very low shear rates are used for viscosity measurements, non-Newtonian characteristics will not be detected for any fluid. Some typical values for model parameters which have been reported in the literature for various types of broths are shown in Table 1. The ranges of parameter values occur, because the cell mass varies with broth age. Roels et al.(16) used a turbine impeller to measure viscosity and treated the data as described by Bongenaar et al.(22) They found that the Casson model parameters depend on cell concentration and morphology, though the actual values cannot be calculated, because the data are reported only in terms of relative magnitudes.

TABLE 1
Representative Rheological Model Parameters Reported for Mycelial Broths

Mycelium	Non-Newtonian Model	Shear Rate Range (s^{-1})	Model Parameters	Reference
P. Chrysogenum	Bingham	1 to 10	μ_o = 1.5 to 48 cp τ_o = 0.07 to 11 dym/cm^2	Deindoerfer and Gaden Appl Microbiol,3,253(1955)
P. Chrysogenum	Casson	Not reported (0.05<N<1.35)	$3.45<\sqrt{\tau}_c<6.26 (Nm)^{1/2}$ $1.34<K_c<3.13 (Ns)^{1/2}m^{-1}$	Roels et al Biotech Bioeng,14,181(1974)
C. Hellebori	Power Law	0.05 to 0.79	K = 2.3 gcm^{-1}s^{n-2} n = 0.29	Deindoerfer and West J. Biochem. Microbiol. Tech. and Eng.,2,165(1960)
C. Hellebori	Bingham	2 to 47	μ_o = 4 to 31 cp τ_o = 0.3 to 2.3 dyne/cm^2	
P. Chrysogenum	Power Law	0 to 310	51 < K < 81 gcm^{-1}s^{n-2}; n = 0.04 to 0.05	
S. Griseus	Bingham	2 to 22	μ_o = 18 to 52 cp τ_o = 0 to 8 dyne/cm^2	
S. Noursei	Newtonian	4 to 28	μ = 11.5 to 40 cp	
S. Aureofaciens	Power Law	2 to 58	K = not reported; 0.28 < n <1.0	Tuffile and Pinho. Biotech Bioeng,12,849(1970)

MASS TRANSFER CORRELATIONS

Mass transfer coefficient correlations can be developed from the governing differential equations according to the methods described by Bird et al.(23), but it is difficult to include all variables deemed to be pertinent. However, the correlations developed this way are firmly connected to the fundamental physics of the mass transfer process. These correlations are presented in terms of dimensionless groups. Dimensionless groups can be deduced also from any list of variables by the Buckingham Pi method as outlined by Langhaar(24). Sideman et al.(25) begin their review paper with such a formal dimensionless correlating equation for gas-liquid mass transfer in agitated vessels.

$$N_{Sh}' = A\,(N_{Sc})^a\,(\mu V_s/\sigma)^b \times (N_{Re})^c\,(\mu_d/\mu)^d \qquad (13)$$

This equation is to apply for any combination of gas and liquid. There is a difficulty here, because this equation contains only five dimensionless groups. The list of variables used is--

$$(k_L a, D_i, \mathcal{D}, \mu, \rho, V_s, \sigma, N, \mu_d) \qquad (14)$$

There are nine members of this set, and if the fundamental units mass, length, and time are chosen, a strict application of the Buckingham Pi method shows that there should be six dimensionless groups rather than five in a complete set from which a mass transfer coefficient correlating equation can be developed. One such complete set which contains all the dimensionless groups suggested by Sideman et al. is--

$$(N_{Sh}', N_{Sc}, \mu V_s/\sigma, N_{Re}, \mu_d/\mu, V_s/D_i N) \qquad (15)$$

A complete set which contains the surface tension, σ, in a familiar "named" dimensionless group is as follows.

$$(N_{Sh}', N_{Sc}, N_{We}, N_{Re}, \mu_d/\mu, V_s/D_i N) \qquad (16)$$

The Weber number, $N_{We} = D_i^3 N^2 \rho/\sigma$, is obtained by combining $(\mu V_s/\sigma)$, $(V_s/D_i N)$, and $(D_i^2 N \rho/\mu)$. The general form of the Weber number is $(L u^2 \rho/\sigma)$ where L and u are the characteristic length and velocity for any process being considered. For mixing in impeller agitated tanks, $L = D_i$ and $u = ND_i$. The same characteristic velocity appearing in the Reynolds number, $D_i^2 N \rho/\mu$, should appear in the Weber number. A correlating equation of the form proposed by Sideman et al. but based on the complete set of dimensionless groups listed in equation (16) would be--

$$N_{Sh}' = A\,(N_{Sc})^a\,(N_{We})^b\,(N_{Re})^c \times (\mu_d/\mu)^d\,(V_s/D_i N)^e \qquad (17)$$

A and the exponents a, b, c, d, and e must be evaluated from experimental data. Dimensional analysis using either the governing equations or the Buckingham Pi method gives only a set of dimensionless groups. The form of any correlating equation is not specified. Equations of the form of equation (17) have been found useful in the past, but the form is strictly empirical.

With the restriction that a given gas-liquid pair is selected, for any particular mixing system, equation (13) can rewritten as--

$$k_L a = A' V_s^b N^c \qquad (18)$$

If the impeller Reynolds number is high enough, $(P/\ N^3 D_i^5)$ is constant, and if (P_g/P_0) is also constant, equation (18) becomes --

$$k_L a = A'' V_s^b (P/V)^{c/3} \qquad (19)$$

Experience shows that correlations of the form of equation (19) are effective for predicting $k_L a$ even though much of the fundamental fluid mechanics information does not appear explicitly. The P/V factor probably includes nearly all the effects of N_{Re} and viscosity, since the power input is different for different fluids. This dimensional form for mass transfer coefficient correlations is the form frequently used for correlating oxygen transfer data for fermenters. Cooper et al.(26) presented early results in this form,

Sideman et al.(25) give an extensive review of data reported through 1965, and Van't Riet(27) has reviewed more recent results. He finds that for oxygen transfer to pure water in vessels having volumes from 2 to 2600 liters, k_La can be predicted within 20-40% by--

$$k_La = 0.026\ (P/V)^{0.4}\ V_s^{0.5} \quad (20)$$

where the units are --k_La, s^{-1}; (P/V), W/m^3; V_s, m/s.

For oxygen transfer in strong electrolyte--non-coalescing--solutions in vessels having volumes from 2 to 4400 liters, k_La can be predicted within 20-40% by--

$$k_La = 0.0020\ (P/V)^{0.7}\ V_s^{0.2} \quad (21)$$

where the units are the same as for equation (20).

Zlokarnik(28) has also discussed the effects of electrolytes on gas-liquid mass transfer in agitated vessels. His work is related to the differences observed between coalescing and non-coalescing liquids. Humphrey(29) reports that the exponents in equation (19) depend on tank volume and suggests--

$$k_La = A''\ (P/V)^{0.95}\ V_s^{0.67} \quad \text{for pilot plants}$$

$$k_La = A''\ (P/V)^{0.4}\ V_s^{0.5} \quad \text{for production plants} \quad (22)$$

Lopes de Figueiredo and Calderbank(30) include the effect of tank volume in a k_La correlation. They found that for vessels having volumes of 43 and 590 liters --

$$k_La = 0.0010(D_T/V^{0.42}) \times P/V)^{0.58}\ V_s^{0.75} \quad (23)$$

if no effects of surface transfer are present.

Miller(31) suggests a more complicated method for predicting k_L separately from a. The method is based on an empirical relationship for $(k_L\ /\ k_{L,calc})$ determined for bubbles in agitated tanks having volumes of 2.5, 25, and 250 liters. $k_{L,calc}$ is calculated from --

$$N_{Sh} = \sqrt{4/\pi}\ \ N_{Pe}^{0.5}$$

$$\text{or} \quad k_{L,calc} = \sqrt{4/\pi}\ (\mathcal{D} u_t/D_{BM})^{0.5} \quad (24)$$

and the empirical correlation is--

$$k_L\ /\ k_{L,calc} = 683\ D_{BM}^{1.38} \quad (25)$$

This gives--

$$k_L = 683\ \sqrt{4/\pi}\ (\mathcal{D} u_t)^{0.5}\ D_{BM}^{0.876} \quad (26)$$

The units for the various factors are $\mathcal{D}$, m^2/s; D_{BM}, m; u_t, m/s; k_L, m/s. D_{BM}, a, and u_t are calculated by the methods described by Calderbank(32,33) and Mendelson(34).

The correlating equations cited above which have the form of equation (19) apply to turbulent mixing conditions in which the liquid is a low viscosity, Newtonian fluid and the species being transferred from the gas to the liquid is oxygen. Using such correlations to predict k_La requires being able to predict the power input per unit volume. Power input, P_g, for a gas-liquid system is related to the power input, P_0, for ungassed conditions by --

$$P_g/P_0 = f_1(Q/ND_i^3) \quad (27)$$

with the function f_1 usually given graphically. Examples are given by Van't Riet(35) and by Oyama and Endoh (36) among others. Michel and Miller (37) present a correlating equation for gassed power per unit volume.

$$P_g/V = (0.796/V) \times [(P_0^2ND_i^3)/Q^{0.56}]^{0.45} \quad (28)$$

The ungassed power is calculated from-

$$(P_0/\rho N^3D_i^5) = f_2(D_i^2N\rho/\mu) \quad (29)$$

with f_2 also given graphically. An example is given by Rushton et al. (38). Nagata(39) has developed algebraic functions which represent the power input data for mixing with no gas present. For turbulent mixing, $(P/\rho N^3D_i^5)$ is constant. P_g/P_0 decreases from 1.0 to approximately

0.4 as Q/ND_i^3 increases from 0 to 0.1. Chapman et al.(40,41) show that impeller geometry and off-bottom position have important effects on gas dispersion and power consumption in gas-liquid mixing. Based on extensive data for vessels having volumes from 19 to 4410 liters, they find that a flat blade disk turbine is a good choice for dispersing gas in water. Predicting k_La requires selecting N and Q, determining P_g/P_0 from the empirical correlations, and then calculating k_La from the correlating equation.

Machon et al.(42) show gassed power input data for aqueous CMC solutions which contained 0.015 to 0.2 weight per cent polymer and which had power law parameters in the following ranges: $0.0031 < K < 0.079$ kg m^{-1} s^{n-2} and $0.78 < n < 0.92$. These solutions were said to have zero visco-elasticity. The data are scattered ±13 per cent but are correlated by--

$$[1 - (P_g/P_0)]^{-1} = 1.26 + 0.041[(P_g/P_0)/N_Q^2] \quad (30)$$

For aqueous glycerol solutions, the values of the constants in equation (30) are 1.76 and 0.038. Nienow et al.(43) report data for aqueous CMC solutions containing 0.3 to 4.0 weight per cent polymer--solutions which were weakly viscoelastic and which had power law parameters $0.003 < K < 1.7$ kg m^{-1} s^{n-2} and $0.7 < n < 1.0$. Superimposed on the expected $P_g/P_0(N_Q)$ dependence was a complicated dependence on gas flow rate and impeller Reynolds number. Data for some polymer solutions which had appreciable viscoelastic behavior were also discussed. In every case, there was a minimum value of P_g/P_0 at some value of N_Q. The effect of viscoelasticity on this minimum was discussed briefly. $(P_g/P_0)_{min}$ decreases from 0.4 to 0.1 as the first normal stress difference increases from 10 t0 100 N m^{-2}. The authors say it is difficult to prepare batches of polymer solutions so that the fluid properties are reproducible. This is especially true for viscoelastic properties, so not much progress has been made in correlating power data with viscoelastic parameters. The only conclusion is that viscoelasticity causes a reduction in power input compared to the power input for fluids which are purely viscous.

Some data for gas-liquid mass transfer with non-Newtonian liquids have been reported in the literature. In most cases, the flow in these systems is not turbulent, so the dimensionless power is not constant. Non-Newtonian mass transfer data have been correlated using dimensionless equations similar to equation (17). Perez and Sandall(44) studied carbon dioxide absorption into aqueous carboxy polymethylene solutions in a baffled tank 0.152 m in diameter. The liquids used were described by the power law model with $0.92 < n < 1.0$ and $0.009 < K < 0.04$ g/(cm s^{2-n}). Absorption with no bubbling--transfer at the liquid surface only--and absorption with bubbling were studied. The effect of non-Newtonian behavior on the mixing process was introduced by using the effective viscosity developed by Metzner and Otto(45) and Calderbank and Moo-Young(46). The effective viscosity is --

$$\mu_e = K(11.0\ N)^{n-1} \times (0.75 + 0.25\ n^{-1})^n \quad (31)$$

The correlating equations developed are--

$$N_{Sh}' = 21.2(N_{Re})^{1.11}\ (N_{Sc})^{0.5} \times (\mu_e V_s/\sigma)^{0.45}\ (\mu_d/\mu_e)^{0.7} \quad (32)$$

for bubbling,and for no bubbling.

$$N_{Sh} = 0.00511(N_{Re})^{0.93}\ (N_{Sc})^{0.5} \quad (33)$$

There are some errors in the paper. The group $(\mu_e V_s/\sigma)$ is given in the paper as $(D_i V_s/\sigma)$ which is not dimensionless, but the correct dimensionless groups have been used in the calculations as reported in the thesis by Perez(47). There are too few groups to form a complete set. Some other group such as $(D_i V_s/N)$ should appear. The effect of such a group may be small, but this is not discussed, and the effect cannot be determined from the results presented.

Yagi and Yoshida(48) studied oxygen desorption from aqueous solutions of glycerol, millet jelly, carboxy methylcellulose, and polyacrylamide. The polymer solutions were described by the power law model, and the non-Newtonian properties were included in the mass transfer correlation by using the Metzner and Otto method. The apparent viscosity for particular mixing conditions was calculated by--

$$\mu_e = \tau/(dv/dr)_{avg} = \tau/(11.5\ N)$$

For $\tau = K\ (dv/dr)^n$,

$$\mu_e = K\ (11.5\ N)^{n-1} \qquad (34)$$

The polymer solutions also had viscoelastic properties, and the effects of viscoelasticity are included in the mass transfer coefficient correlation by using the Deborah number--the ratio of the characteristic relaxation time for the polymer solution to the characteristic time for the mixing process. The characteristic times for the polymer solutions were estimated as the inverse of the shear rate at which $\mu_e/\mu_0 = 0.67$ where μ_0 is the zero shear viscosity. The characteristic time for the process is 1/N. Mass transfer coefficient correlations take the form of equation (15) but with viscosity ratio omitted and with the effect of gravity included. For Newtonian fluids--Deborah number equal to zero--

$$N_{Sh}' = 0.060\ N_{Re}^{1.5}\ N_{Fr}^{0.19}\ N_{Sc}^{0.5} \times (\mu V_s/\sigma)^{0.6}\ (ND_i/V_s)^{0.32} \qquad (35)$$

For non-Newtonian fluids which are also viscoelastic --

$$N_{Sh}' = 0.060\ N_{Re}^{1.5}\ N_{Fr}^{0.19}\ N_{Sc}^{0.5} \times (\mu_e V_s/\sigma)^{0.6}\ (ND_i/V_s)^{0.32} \times [1 + 2.0\sqrt{\lambda N}\ ]^{-0.67} \qquad (36)$$

The Newtonian and non-Newtonian correlations are made to coincide by using the equivalent viscosity for non-Newtonian mixing and by including the Deborah number in the factor $[1+ 2.0\sqrt{\lambda N}]^{-0.67}$ which goes to unity as $\lambda \rightarrow 0$. This factor causes k_La for viscoelastic solutions to be lower than for purely viscous solutions being processed under the same conditions. The Yagi and Yoshida correlations include the members of a complete set of dimensionless groups which would be obtained by following the Buckingham Pi method. In particular, the effect of the group (D_iN/V_s) is included, and this group is shown to be nearly as important as N_{Sc} or $(\mu_e V_s/\sigma)$. Neglecting the viscosity ratio which Perez and Sandall found to appear to the 0.7 power has not been discussed by Yagi and Yoshida. Perez and Sandall did not include any effect of N_{Fr} in their correlation, and Yagi and Yoshida show that it appears only to the 0.2 power. This indicates that the effect of N_{Fr} is truly small for the baffle arrangement--$W/D_T = 0.1$-- used in both studies.

Ranade and Ulbrecht(20) found a reduction in k_La when small--100 to 1000 ppm--amounts of carboxy methylcellulose (CMC) or polyacrylamide(PAA) were added to water being aerated and stirred. The reduction was by a factor of 2 or 3 for CMC and by a factor of 12 for PAA. The PAA solutions showed more pronounced viscoelasticity than the CMC solutions.

Fermentation broths may contain suspended solids which may affect mass transfer. This question has been studied by Chapman et al.(49). For constant gas flow rate, k_La for transfer from gas bubbles to water containing suspended soda glass ballotini could be correlated using the power input per unit mass by an equation equivalent to equation (19). There is also a considerable effect of gas flow rate and an effect of solids concentration as shown in Table 2 for mixing with a disk turbine.

Mills et al.(50) measured oxygen transfer to slurries of glass beads suspended in 0.1 molar aqueous sodium chloride solutions containing up to 0.4 volume fraction of solids. They determined that k_La decreases with concentration according to--

$$k_La = (0.10 - 0.0018\,\phi) \times (P_T/V)^{0.67}\ V_s^{0.31} \qquad (37)$$

TABLE 2
Effect of Solid Concentration on k_La for Mixing with a Disk Turbine

Q/V	ϵ_T	$k_La\vert_{0\%}$	$k_La\vert_{3\%}$	$k_La\vert_{20\%}$	per cent reduction
(min^{-1})	(W/kg)	(s^{-1})	(s^{-1})	(s^{-1})	
0.25	1.9	0.074	0.074	0.052	30
1.0	2.0	0.21	0.21	0.079	62

where the units are kW/m^3 and cm/s. The glass particles used were rigid, so they did not simulate exactly the cells which would be found in a fermentation broth. A reduction in k_La when solids are present has also been reported by Joosten et al.(51).

Loucaides and McManamey(52) studied oxygen transfer in a suspension containing 16 kg/m^3 of paper pulp in water. The paper pulp is a non-rigid, filamentous material, and the suspension was non-Newtonian. The power law model ($K = 0.254\ N\ s^n\ m^{-2}$ and $n = 0.57$) was used to describe it. They found k_La values to be lower with paper pulp present than k_La values measured under the same conditions and by the same method but without paper pulp present. They present a complicated correlation for the specific oxygen transfer coefficient, (k_La/H), which involves determining an extrapolated value, $(k_La/H)_0$, from the intercept of (k_La/H) plotted versus (P_g/V) at $(P_g/V) = 0$. This extrapolated value was correlated by--

$$(k_La/H)_0 = (2.4\text{x}10^{-4})\ N\ D_iD_T^{-0.5} + 2.1\text{x}10^{-4} \quad (38)$$

with the dimensions being in m and s. The correlation for k_La/H is written in terms of

$$\Delta(k_La/H) = (k_La/H)-(k_La/H)_0.$$

$$\Delta(k_La/H) = (3.76\text{x}10^{-6})(P_g/V)V_s^{0.3} \quad \text{for } (P_g/V) < (P_g/V)_b \quad (39a)$$

and

$$\Delta(k_La/H)=[\Delta(k_La/H)_b/(P_g/V)_b^{0.5}](P_g/V)^{0.5} \quad \text{for } (P_g/V) > (P_g/V)_b \quad (39b)$$

where

$$(P_g/V)_b=200[(ND_i)^2 -0.185]/(D_T^{0.45}H_L)$$

and

$$\Delta(k_La/H)_b = (3.76\text{x}10^{-6})(P_g/V)_b)\ V_s^{0.3}$$

with all dimensions in W, m, and s. The data used for developing these correlations were obtained in vessels having diameters of 0.187, 0.291, and 0.451 m. The sodium sulfite solution used to promote oxygen transfer was a non-coalescing system, so the results may be applicable only to salt solutions.

FERMENTATION PROCESS SCALE-UP

Several papers and chapters in books and monographs mention scale-up, but most authors simply discuss the consequences of scaling-up a process. They make frequent reference to the work of Oldshue(4) in which it is shown that when a particular scale-up strategy is carried out by maintaining a specific parameter constant for laboratory and plant scale fermenters, other parameters cannot be controlled and may take on much different values as the system volume changes. Bailey and Ollis(8) emphasize that this is particularly important for bioreactor scale-up. If a successful fermentation is carried out on the laboratory scale and scale-up is carried out by a particular strategy, the variables not controlled may change in unexpected ways as volume is increased. This may have unexpected effects on the yield, because so many factors affect microbe growth. Once the scale-up strategy is selected, the changes in the uncontrolled variables should be predicted and the consequences should be studied as part of the overall process design.

Cooney(53) makes the point that bioreactor performance is usually limited by the heat and mass transfer capabilities of the system. Heat transfer and mass transfer are separate issues from a system design standpoint, because mass transfer takes place throughout the broth while heat transfer takes place only at the boundary surfaces. The two factors can be considered independently with mass transfer being considered most strongly in the scale-up strategy and the required heat transfer capability being provided separately.

A few authors make some specific recommendations about how scale-up should be accomplished for fermentation systems. Aiba et al.(3) focus on the strategy of maintaining mass transfer coefficients constant in scaling-up processes involving aeration. The method suggested is based on having turbulent mixing in a broth which has properties not much different from water. The gas flow rate is scaled-up based on the k_La correlation for bubble aeration without impeller agitation.

$$k_La \sim (Q/V)(H_L/d_b^{1.5}v_b^{0.5}) \quad (40)$$

If bubble size and velocity do not change in scale-up and $(k_La)_1 = (k_La)_2$, then--

$$\frac{(Q/V)_1}{(Q/V)_2} = \frac{H_{L2}}{H_{L1}} \quad (41)$$

The impeller speed is scaled-up using the strategy that $(K_vp_t)_1 = (K_vp_t)_2$. The mass transfer coefficient correlation for K_v in impeller agitated fermenters is--

$$K_v = 0.0635\ (P_g/V)^{0.95}\ V_s^{0.67} \quad (42)$$

This is used with $N_{Pow}(N_{Re})$ = constant and (P_g/P_0) data. (K_vp_t) for the laboratory fermenter can be calculated from equation (42) or some similar correlation provided the power density can be measured or determined in some other way. The method also requires knowing that the mixing is turbulent and that the gas bubble size and velocity are not functions of the tank volume. This scale-up method has some weak points. One discrepancy is using two different mass transfer coefficient correlations--one to establish the gas flow rate and one to establish to impeller speed for the large system. One of the correlations is clearly not applicable to impeller agitated systems. An alternative procedure for selecting the gas flow rate for the plant sized system would be to keep (Q/V), (Q/ND_i^3), or V_s constant in scale-up. Another difficulty with the method is that it is based on mass transfer coefficients for turbulent mixing of air and water. Fermentation broths contain other materials which can affect the viscosity and surface tension which in turn can affect the mass transfer from the gas to the liquid. To avoid this difficulty, mass transfer coefficient correlations which include non-Newtonian properties could be used while retaining the strategy that $(k_La)_1 = (k_La)_2$.

Moo-Young and Blanch(9) make general comments about scale-up using the strategy of maintaining k_La constant. They recommend scaling up the gas flow rate by keeping VVM (or Q/V) constant. They mention organism damage, liquid blow-out, and oxygen poisoning as limitations on the gas flow rate and impeller speed. For turbulent mixing with Newtonian broths, they recommend using a dimensional k_La correlation such as that presented by Van't Riet--equations (20) and (21). For mixing viscous, non-Newtonian broths, they recommend using dimensionless k_La correlations such as presented be Perez and Sandall--equation (32)--or by Yagi and Yoshida--equation (36).

Humphrey(29) suggests finding optimum operating conditions on the laboratory scale. The optimum conditions are those which give the maximum yield of cells or extracellular product. Scale-up should be accomplished by achieving the same oxygen transfer capability on the plant scale as was observed on the laboratory scale. He discusses typical values of k_La, P/V, and Q/V needed to attain a biomass productivity of 2 to 5 kg/m^3hr in a bioreactor of 500 m^3. The mass transfer coefficient correlation of Miller--equations (24), (25), and (26)--is suggested with the effective power input being given by the sum of the mechanical power from the impeller and the power input due to gas injection. As an alternate k_La correlation, equation (22) is suggested. This is to be used along with maintaining impeller tip speed constant. Humphrey says one should determine what processing conditions can be achieved in a large plant and then scale-down these conditions to the laboratory scale in order to have an effective scale-up strategy.

Bailey and Ollis(8) make specific comments about scale-up strategy. They point out that power input per unit volume influences mass transfer coefficients, bubble and droplet sizes, and other parameters which are important in bioreactor operation, but using constant power per unit volume as a scale-up strategy seems to fail unless P/V = 1.0 HP/100 gal. This point is illustrated by quoting the data of Bartholomew(54) for penicillin production. They also say that the data of Bartholomew for vitamin B_{12} production show that the strategy of maintaining k_La constant is an effective--though not perfect--strategy for scaling-up fermentation systems.

Oldshue(55,56) makes general comments related to scale-up strategy. He says that maintaining power per unit volume constant gives conservative scale-up. His discussion focusses on including the effect of the shear rate to flow rate ratio for the impeller zone in scale-up calculations. This ratio can be matched for the laboratory and plant sized systems by relaxing the requirement for geometric similarity to the extent of changing D_i/D_T with scale. Oldshue observes that the range of shear rates in a large tank will be greater than the range of shear rates in a small tank. The basic scale-up strategy recommended is to maintain k_La constant. This includes estimating the shear stress at different points in the laboratory system in which a successful fermentation has been accomplished and using these data to estimate the effective viscosity at different points in the system. These results are helpful for estimating the mass transfer coefficient. Scale-up is carried out by choosing operating conditions for the plant size system which will duplicate the conditions observed in the laboratory.

Humphrey(57) summarizes the uncertainties about scale-up which persist by noting that in the correlating equations relating k_La to (P/V) and V_s, it is not yet understood why the exponents on these factors seem to depend on scale.

Charles(12) reviews many of the factors which affect fermenter scale-up and makes comments about the reliability of various data correlations used for fermenter design. He has a rather gloomy outlook toward the state of development of scale-up methods saying that not nearly enough attention is paid to the effects of non-Newtonian behavior in correlating mixing and mass transfer data in forms useful for scale-up calculations. He makes comments about scale-up for non-Newtonian broths pointing out that the non-Newtonian model parameters must not change with scale if any scale-up method is to be effective. The k_La correlation he uses for his example is one which has not been reported before, and no reference or supporting data are given.

RECOMMENDATIONS ABOUT SCALE-UP PROCEDURES

Once a fermentation has been accomplished successfully on the laboratory scale and the criterion for success has been established, scale-up calculations can be made. The procedures recommended are as follows.

For the laboratory scale fermenter --

* Measure Q, N, yield, fermenter dimensions
* Determine broth properties: ρ, $\mathcal{D}$, σ, and μ or μ_e
* Calculate VVM = (Q/V), (Q/ND_i^3), πND_i, N_{Re}, etc.
* Predict $P_0/\rho N^3D_i^5$, P_g/P_0, P_g/V, and, k_La

For the plant scale fermenter--

* Choose the volume required based on yield and required capacity

* Calculate the dimensions based on geometric similarity

* Establish the scale-up strategy to be used

$$(k_La)_{plant} = (k_La)_{lab}$$

* Calculate Q and N

METHOD 1

a. Determine Q using
VVM = constant
or Q/ND_i^3 = constant
or V_s = constant

b. Calculate N from power and k_La correlations

METHOD 2

a. Determine N using
πND_i = constant

b. Calculate Q from power and k_La correlations

* Estimate the power consumption

The scale-up methods described involve measuring only the rheological properties of the broth, the system dimensions, and the directly controlled variables Q and N. k_La is not a directly controlled variable even though it is crucial to the successful operation of the fermenter. Applying the methods is not completely straight-forward. k_La can be calculated directly for the laboratory scale fermenter. Calculating Q and N for the plant scale fermenter usually involves some iteration, because Q and N appear in more than one factor in the mass transfer coefficient correlations. If the impeller speed is to be maintained constant as in Method 2, N for the plant scale can be calculated directly, and μ_e can be estimated from equation (34) for broths behaving according to the power law model. Then Q can be calculated from the appropriate correlation for k_La. Since Q appears implicitly, iterative calculations are required. If either VVM or V_s is to be maintained constant as in Method 1, Q for the plant scale can be calculated directly. Then N and μ_e for the plant must be determined simultaneously from equation (34) and the appropriate k_La correlation. If N_Q is to be maintained constant(Method I), a trial value for Q for the plant scale must be chosen. Then the calculations can proceed as in Method 2. The value of Q calculated is compared to the trial value, and the trial value is modified until convergence is attained.

EFFECT OF SCALE-UP ON POWER CONSUMPTION

Einsele(58) surveyed European fermentation plants and was able to correlate power input data for fermenters having volumes from 0.5 to 300 m^3 by--

$$(P/V) \sim V^{-0.37} \tag{43}$$

This shows that in actual practice, the power per unit volume used becomes smaller as the system volume becomes larger. Equation (43) leads to a general expression useful for scaling-up power per unit volume.

$$\frac{(P/V)_{plant}}{(P/V)_{lab}} = \left(\frac{V_{plant}}{V_{lab}}\right)^s \tag{44}$$

where the value of the exponent, s, depends on the scale-up strategy selected. Equations of this form can be developed for any scale-up strategy following the methods described by Penney and Tatterson(59). If a particular strategy gives s > 0, that strategy is probably not useful, since P/V will be greater on the large scale than on the small scale. This does not correspond to actual practice. For example, the strategy--maintaining k_La and V_s constant--suggested in the work of Lopes de Figueiredo and Calderbank(30) leads to s = 0.16. One might reject this strategy, because it could lead to excessive power consumption in large fermenters. The value of the scale-up exponent, s, for different scale-up strategies is given in Table 3.

The actual power consumption can be determined using equation (44) rearranged as follows.

$$\frac{P_{plant}}{P_{lab}} = \left(\frac{V_{plant}}{V_{lab}}\right)^{s+1} \tag{45}$$

Equation (45) is plotted in Figure 1 to show the effect of different scale-up strategies. The data of Connolly and Winter(60) for suspending sand in water are also plotted. The results illustrated in this figure can be achieved if each part of the fermenter volume is as productive as every other part.

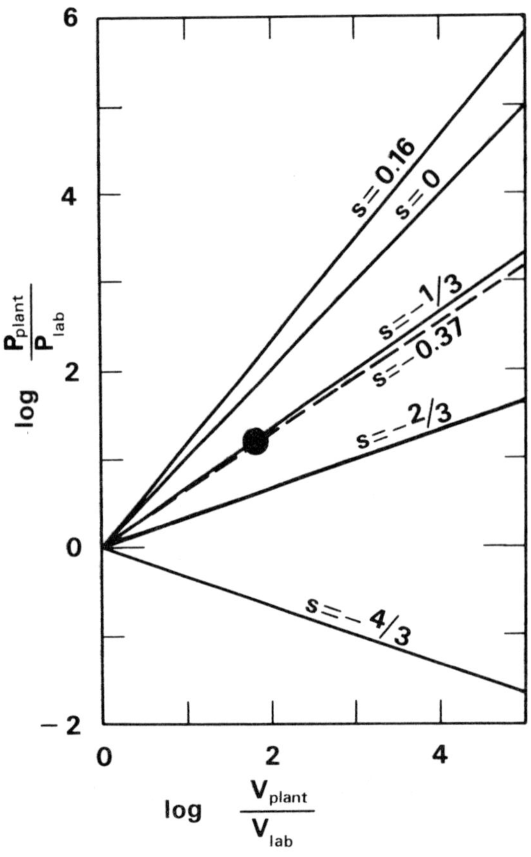

FIGURE 1. Effect of Scale-Up on Power Consumption for Newtonian Mixing [---- Einsele (*58*), ● Connolly and Winter (*60*)].

TABLE 3
Effect of Scale-up on Power Consumption for Mixing Newtonian Fluids

Scale-up Strategy	s
Maintain k_La and V_s constant	0.16
Maintain P/V constant	0
Maintain πD_iN constant (turbulent mixing)	- 1/3
Maintain πD_iN constant (laminar mixing)	- 2/3
Maintain N_{Re} constant	- 4/3

NOMENCLATURE

c^* oxygen concentration at the gas-liquid interface

c_L oxygen concentration in the broth

$\mathcal{D}$ molecular diffusion coefficient

D_i impeller diameter

D_T fermenter diameter

g gravitational acceleration

H Henry's law constant

K consistency index

k_La mass transfer coefficient per unit volume

$K_V = k_La/H$

n flow behavior index

N impeller speed

N_1 first normal stress difference

$N_{Fr} = D_iN^2/g$

$N_{Re} = D_i^2N\rho/\mu$ or $D_i^2N\rho/\mu_e$

$N_{Sc} = \mu/\rho\mathcal{D}$ or $\mu_e/\rho\mathcal{D}$

$N_{Sh} = k_LD_i/\mathcal{D}$

$N_{Sh}' = k_LaD_i^2/\mathcal{D}$

OTR oxygen transfer rate

OUR oxygen uptake rate

P power input

P_g power input with gas present

P_0 power input without gas present

P_T sum of gassed power input and gas expansion power input

p_t total pressure

Q volumetric gas flow rate

r radial coordinate

v velocity

V fermenter volume

V_s superficial gas velocity

v_z velocity component in the z-direction

VVM = Q/V

W baffle width

x coordinate direction

ϵ_T power input per unit mass

λ relaxation time

μ broth viscosity

μ_{app} apparent viscosity

μ_d dispersed phase(gas) viscosity

μ_e effective viscosity

ρ broth density

σ surface tension

τ shear stress

τ_{xz} xz component of the shear stress tensor

ϕ volume fraction of solids

LITERATURE CITED

1. Anderson, C., G.A. LeGrys, and G.I. Solomons, The Chem Eng, No. 377, 43 (1982) .

2. Aiba, S., A.E. Humphrey, and N.F. Millis, Biochemical Engineering, pp. 163-185, Academic Press, New York (1965) .

3. Aiba, S., A.E. Humphrey, and N.F. Millis, Biochemical Engineering, 2nd Ed., pp. 195-217, Academic Press, New York (1973) .

4. Oldshue, J.Y., Biotech Bioeng, 8, 3 (1966) .

5. Oldshue, J.Y., Genetic Engineering News, 3(6), 46 (1983) .

6. Oldshue, J.Y., "Agitation", in Fermentation and Biochemical Engineering Handbook, Vogel, H.C., ed., Noyes Publications, Park Ridge, N.J. (1983) pp. 142-161 .

7. Blanch, H.W. and S.M. Bhavaraju, Biotech Bioeng, 18, 745 (1976) .

8. Bailey, J.E. and D.F. Ollis, Biochemical Engineering Fundamentals, pp. 462-464 McGraw-Hill, New York (1977) .

9. Moo-Young, M. and H.W. Blanch, "Design of Biochemical Reactors: Mass Transfer Criteria for Simple and Complex Systems," in Advances in Biochem Eng, Vol. 19. Ghose, T.K., A. Fiechter, and N. Blakebrough, eds., Springer-Verlag, Berlin (1981) pp. 2-69 .

10. Moo-Young, M. and H.W. Blanch, "Kinetics and Transport Phenomena in Biological Reactor Design," Foundations of Biochemical Engineering, Blanch H. W., E. T. Papoutsakis, and G. Stephanopoulos, eds., ACS Symposium Series 207, Am. Chem. Soc., Washington, D. C. (1983) pp. 335-354 .

11. Prokop, A., "Reactor Design Fundamentals," in Foundations of Biochemical Engineering, Blanch, H.W., E.T. Papoutsakis, and G. Stephanopoulos, eds., ACS Symposium Series No. 207, Am Chem Soc, Washington, D. C. (1983) pp. 370-374 .

12. Charles, M. . "Fermenter Design and Scale-Up," in Comprehensive Biotechnology, Vol. 2, Moo-Young, M. ed., Pergamon Press, Oxford (1985) pp. 57-75 .

13. Richards, J.W., Progress in Industrial Microbiology, 3, 143 (1961) .

14. Finn, R.K., "Aeration and Agitation," in Biochemical and Biological Engineering Science, Vol. 1, Academic Press, New York (1967) pp.69-99 .

15. Bird, R.B., W.E. Stewart, and E.N. Lightfoot, Transport Phenomena, pp.3-15, John Wiley and Sons, New York (1960) .

16. Roels, J.A., J. Van Den Berg, and R.M. Voncken, Biotech Bioeng, 16, 181 (1974) .

17. Carreau, P.J., Trans Soc Rheol, 16, 99 (1972) .

18. Bird, R.B., R.C. Armstrong, and O. Hassager, Dynamics of Polymeric Liquids: Fluid Mechanics, pp. 445-449, John Wiley and Sons, New York (1977) .

19. Ulbrecht, J., The Chem Eng, No. 286, 374 (1974) .

20. Ranade, V.R. and J.J. Ulbrecht, AIChE Journal, 24, 796 (1978) .

21. Charles, M., "Technical Aspects of the Rheological Properties of Microbial Cultures," in Adv in Biochem Eng, Vol. 8: Mass Transfer in Biotechnology, Ghose, T.H., A. Fiechter, and N. Blakebrough, editors, Springer-Verlag, Berlin (1978) pp. 1-62 .

22. Bongenaar, J.J.T.M., N.W.F. Kossen, B. Metz, and F.W. Meijboom, Biotech Bioeng, 15, 201 (1973) .

23. Bird, R.B. et al, Transport Phenomena, pp. 107-111, 580-583 .

24. Langhaar, H.L., Dimensional Analysis and Theory of Models, pp. 13-58, John Wiley and Sons, New York (1951) .

25. Sideman, S., O. Hortacsu, and J. W. Fulton, Ind Eng Chem, 58(7),32 (1966) .

26. Cooper, C.M., G.A. Fernstrom, and S.A. Miller, Ind Eng Chem, 36, 504 (1944) .

27. Van't Riet, K., Ind Eng Chem Process Des Dev, 18, 357 (1979) .

28. Zlokarnik, M., "Sorption Characteristics for Gas-liquid Contacting in Mixing Vessels," in Adv in Biochemical Engineering No. 8, Mass Transfer in Biotechnology, Ghose, T.K., A. Fiechter, and N. Blakebrough, eds., Springer-Verlag, Berlin. (1978) pp. 133-151.

29. Humphrey, A.E., "Biochemical Engineering," in Encyclopedia of Chemical Processing and Design. Vol. 4., McKetta, J.J., ed., Marcel Dekker, New York (1977) pp. 359-394 .

30. Lopes de Figueiredo, M.M. and P.H. Calderbank, Chem Eng Science, 34, 1333 (1979) .

31. Miller, D.N., AIChE Journal, 20, 445 (1974) .

32 .Calderbank, P.H., Trans Inst Chem Eng, 36, 443 (1958) .

33. Calderbank, P.H., Trans Inst Chem Eng, 37, 173 (1959) .

34. Mendelson, H.D., AIChE Journal, 13, 250 (1967) .

35. Van't Riet, K, "Turbine Agitator Hydrodynamics and Dispersion Performance," Thesis. Delft Technical University, Delft, The Netherlands (1975) .

36. Oyama, Y. and K. Endoh, Kagaku Kogaku, 19, 2 (1955) .

37. Michel, B.J. and S.A. Miller, AIChE Journal, 8, 262 (1962) .

38. Rushton, J.H., E.W. Costich, and H.S. Everett, Chem Eng Prog, 46, 467 (1950) .

39. Nagata, S., Mixing, pp. 24-59, Kodansha, Ltd., Tokyo (1975) .

40. Chapman, C.M., A.W. Nienow, M. Cooke, and J.C. Middleton., Chem Eng Res Des, 61, 71 (1983) .

41. Chapman, C.M., A.W. Nienow, M. Cooke, and J.C. Middleton., Chem Eng Res Des, 61, 82 (1983) .

42. Machon, V., J. Vlcek, A. W. Nienow, and J. Solomon, The Chem Eng Journal, 19, 67 (1980) .

43. Nienow, A.W., D.J. Wisdom, J. Solomon, V. Machon, and J. Vlcek, Chem Eng Comm, 19, 273 (1983) .

44. Perez, J.F. and O.C. Sandall, AIChE Journal, 20, 770 (1974) .

45. Metzner, A.B. and R.E. Otto, AIChE Journal, 3, 3 (1957) .

46. Calderbank, P.H. and M.B. Moo-Young., Trans Inst Chem Eng, 37, 26 (1959) .

47. Perez, J.F., "Gas Absorption by Non-Newtonian Fluids in Agitated Vessels," M.S. Thesis, Univ. of Cal-Santa Barbara, Santa Barbara (1973) .

48. Yagi, H. and F. Yoshida, Ind Eng Chem Process Des Dev, 14, 488 (1975) .

49. Chapman, C.M., A.W. Nienow, M. Cooke, and J.C. Middleton., Chem Eng Res Des, 61, 167,182 (1983) .

50. Mills, D.B., R. Bar, and D.J. Kirwan, "Effect of High Solids Concentration on Oxygen Transfer in Agitated Three-Phase Systems," Paper 8f presented at the A.I.Ch.E. 1986 Annual Meeting, Miami Beach (1986) .

51. Joosten, G. E. H., J. G. M. Schiler, and J. J. Janssen, Chem Eng Science, 32, 563 (1977) .

52. Loucaides, R. and W.J. McManamey, Chem Eng Science, 28, 2165 (1973).

53. Cooney, C.L., Science, 219, 728 (1983) .

54. Bartholomew, W.H., Adv. in Appl. Microbiol., 2, 289 (1960) .

55. Oldshue, J.Y., Chem Eng, 90(12), 82 (1983) .

56. Oldshue, J.Y., Fluid Mixing Technology, pp. 89-92, 192-215, McGraw-Hill, New York (1983) .

57. Humphrey, A.E., "Introduction" (to the section about "Bioreactor Design, Operation, and Control") in Comprehensive Biotechnology, Vol. 2., Moo-Young,M., ed., Pergamon Press, Oxford (1985) pp. 3-4 .

58. Einsele, A., "Scaling of Bioreactors, Theory and Reality,". Paper 4.13 presented at the 5th Int Ferm Symp, Berlin (1976) .

59. Penney, W.R. and G.B. Tatterson, Food Technology, 37(2), 62 (1983).

60. Connolly, J.R. and R.L. Winter, Chem Eng Prog, 65(8), 70 (1969) .

Scale-up Using a Biochemical Process Simulator

HERBERT E. KLEI
ROBIN D. BAENA
THOMAS F. ANDERSON
DONALD W. SUNDSTROM
ALBERTO BERTUCCO
Department of Chemical Engineering
University of Connecticut
Box U-139
191 Auditorium Road
Storrs, Connecticut 06268

A software program has been developed for use on IBM PC or IBM-compatible personal computers, to simulate a variety of free cell or immobilized cell biochemical reactors to allow sizing and scale-up. The program is useful for simulation of fermentors to produce biochemicals. Several examples using the BPS simulation program are presented.

Biochemical reactors come in a variety of configurations, using either immobilized or free cells and employing many possible reaction kinetic expressions.

In order to scale-up laboratory data, the laboratory reactor must be simulated to determine kinetic constants, and to verify the applicability of certain assumptions as to mixing, particle effectiveness factors, and kinetic expressions.

Since the modeling process involves repeated solution of many different trial models, the task of programming to solve the many simultaneous nonlinear differential equations can be time consuming, especially if recycle streams are involved. For this reason, a Biochemical Process Simulator (BPS) was developed for use on the IBM-PC to allow the user to select his mixing model, kinetic expression, and reactor configuration from a series of graphics driven menus. The appropriate constants are entered into the graphics screens for data input and the output is displayed and summarized on the screen and is available in detailed form for use with a spreadsheet and graphical plotting package such as Lotus 123®. The program can be run on color or monochrome monitors without graphics hardware.

Alberto Bertucco is now at the Universita di Padova, Padova, Italy.

PROGRAM DESCRIPTION

A flowsheet summarizing the screen and the data or decisions to be made on them is given in Fig. 1. The Main Menu screen is entered first, which consists of a series of menu choices: file specify, configure, run, results, and exit. When file specify is chosen, the bottom of the screen becomes active allowing the user to specify drive (A,B,C) and an eight character name with the extension .BPS automatically added. When the configure cell is chosen, a second screen appears allowing the user to specify one of the following: free cell, activated sludge, and immobilized cell. The activated sludge cell is included separately because it is an important biological reactor used in wastewater processing and employs solids settlers and solids recycle both of which are becoming more common in biochemical reactor design. Once a configuration has been chosen, a screen will appear allowing for the specification of inlet stream concentrations, nutrients, etc., as well as reactor kinetics expression, type of reactor hydraulics (CSTR, Plug flow, etc.), and type of settler model if solids recycle is used. Once these selections are made, the appropriate kinetics screen will appear allowing input of values for kinetic constants or the choice default values from a menu of common substrates. After the kinetics selections are made, the appropriate reactor/settler screen will appear allowing data input for such

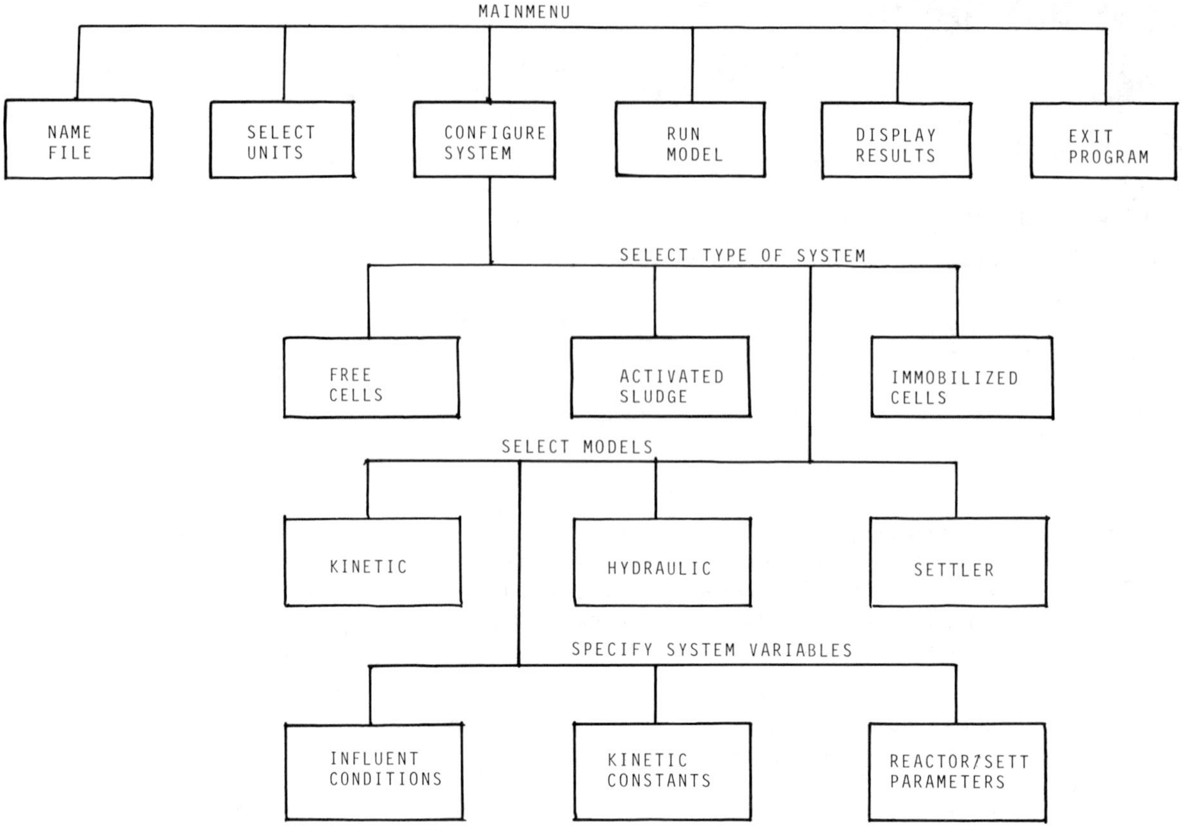

FIGURE 1. Menu flowsheet.

TABLE 1
Kinetic Models Equation for $(-r_S)$

A. Free Cells or Activated Sludge

1) Michaelis Menten/Monod

$$\frac{k_o X S}{Y_X(K_m + S)}$$

2) Competitive Inhibition

$$\frac{k_o X S}{Y_X(K_m + S + I\ K_m/K_I)}$$

3) Non-Competitive Inhibition

$$\frac{k_o X S}{Y_X(K_m+S)(1+I/K_I)}$$

4) Substrate Inhibition

$$\frac{k_o X S}{Y_X(K_m+S + S^2/K_I)}$$

5) Product Inhibition*

$$\frac{k_o X S}{Y_X(K_m+S)} \cdot \frac{K_p}{K_p+P}$$

$$P + P_o + Y_p\ (S_o - S)$$

*not a choice under activated sludge

Nomenclature

$k_o = [\text{time}^{-1}]$

$K_m = \left[\frac{\text{mass substrate}}{\text{vol. reactor}}\right] \text{or} \left[\frac{\text{moles substrate}}{\text{vol. reactor}}\right]$

$K_I = \left[\frac{\text{mass inhibitor}}{\text{vol. reactor}}\right] \text{or} \left[\frac{\text{moles inhibition}}{\text{vol. reactor}}\right]$

$I = \left[\frac{\text{mass inhibitor}}{\text{vol. reactor}}\right] \text{or} \left[\frac{\text{moles inhibition}}{\text{vol. reactor}}\right]$

$Y_X = \left[\frac{\text{mass biomass}}{\text{mass substrate}}\right] \text{or} \left[\frac{\text{moles biomass}}{\text{moles substrate}}\right]$

$k_d = [\text{time}^{-1}]$

$Y_p = \left[\frac{\text{mass product}}{\text{mass substrate}}\right] \text{or} \left[\frac{\text{moles product}}{\text{moles substrate}}\right]$

$P = \left[\frac{\text{mass product}}{\text{vol. reactor}}\right]$

B. Immobilized Cells

1) Michaelis Menten

$$\frac{k_o E S}{K_m + S};\ E = E_o e^{-k_d t}$$

2) Substrate Inhibition

$$\frac{k_o E S}{K_m + S + S^2/K_I};\ E = E_o e^{-k_d t}$$

TABLE 1 *(Continued)*

$k_o = \left[\frac{\text{mole substrate}}{\text{g enzyme-time}}\right] \text{or} \left[\frac{\text{mole substrate}}{\text{g cell-time}}\right]$

$E = \left[\frac{\text{g enzyme}}{\text{vol. reactor}}\right] \text{or} \left[\frac{\text{g cell}}{\text{vol. reactor}}\right]$

$K_m = \left[\frac{\text{mole substrate}}{\text{vol. reactor}}\right]$

$k_d = [\text{time}^{-1}]$

3) Product Inhibition

$$\frac{k_o E S}{K_m + S}\ \frac{1}{1 + P/K_p};\ E = E_o e^{-k_d t}$$

$P = P_o + Y_p\ (S_o - S)$

$K_p = \left[\frac{\text{mole product}}{\text{vol. reactor}}\right] \text{or} \left[\frac{\text{mass product}}{\text{vol. reactor}}\right]$

$Y_p = \left[\frac{\text{mole product}}{\text{mole substrate}}\right] \text{or} \left[\frac{\text{mass product}}{\text{mass substrate}}\right]$

$k_d = [\text{time}^{-1}]$

4) Equilibrium

$$\frac{k_o' E(S-S_e)}{K_M' + (S-S_e)}$$

S_e = equilibrium substrate concentration, $\left[\frac{\text{mass substrate}}{\text{vol. reactor}}\right]$

K_M' = f(S_e, elementary reaction rate constants)

= constant for a given value of S_e, $\left[\frac{\text{mass substrate}}{\text{vol. reactor}}\right]$

k_o' = f(elementary kinetic constants) $\left[\frac{\text{mass substrate}}{\text{gm cells-time}}\right]$

[See Kikkert, A., et al., Biotech. and Bioeng., 23, 1087-1101 (1981)] variables as, recycle ratio, number of tanks in series (if needed), Peclet number and reactor length (for axial dispersion), sludge age, solids settling constants (if needed for the flux model), reactor residence time, final time and reactor volume (for batch reactor), and volumetric flow rate.

Once this data is entered, all parameters are fixed and the solution can be generated by returning to Main Menu and choosing the cell marked RUN. A screen will appear giving the status of the solution and will tell when the solution is complete. The results are obtained by either going to Main Menu and choosing cell marked RESULTS or by exiting the program and calling up the file with the chosen filename and the extension (.PRN). Custom tailored graphs can be created by importing this (.PRN) file into standard spread sheet programs such as Lotus 123®. If any changes need to be made in one of the parameters and the solution repeated, the chosen variables can be changed and the solution re-run without respecifying all the remaining parameters.

KINETIC MODELS

The choices of kinetic expressions for substrate removal ($-r_S$) under the free cell, activated sludge, or immobilized cell configurations are given in Table I. The net rate of cell growth, (r_x) is taken to be,

$$r_x = (-r_s)Y_x - k_d X \tag{1}$$

The rate of product formation kinetics is taken to be,

$$r_p = (-r_s)Yp \tag{2}$$

The units associated with the values of the kinetic constants fed into the program determine the units associated with output variables. For example, if a value of K_M = 100 is placed into the screen asking for K_M and the units of K_M are mg substrate/liter of reactor, then the units of S from the program output will have mg substrate/liter of reactor. As a result, if Y_x = mg biomass produced/mg substrate consumed, then X will have units of mg biomass/liter of reactor. The units for the problem are selected on a Units Select Screen prior to the kinetics screens. However, when entering numbers for kinetic constants, the associated units of the constants should be consistent with those previously chosen. If the substrate concentration unit was previously chosen as lb/gallon, then K_M would be entered as 100 x (1lb/454,000 mg) x (3.78 liters/gallon) = 8.32×10^{-4} lb/gallons.

The units became very critical in immobilized reactors, where enzyme or cell concentrations are quoted as dry mass or mg of protein per volume of particle or per volume of reactor. By convention this program has taken the enzyme or cell concentration, E, to be dry mass of cells per volume of reactor, or gms of enzyme per volume of reactor. If enzyme units per volume of reactor are used, then it is entered as the product, k^oE, and time must be in minutes. For convenience, enter k_o = 1.00, and E = EU x 10^{-6}. The immobilized cell kinetic equations assume that cell synthesis in the reactor is negligible and therefore Y_x is taken to be zero. Cell death or enzyme deactivation, however, is allowed by specifying the first order coefficient, k_d.

EFFECTIVENESS FACTOR

In immobilized cell systems as well as in cell flocs, substrate and product diffusion within the particle can be the rate limiting step, resulting in reaction rates less than predicted based upon only bulk liquid concentrations (6,7). An additional diffusion resistance can also occur across a stagnant external liquid film around the particles, but this resistance is taken to be negligible compared to the resistance to diffusion within the particle.

The rate of consumption for substrate, ($-r_s$) under conditions of intraparticle substrate diffusion is represented by

$$(-r_s) = \eta \text{ x (reaction rate under bulk liquid concentrations)}, \tag{3}$$

and

η = effectiveness factor
= observed reaction rate/reaction rate under bulk liquid concentrations (4)

At steady state, the observed reaction rate, which is the numerator of Equation (4), is equal to the diffusion rate within the particle when evaluated at the particle surface. Therefore if a substrate concentration profile could be calculated within the particle, and its slope evaluated at the surface, then η and the reaction rate could be calculated in the reactor.

In order to evaluate the concentration profile within the particle, a steady state balance on substrate and product in a spherical shell of the particle is written and yields,

$$D_s \frac{d^2S}{dr^2} + \frac{2}{r}\frac{dS}{dr} + r_s = 0 \qquad (5)$$

$$D_p \frac{d^2P}{dr^2} + \frac{2}{r}\frac{dP}{dr} + r_p = 0 \qquad (6)$$

The boundary conditions are,

$$\text{at } r = 0 \quad \frac{dS}{dr} = 0, \ \frac{dP}{dr} = 0 \qquad (7)$$

$$\text{at } r = R \quad k_L(S_o - S_R) = D_s \left.\frac{dS}{dr}\right|_{r=R} \qquad (8)$$

where k_L = external mass transfer coefficient and S_R is substrate concentration at the particle surface. This system of differential equations and boundary conditions can be reduced into a single differential equation by adding the two differential equations, together with the equations (1) and (2) to yield,

$$P = P_o + (S_o - S)\, Y_p \, \frac{D_R}{D_p} \qquad (9)$$

Therefore only the first differential equation needs to be solved, which can be rewritten in terms of a dimensionless radius, $\bar{r} = r/R$,

$$\frac{d^2S}{d\bar{r}^2} + \frac{2}{\bar{r}}\frac{dS}{d\bar{r}} + \frac{R^2 r_S}{D_s} = 0 \qquad (10)$$

Equation (10) with boundary conditions (7) and (8) are solved using the method of orthogonal collocation (15) which converges fast for this system. The constants in the collocation equations are solved by a Newton-Raphson technique to give point values of substrate within the particle as a function of dimensionless radius. The derivative at $\bar{r}$ = 1 is evaluated and substituted into Equation (4) to allow the calculation of η for a given value of external substrate concentration, S_o. The solution process is repeated many times for different values of S_o ranging from the feed concentration to near zero, thus giving a table of value of η as a function of the external particle substrate concentration. This generated table allows the effectiveness factor to change along the length of the reactor as the bulk substrate concentration decreases.

REACTOR MIXING MODELS

The type of liquid mixing may be modeled according to batch, a single CSTR, N equal volume CSTR in series, plug flow, and axial dispersion equations given in Table II. The biological reactor for free cells and activated sludge configurations, with the exception of the batch case, can have continuous feed, cell settling, and solids recycle. The reactor for such a case would fit into a flow sheet given in Fig. 2.

For free cell flow reactors using no solids separation or recycle, the recycle rate R would be set to 0.0 and the settler system and recycle calculations are bypassed. For immobilized systems, there are no solids leaving the reactor and hence no settler or solids recycle are needed. The immobilized reactor choices are: batch, CSTR, column in plug flow, and column in axial dispersion flow. For immobilized systems, the biomass or enzyme concentration in the reactor is designated by replacing X with E in the product and substrate equations of Table II. This change allows consistency with the kinetic equations.

In systems utilizing solids recycle, an index of solids residence time must be specified. An approximation to a solids residence time, called sludge age, is often used and is designated as θ_c. Sludge age and its approximation to a mean solids residence time are discussed elsewhere (1). By making a solids balance around the settler, the sludge age is related to other system parameters by the relationships given in Table II. The sludge age has significance in predicting solids settling characteristics and is one of the independent variables which must be specified. Typical values of sludge age for activated sludge systems lie around 3-5 days.

SETTLER MODELS

One principal factor affecting the efficiency of the activated sludge process

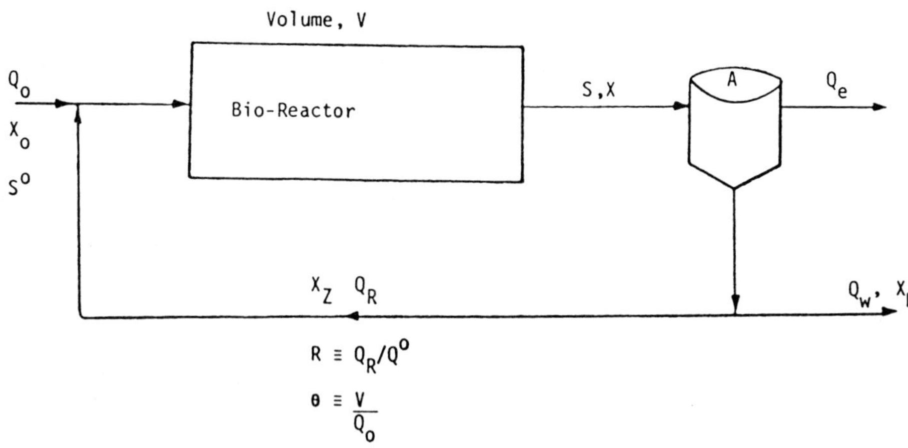

FIGURE 2. Biological reactor location in flowsheet for free cell and activated sludge case.

TABLE II
Reactor Mixing Models

* Equation not used in Immobilized Systems

1) Batch	$\frac{dS}{dt} + (-r_s) = 0$	
	* $\frac{dX}{dt} + (-r_x) = 0$	
	$\frac{dP}{dt} + (-r_p) = 0$	
	t = time	
2) C.S.T.R.	$S-S_o + (-r_s)\,\theta = 0$	$\theta \equiv \frac{\text{Reactor Volume}}{\text{Fresh Feed rate, Vol/time}}$
	* $-(1+R)X+X_o+RX_R+(r_x)\theta=0$	R = recycle ratio = $\frac{Q_R}{Q_o}$
	$P_o-P + (r_p)\,\theta = 0$	$X_R = [\frac{\text{g biomass}}{\text{volume}}]$
	* $\theta_c = \frac{\theta}{1+R(1-X_{R/X})}$	θ_c = sludge age, [time^{-1}]
		i = reactor index (1...N)
3) N CSTR in series for i=2....N	$S_i-S_{i-1} + (-r_s)\frac{\theta}{R+1}=0;$	$\bar{X} = \frac{\Sigma_i X_i}{N}$
	* $X_i-X_{i-1} + (-r_x)\frac{\theta}{R+1}=0$	= average biomass concentration in tank system
	$P_i-P_{i-1} + (-r_p)\frac{\theta}{R+1}=0$	
for i = 1	$(1+R)S_1-S_o-RS_N + (-r_s)\theta=0$	
	* $(1+R)X_1-X_o-RX_r+(-r_x)\theta=0$	
	$(1+R)P_1-P_o-RP_N+(-r_p)\theta=0$	
	* $\theta_c = \frac{\theta\bar{X}N}{(R+1)X_N-RX_r}$	

TABLE II (*Continued*)

4) Plug Flow	$\frac{dS}{d\theta} + (\frac{-r_s}{R+1}) = 0$	
	* $\frac{dX}{d\theta} + (\frac{-r_x}{R+1}) = 0$	$Z' = Z/L$
	$\frac{dP}{d\theta} + (\frac{-r_p}{R+1}) = 0$	L = reactor length [m]
	or: $\frac{dS}{dZ'} + (-r_s)\frac{\theta}{R+1} = 0$	Z = reactor axial coord [m]
	* $\frac{dX}{dZ'} + (-r_x)\frac{\theta}{R+1} = 0$	$\bar{X} = \int_0^1 X dZ'$, $[\frac{\text{g biomass}}{\text{l reactor}}]$
	$\frac{dP}{dZ'} + (-r_p)\frac{\theta}{R+1} = 0$	
5) Axial Dispersion	$-\frac{1}{P_{e_s}}\frac{d^2S}{dZ'^2} + \frac{dS}{dZ'} + (-r_s)\frac{\theta}{R+1} = 0$	$Z' = Z/L$
	* $-\frac{1}{P_{e_x}}\frac{d^2X}{dZ'^2} + \frac{dX}{dZ'} + (-r_x)\frac{\theta}{R+1} = 0$	L = reactor length [distance]
	$-\frac{1}{P_{e_p}}\frac{d^2P}{dZ'^2} + \frac{dP}{dZ'} + (-r_p)\frac{\theta}{R+1} = 0$	Z = reactor axial coord [dist]
		Pe = Peclet No. $\equiv \frac{\bar{u}L}{Z} = \frac{L^2}{Z}\frac{(1+R)}{Z}$
	* $\theta_c = \frac{\theta\bar{X}}{(R+1)X_{out}-RX_R}$	$\bar{u}$ = average axial velocity, $\frac{\text{distance}}{\text{time}}$
		$\bar{X} = \int_0^1 X dZ'$ $[\frac{\text{g biomass}}{\text{l reactor}}]$
		X_{out} = X at Z=L

or other flocculating cell systems is the performance of the secondary settler. High effluent concentrations are often the result of problems with floc formation and sludge settlability rather than from difficulties encountered with biodegradation. The effluent quality as well as the size and cost of the facilities often is controlled by the sludge settling characteristics. Failure to consider the settler in the overall design can lead to unsatisfactory performance of the entire process. None the less, a good quality effluent can be produced if the settler is properly designed and operated.

The clarifier model in this Program allows two choices: Perfect settler or a Flux limiting settler. The perfect settler assumes that all solids entering the clarifer will settle and leave the bottom, for all inlet conditions. The flux limiting settler uses the settling flux theory to predict when the perfect settler will fail due to excessive solids loading. Both models assume that no solids will appear in the overflow effluent and that the bottom solids concentration can be determined by a material balance around the settler.

The models for particle removal are based on the assumption of ideal flow. The model assumes 1) quiescent or non-turbulent flow, 2) uniform distribution of velocity over all sections normal to the flow direction, and 3) no scour or resuspension of particles from the tank bottom. Under these assumptions, performance is a function of surface overflow rates and settling rates only.

For clarification, sufficient surface area must be provided so that the hydraulic loading per unit area does not exceed the settling velocity of the slowest settling material which is to be completely removed. That is, the upward flow rate must be less than the solids' settling velocity:

$$Q_e/A_c < V_s$$

where Q_e = overflow volumetric rate of clarified liquid

A_c = clarifier surface area
V_s = initial settling rate of the suspension at the feed concentration

Improper consideration of the thickening function can lead to failure in two ways. First, it can lead to direct deterioration of the effluent quality by loss of suspended solids over the weir. Secondly, the inability of the settler to produce a concentrated underflow affects the performance in the aeration tank since solids are needed to sustain the process. Furthermore, increasing the organic loading e.g., (mg BOD/mg solids) may also adversely affect the flocculating and settling characteristics of the sludge (1), making it even more difficult to achieve the desired underflow concentration.

Coe and Clevenger (8) and Kynch (9) established the importance of the flux theory in the design of settling tanks. Solids transport is assumed to be dependent on the settling characteristics of the solids and the tank underflow rate only.

For thickening, sufficient surface area must be provided so that solids are applied at a rate which is not in excess of the rate at which solids are able to reach the bottom of the tank.

In continuous thickeners, solids are transported downward by two mechanisms: 1) gravity settling and 2) bulk transport due to solids removal. The flux due to gravity transport, G_g, depends on the settling characteristics of the sludge and is defined by:

$$G_g = X_i * V_i$$

The flux due to solids withdrawal, G_u, can be controlled by the operator and is defined by:

$$G_u + X_i * U_b$$

where X_i = the local concentration of solids

V_i = the settling velocity of solids at concentration X_i
U_b = Q_u/A is the bulk downward velocity of the suspension.

The total flux is the sum of the gravity and bulk fluxes:

$$G = G_g + G_u = (V_i + U_b) * X_i$$

as illustrated in Figure 3. The minimum in the total flux curve corresponds to the maximum solids flux (commonly called the

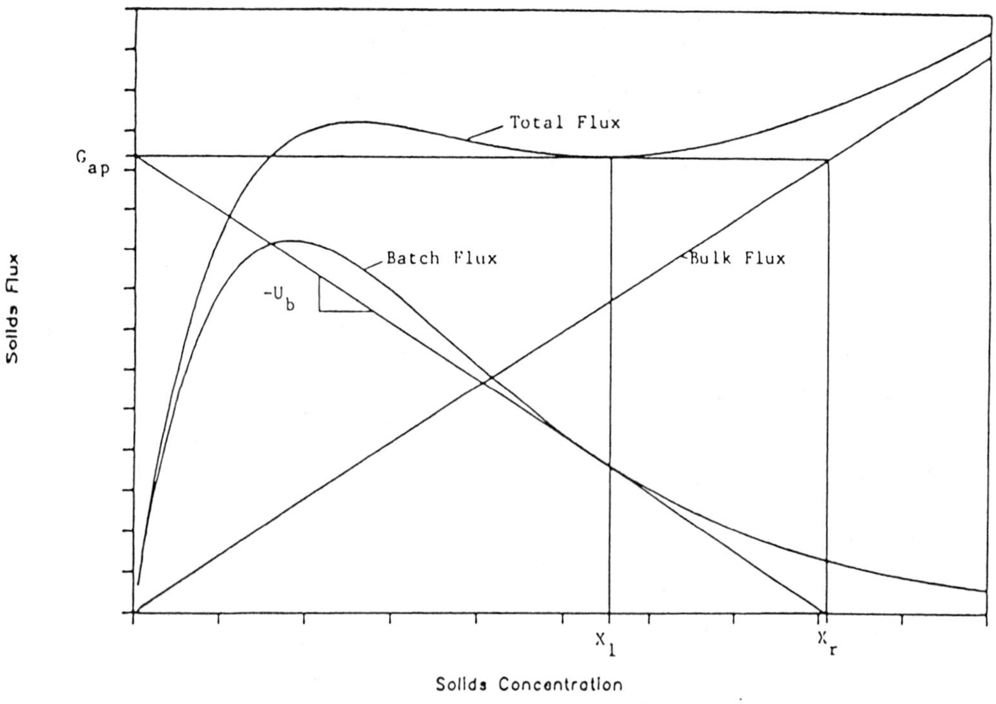

FIGURE 3. Flux analysis in settler operation.

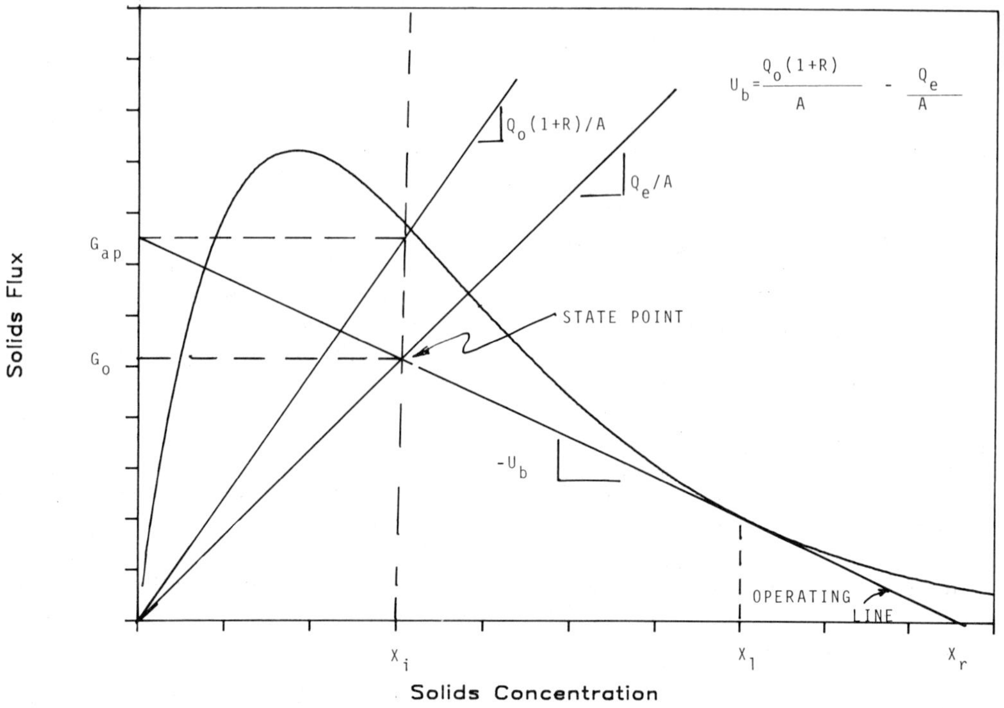

FIGURE 4. Design lines for predicting satisfactory settler operation.

limiting flux) that can be transported to the bottom of the basin and concentrated to the given underflow concentration. The flux due to gravity is zero at the bottom of the tank. Since the solids are removed by bulk flow, the underflow concentration is the x-coordinate on the bulk flux line where the y-coordinate equals the limiting flux.

The batch flux technique developed by Yoshioka et al. (10) is a useful method for analyzing settler thickening performance. This method is simpler and more versatile than the total flux approach, although it produces the same results. In addition, the effects of operational changes can also be analyzed using the batch flux curves.

The graphical batch flux concent is illustrated in Fig. 4. A line from the origin with a slope equal to the overflow rate, Q_e/A, is superimposed on the batch flux curve. The point on this line with an x-coordinate equal to the solid concentration into the settler is defined as the state point. The state point has a y-coordinate of G_o, the applied flux at zero recycle.

The operating line has a slope of $-U_b$ and passes through the state point. This line intersects the y-axis at the applied flux, $G_{ap}=X_oQ_o(1+R)/A$, and the x-axis at the underflow concentration (except in overloaded case). The state point serves as a pivot point of the operating line for changing recycle flow rates as shown in Figure 5.

When the operating line is tangent to the batch flux curve, the system is critically loaded with respect to the thickening function. This line represents the minimum recycle ratio or maximum flux possible to obtain a desired underflow concentration. If the recycle is decreased so that the operating line cuts the flux curve at three points (broken operating line in Figure 5), overloaded conditions prevail with respect to the thickening capacity. That is, the applied flux is greater than the maximum flux that can be transmitted to the bottom of the tank and concentrated to the desired underflow concentration. A sludge layer of limiting concentration, X_1, will build up and if the overloaded conditions continue, the clarifier will fail, spilling solids over the weir.

Changes in the influent flow rate change the position of the state point (see Fig. 5). Critical conditions with respect to clarification prevail when the state point is on the gravity flux curve. If the state point is outside the envelope of the flux curve, the overflow rate Q_e/A is greater than the settling velocity at the feed concentration. Hence solids will be carried over the weir and settler failure will occur.

Changes in settling characteristics of the sludge result in a change in the shape of the batch flux curve. Settler failure will occur if the change causes the operating line to cut the flux curve in three places or the state point to fall outside of the envelope of the flux curve.

The settling flux approach can be used for system design and process operational analysis. It can help the operator make control decisions to prevent system failure and maximize overall treatment efficiency. This method is easily adopted for computer solutions. However, a mathematical relationship between local solids concentration and settling velocity is needed. The semi-log expression proposed by Vesilind (11) and tested by Smollen and Ekama (12) was used. Vesilind's model is given by the following equation:

$$V_i = V_o \exp(-b * X_i)$$

where b and V_o are experimentally determined constants. These constants can be related to the Sludge Volume Index as done by Daigger and Roper (13) and Koopman and Cadee (14). Typical values for municipal sludge are b = $0.55 m^3/kg$, v_o = 138 m/day, (12).

The program first checks for settler failure with respect to the clarification function by determining if the state point is within the envelope of the batch flux curve. The applied flux at zero recycle, G_o, which is the y-coordinate of the state point and the gravity flux, G_g, at the inlet concentration, X_i, are calculated and compared. If G_o is greater than G_g at X_i, the state point is above the batch flux curve and settler failure has occurred. The program gives the message: "SETTLER FAILURE... SOLIDS CARRY-OVER DUE TO EXCESSIVE OVERFLOW RATE". If the fluxes are equal, the state point is on the batch flux curve. The program gives the message: "SETTLER OPERATING AT CRITICAL CONDITIONS WITH RESPECT TO SOLIDS HANDLING CRITERION". If G_o is less than G_g at X_i, the state

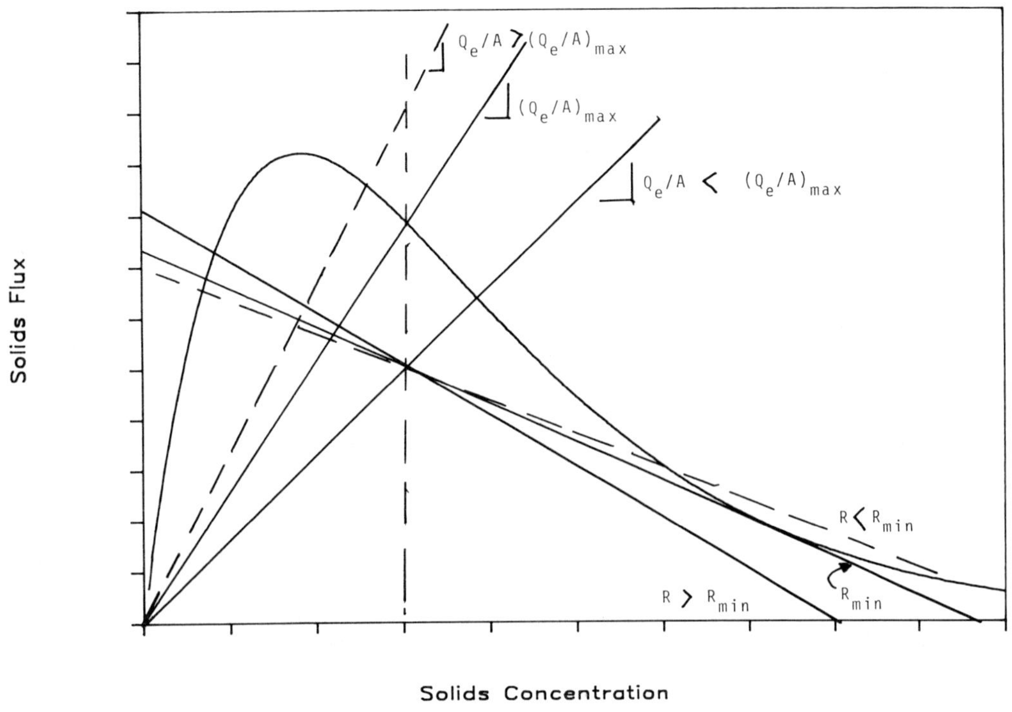

FIGURE 5. Settler operating lines under failure (----).

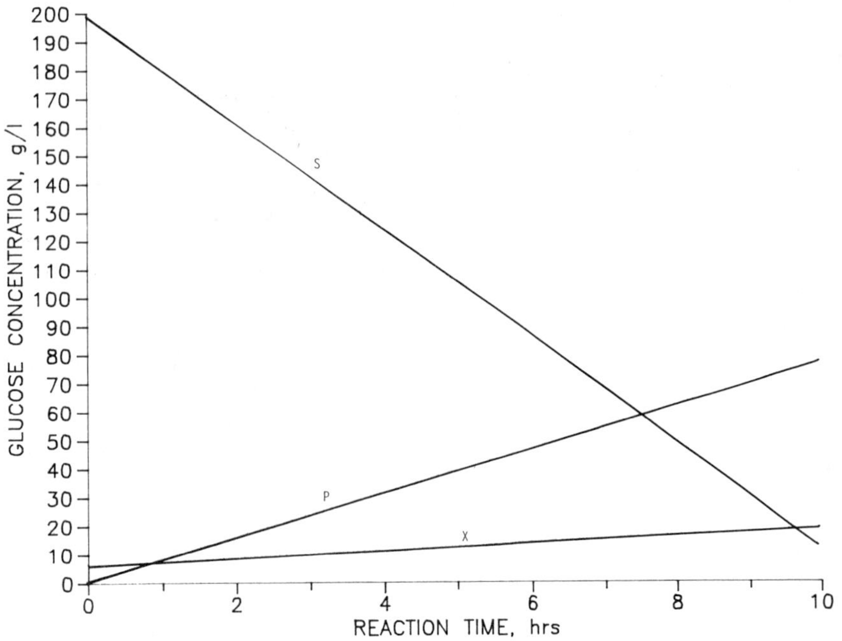

FIGURE 6. Example 1. Batch free cell conversion of glucose to ethanol. S_0 = 200 gm glucose/liter, X_{in} = 3 gm yeast/liter, t_f = 10 hours. Product inhibition kinetics.

point is within the envelope of the flux curve and no error message is given. In all three cases, the minimum area required for safe operation is calculated.

If the clarification criterion are met, the program checks for failure with respect to the thickening function. The minimum recycle ratio, the recycle ratio required for the operating line to be tangent to the batch flux curve, is determined. If the desired recycle ratio is less than the minimum, settler failure has occurred. The program gives the message: "SETTLER FAILURE... SOLIDS CARRY-OVER DUE TO INSUFFICIENT RECYCLE". If the desired recycle ratio is equal to the minimum, the operating line is tangent to the batch flux curve. The program gives the message: "SETTLER OPERATING AT AT CRITICALLY LOADED CONDITIONS WITH RESPECT TO THE THICKENING CRITERION". If the desired recycle ratio is greater than the minimum, safe operating conditions prevail and no error message is given.

EXAMPLES

In order to illustrate the type of problems solved with this program, several examples will be simulated.

I - Batch Free Cell Fermentation of Glucose to Ethanol

To simulate this system, the Free Cell routine was selected and on the Main Menu the Configure routine selected was Batch. The initial substrate concentration S_o was entered as 200 gm glucose/liter, $X_{initial}$ = 3 gm yeast/liter, and t_{final} = 10 hours. The kinetic mechanism selected was product inhibition with the following literature values for kinetic constants (2),

k_o = 0.45 hr^{-1}
Y_p = 0.46 gm ethanol/gm glucose
K_M = 1.0 gm glucose/liter
Y_x = 0.1 gms cells/gm glucose
K_p = 23 gm ethanol/liter

After specifying the kinetic constants, the Run selection was made in the Main Menu with the results going to a file with the .PRN extension name. The results are plotted using Lotus in Fig. 6, showing the concentrations of glucose (S),cells (X) and ethanol (P) with time.

II - Ethanol Production with Immobilized Cells in a Plug Flow Column

This simulation is begun by entering the Immobilized Cell routine on the first menu and under the Configure choice of the Main Menu select Plug Flow Column as the type of reactor. When the reactor screen appeared, values of θ = 5 hrs residence time, E_o = 20 or 40 gms cells/liter of reactor and S_o = 200 gm glucose/liter were entered. The kinetic model chosen was product inhibition with the same kinetic constants as Example I. The particle diameter was selected to be 0.3 cm, and the glucose diffusion coefficient in the particle was set at D_s = $5x10^{-6}$ cm^2/sec, which is consistent with literature values (3,4). A cell concentration of E = 20 gm/liter of reactor was entered on the reactor screen, resulting in the solution shown in Fig. 7. Changing E to 40 gm cells/liter of reactor gave almost complete conversion of the glucose at a distance of 80% of the reactor length.

III- Ethanol Production With Immobilized Cells in a Fixed Bed Reactor Using Axial Dispersion Model.

The example II was changed to axial dispersion by entering the configure screen and changing the choice of reactor from plug flow to axial dispersion column, with all other column values for kinetic model and kinetic constants remaining the same. At the reactor screen, values of θ were entered from 1 to 6 hours and the Peclet number was varied from 0 to ∞ to simulate mixing from plug flow to completely mixed. The change in outlet concentration with residence time for several constant Peclet Numbers is shown in Fig. 8, indicating the increased substrate conversion as the mixing shifts towards plug flow.

IV - Isomerization of Glucose to Fructose in an Immobilized Fixed Bed System

A. Effect of particle size on conversion.

This simulation was entered by selecting the Immobilized Cell system and under the configure selection of the Main Menu choosing a plug flow column. The kinetic model from the Menu choice was Equilibrium, and the values of constants entered were k_o, = 0.161 moles glucose/gm cell-hr, K_M = 21.57 moles glucose/liter of reactor and an equilibrium glucose concentration of 0.54 times the feed

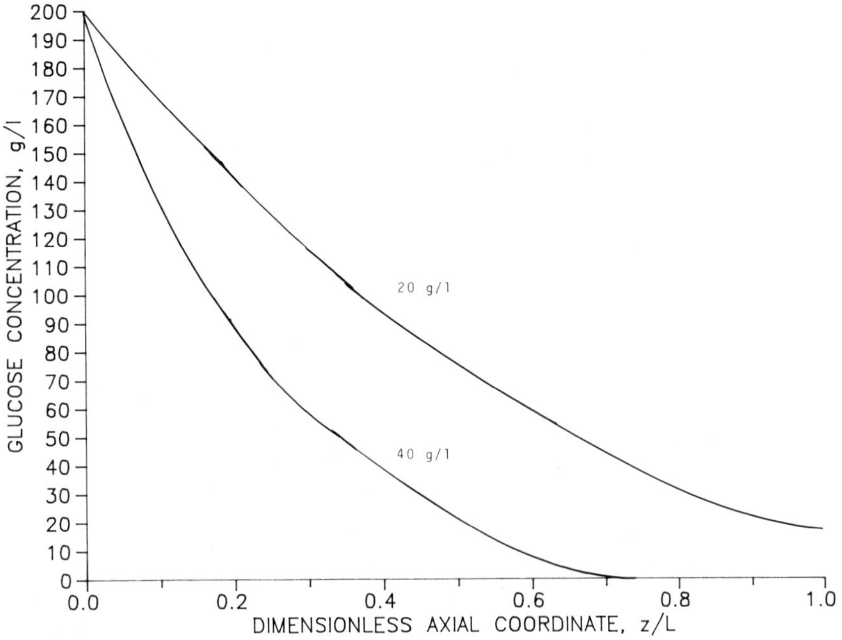

FIGURE 7. Example 2. Ethanol production with immobilized cells in a plug flow column for two different cell concentrations. θ = 5 hours, D_p = 0.15 cm; Product inhibition.

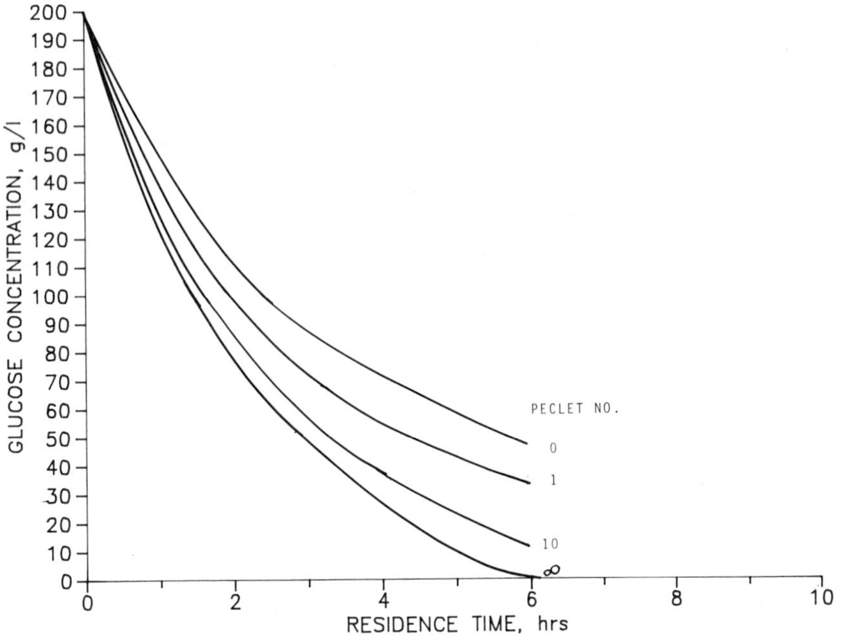

FIGURE 8. Example 3. Ethanol production with immobilized cells in a fixed bed using axial dispersion model.

concentration (5). The residence time, θ, was chosen to be 2.5 hours, S_o = 2.78 moles glucose/l, and E = 102 gms cells/liter of reactor. Three runs were made at different particle diameters: D_p = 0.07, 0.28, and 0.56 cm. The effectiveness factors for these particles corresponded to 1.0, 0.945, and 0.834 respectively. The function $(S_o-S)/(S_o-s_e)$ changes along the length of the reactor as shown in Fig. 9, and decreases as the effectiveness factor decreases.

B. Effect of Mixing

The simulation in IV-A was changed to S_o = 0.0556 mols/glucose liter, and θ varied from .073 to .037 hours corresponding to a superficial mean liquid velocity range of 0.37 to 0.78 cm/seconds to correspond to the runs Ching, Ho, and Rather (5). The decrease in $(S_o-S)/(S_o-S_e)$ with increased velocity is shown in Fig. 10. The runs were repeated by changing the choice of reactor from plug flow to CSTR, and the corresponding results were shown also on Fig. 10. The data points are those of Ching, Ho, and Rathor (5), showing that their conversion in the fluidized bed reactor lies between these mixing extremes. Their data could have been simulated by the axial dispersion model, adjusting the Peclet Number to obtain a fit, but the closeness between the plug flow and CSTR did not warrant the effort.

V - Activated Sludge with CSTR

A municipal activated sludge reactor was simulated by entering Activated Sludge on the configure screen. Monod kinetics was chosen with K_M = 120 mg MLSS/l, k_o = 0.55 hr^{-1}, k_d = 0.0025 gr^{-1}, Y = 0.5 mg MLSS/mg BOD (16). The mixing was chosen as a single CSTR with θ = 6 hrs, Q_o = 37854 m^3/day, and R = 0.40. The Flux Settler model was chosen with Q_o/A = 25m^2/m^3-day, v_o = 138m/day and b = 0.55 m^3/kg MLSS. After returning to the Main Menu and selecting RUN, the solution given was S = 3.68 mg BOD/l, X = 2521.36 mg MLSS/l, X_R = 8306 mg MLSS/l. Checks on the settler were performed and specified parameters were found to be feasible. The minimum possible recycle ratio was calculated to be 0.341, and the actual settler area needed was 1514m^2 with the minimum feasible area to be 1071m^2 with the minimum feasible area to be 1071m^2. The solution for a plug flow reactor was easily done by only changing the chosen reactor and leaving all other choices untouched. Likewise 4 tanks in series could also be simulated by only selecting N-Tanks in Series as the reactor model. The results for these cases are shown in Figure 11.

VI - Activated Sludge With Phenol as Substrate

Many biological materials, such as phenol or nitrogen compounds have substrate inhibition kinetics which give in theory multiple solutions and have sharpely broken concentration profiles along the length of the reactor. Phenol oxidation in an activated sludge reactor wassimulated using product inhibition kinetics: K_M = 49 mg BOD/l, k_o = 0.21 hr^{-1}, Y_1 = 1.02 mg MLSS/mg BOD, k_d = .0212 hr^{-1}, K_I = 154 mg BOD/l (17). The same choice of reactors was made as in Example 5 with θ =6 hrs, θ_c = 3 days, R = 0.25 and S_o = 500 mg BOD/l. The results are shown in Fig. 12 where the phenol concentration in the plug flow case shows the sudden drop near the middle of the reactor indicating the necessiry of using sufficient residence time in plug flow mixing.

VII - Activated Sludge With Settler

Example V was modified to show failure by several causes. The recycle ratio was first lowered to 0.25 and the solution repeated. The error message indicating settler failure due to insufficient recycle was obtained and can be graphically seen in Fig. 13 where the line through the state point lies above the tangent line and indicated as "thickening failure". The Q_o/A_2 was next raised from 25 to 40 m^3/m^2-day with R = 0.40. The error message indicating settler failure due to solids carry over from the excessive overflow rate was obtained. Failure for this case can also be seen on Fig. 13 since the state point has been raised above the batch settling envelope and is indicated as "clarification failure".

LITERATURE CITED

1. Sundstrom, D. W., and Klei, H. E., Wastewater Treatment, Prentice Hall, New York, 1979.

2. Kallerup, F., and Daugulis, A., "A Mathematical Model for Ethanol Production by Extractive Fermentation in a Continuous Stirred Tank Fermentor", Biotech. and Bioeng., 27, 1335-1346 (1985).

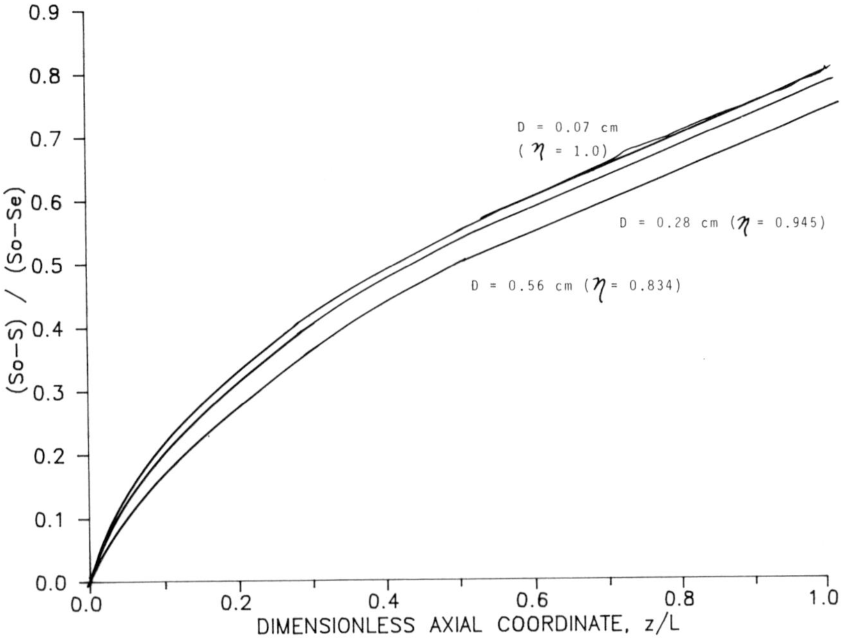

FIGURE 9. Example 4. Conversion of glucose to fructose using plug flow immobilized bed reactor using several diameters of catalyst particle and equilibrium kinetics.

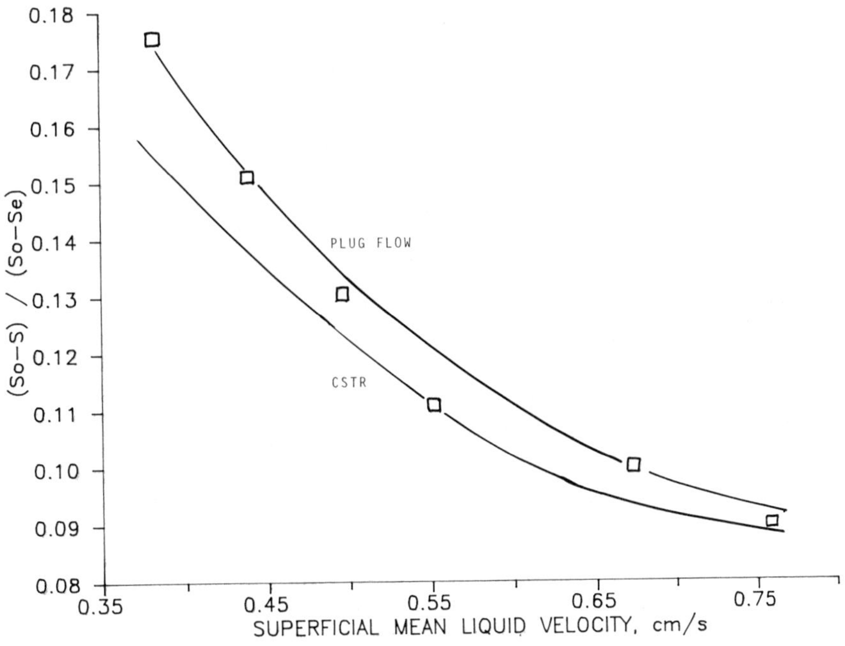

FIGURE 10. Example 4. Conversion of glucose to fructose in both CSTR and Plug Flow Reactors. □ are data points of (5).

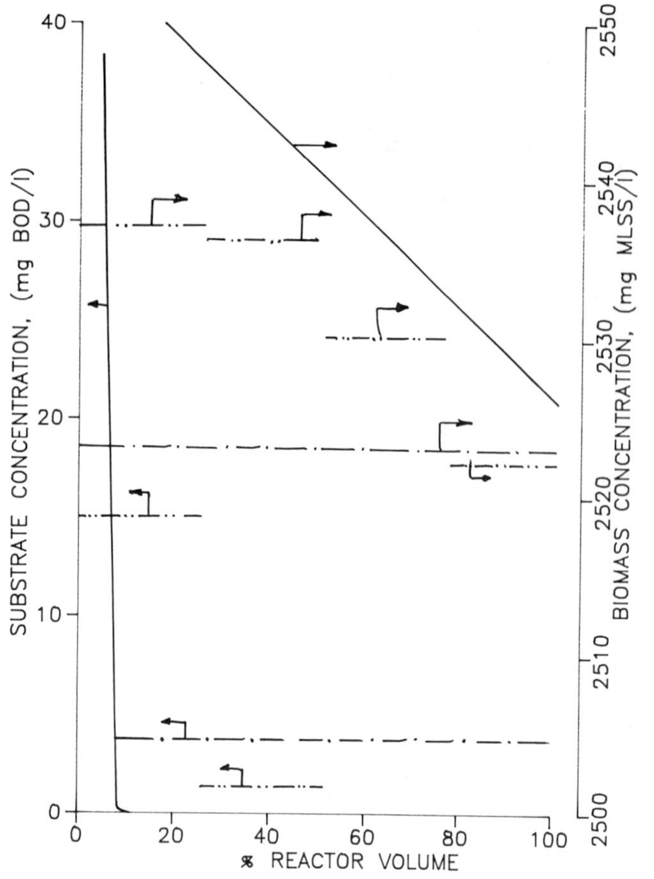

FIGURE 11. Example 5. Conventional activated sludge with municipal wastewater in plug flow (—), single CSTR (—·—) and 4 CSTR in series (—··—). Parameters: S_0 = 500 mg BOD/l, R = 0.40 θ_c = 3 days, θ = 6 hours; Monod kinetics – K_M = 120 mg BOD/l, k_0 = 0.55 hr^{-11}, Y = 0.5 mg MLSS/mg BOD, k_d = 0.0025^{-1}. Flux settler: b = 0.55 m/kg MLSS, v_0 = 138 m/day.

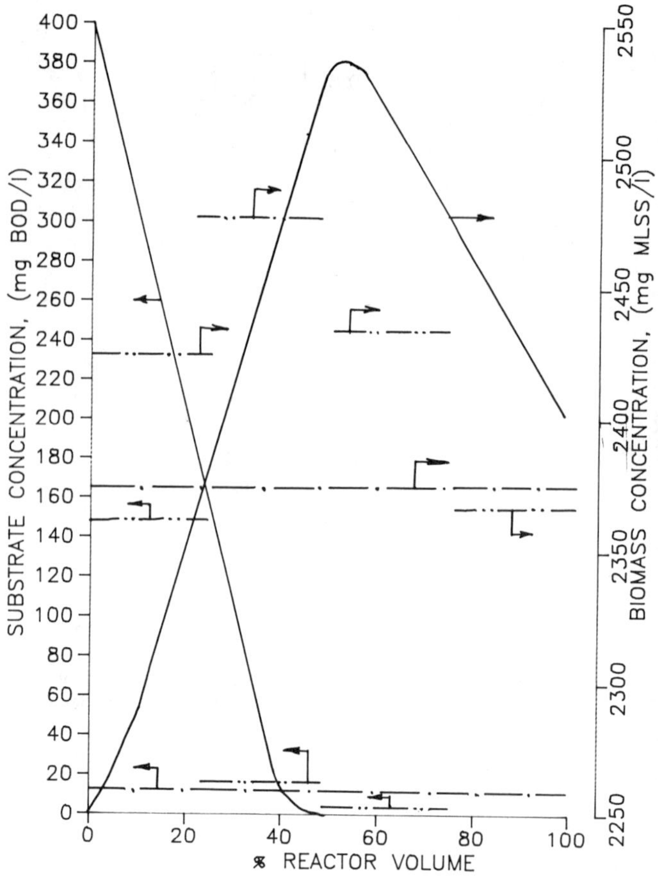

FIGURE 12. Example 6. Phenol degradation in plug flow (—), single CSTR (— · —) and 4 CSTR in series (— · · —). Parameters: S_0 = 500 mg BOD/l, R = 0.25, θ_c = 3 days, θ = 6 hours; substrate inhibition – K_m = 49 mg BOD/l, k_0 = 0.21 hr^{-1}, Y = 1.02 mg MLSS/mg BOD, k_d = .0212 hr^{-1}, K_1 = 154 mg BOD/l. Perfect settler.

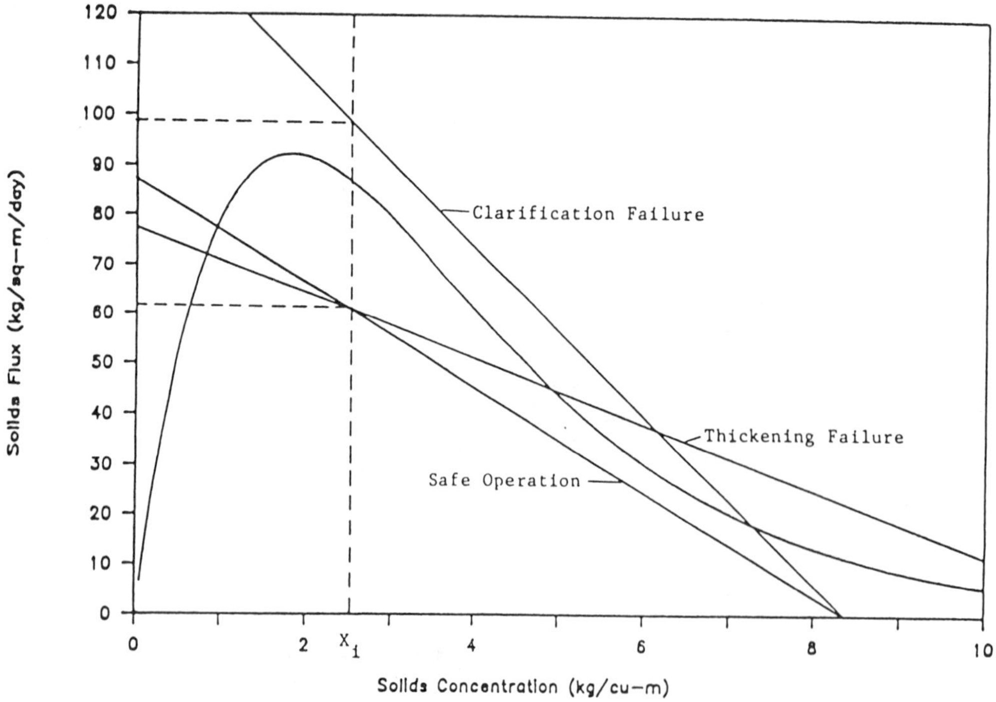

FIGURE 13. Example 7. Settler failure for either too low recycle or too high overflow rate.

3. Brezin, I. V., Klibanov, A., and Martinek, K., "Kinetic and Thermodynamic Aspects of Catalysis by Immobilized Enzymes", Russian Chemical Reviews, 44(1), 9-25 (1975).

4. Hannoun, B. J. M., and Stephanipoulos, G., "Diffusion Coefficients of Glucose and Ethanol in Cell-free and Cell-occupied Calcium Algenate Membranes", Biotech. and Bioeng., 28(6), 829-835 (1986).

5. Ching, C. B., Ho, Y., and Rathor, M., "Isomerization of Glucose to Fructose in a Fluidized Bed Reactor", Biotech. and Bioeng., 26, 820-823 (1984).

6. Wadrak, D. T., and Carbonell, R. G., "Effectiveness Factors for Substrate and Product Inhibition", Biotech. and Bioeng., 17, 1761-1773 (1975).

7. Frouws, M. J.. Vellenga, K., and DeWilt, H. G. J., "Combined External and Internal Mass Transfer Effects in Heterogeneous (Enzyme) Catalysis", Biotech. and Bioeng., 18, 53-62 (1976).

8. Coe, H. S. and Clevenger, G. H., "Methods for Determining the Capacities of Slime Settling Tanks" Trans. AIME, 55, 356, (1916).

9. Kynch, G. J., "A Theory of Sedimentation", Trans. Faraday Soc. 48, 166, (1952).

10. Yoshioka, N., et al., Continuous Thickening of Homogenous Flocculated Slurries", Chemical Engineering (Tokyo) 21, 66, (1957).

11. Vesilind, P. A., "Design of Prototype Thickeners from Batch Settling Tests", Water Sew. Wks., 115, 302, (1968).

12. Smollen, M. and Ekama, G. A., "Comparison of Empirical Settling - Velocity Equations in Flux Theory for Secondary Settling Tanks", Water SA, 10, 175, (1984).

13. Daigger, G. T., and Roper, R. E., "The Relationship Between SVI and Activated Sludge Settling Characteristics", Journal WPCF, 57, 859, (1985).

14. Hoopman, B., and Cadee, K., "Prediction of Thickening Capacity Using Diluted Sludge Volume Index", Water Res. 17, 1427, (1983).

15. Finlayson, B. A., The Method of Weighted Residuals and Variational Principles with Application in Fluid Mechanics and Mass Transfer, Academic Press, NY (1972).

16. Mynhier, M. D., and Grady, C. P., "Design Graphs for Activated Sludge Process", Journal Environmental Engr. Div. ASCE 101, No. EE5, 829, (1975).

17. Rozich, A. F., et al., "Predictive Model for Treatment of Phenolic Wastes by Activated Sludge", Water Research 17, 1453, (1983).

A New Method for Fermentor Scale-up Incorporating Both Mixing and Mass Transfer Effects—I. Theoretical Basis

VIJAY SINGH
R. FUCHS
Schering Corporation
1011 Morris Avenue
Union, New Jersey 07083

A. CONSTANTINIDES
Department of Chemical and Biochemical Engineering
Rutgers University
Piscataway, New Jersey 08854

This paper describes a novel method for conducting fermentor scale-up. Fermentor hydrodynamics are modelled by considering the fermentor as a series of compartments. By solving oxygen balances, in conjunction with correlations for mass transfer and impeller pumping capacity for each compartment, it is possible to estimate the dissolved oxygen concentrations in various parts of the fermentor. The objective of the scale-up method is to determine operating conditions, such as agitation speed and aeration rate, that minimize power consumption while keeping the dissolved oxygen concentrations above some predetermined critical value.

The technique is suitable for use in both Newtonian, as well as non-Newtonian fermentation broths.

Predicting operating conditions in large scale-up fermentors from experiments conducted at small scales is a key problem in biochemical engineering, especially in systems with highly viscous or non-Newtonian rheology. It has become apparent in recent years that mixing is critical to the proper operation of fermentation equipment, and that conventional scale-up methods based on overall mass transfer correlations are not adequate.

This paper introduces a new method for fermentor scale-up that incorporates both mixing and mass transfer effects. The technique utilizes a liquid circulation model in conjunction with mass transfer correlations to predict local dissolved oxygen concentrations in various sections of the fermentor. Aeration and agitation conditions are then selected so that these local dissolved oxygen concentration are kept above some critical value established at small scale. Selection of the operating conditions can be conducted so as to minimize overall power consumption.

LITERATURE SURVEY

This paper will only deal with the scale-up of stirred tank fermentors as these are the type most commonly used in industry. Most scale-up in industrial practice is conducted by keeping selected parameters constant with change in scale. The most common parameters used are:

a. Equal power/volume (P/V) ratio.
b. Equal oxygen transfer (k_La).
c. Equal shear.
d. Equal mixing.
e. Combination of oxygen transfer, shear and mixing.

These methods typically use empirical correlations to estimate the parameters at different scales. Most methods assume geometric similarity, though in practice minor adjustments are made to keep all relevent parameters within desired ranges. These different techniques will be discussed.

Equal power/volume (P/V) ratio:

This method has been used in many antibiotic fermentations as a primary scale-up parameter. Typically a P/V ratio of 1.0 - 2.0 kW/m^3 is used (Aiba et al, (1)). However data on scale-up using equal power is sparse, and further suggests that this criteria may be applicable only to geometrically similar systems, and that other parameters such as impeller diameter need to be taken into account. It is known that power/volume requirements decrease on increase in scale suggesting that this technique may not provide the optimum in energy efficiency.

Equal oxygen transfer (k_La):

This approach to scale-up is based on the fact that most fermentations are oxygen limited, and scale-up is conducted on the basis of providing equal oxygen transfer at various scales of operation. Most correlations to estimate oxygen transfer capacity are based on power input and gas superficial velocity. The general validity of this type of correlation, at least for local mass transfer has been well justified by Van't Riet (2).

In general : $k_La = k_1 \times \left(\frac{P_g}{V}\right)^{k_2} \times (v_s)^{k_3}$

The constant k_1 is characteristic of fermentor geometry, and Bartholomew (3) showed that exponents k_2 and k_3 are dependent on scale.

Though oxygen transfer limited scale-up is done by designing for equal mass transfer coefficient k_La, it is really based on providing equal oxygen transfer rate (OTR). k_La and OTR are related by the following equation:

$$OTR = k_La\ (C^* - C)$$

where C^* is the dissolved oxygen concentration in equilibrium with the gas.
C is the actual dissolved oxygen concentration in the liquid

The difficulty with this technique is that it assumes that the fermentor is sufficiently well mixed so that a local mass transfer rate can provide an estimate of the overall oxygen consumption. In viscous, non-Newtonian fluids, or in very large fermentors, large oxygen gradients are known to exist and fluid mixing may be a major problem (Oosterhuis (4), Van't Riet (2)). In these cases the conventional k_La based technique may significantly underestimate the power input required for adequate performance. Also the values of the constants k_1, k_2 and k_3 must usually be experimentally determined for a given fermentor configuration.

Equal shear:

Scale-up to keep fluid shear constant is traditionally based on maintaining equal impeller tip speed. However, Oldshue (5) pointed out that the impeller tip speed is related to the maximum shear rate, while the average shear rate is a function of impeller speed. Typically on scale-up maximum shear increases and average shear decreases. It is still unclear whether scale-up for shear sensitive organisms should be based on maximum or average shear, or on some other parameter such as power input (van Suijdam and Metz, (6)).

Equal mixing:

To combat the problems of poor mixing in large fermentors, especially with viscous non-Newtonian systems, some researchers such as Fox and Gex (7) have proposed scale-up methods to maintain equal mixing time. However it can be shown that the power required to maintain equal mixing time increases as the cube of the impeller diameter! Experience proves that this is much more than required for practical scale-up.

Combination of parameters:

In most scale-up situations two or more parameters, such as oxygen transfer and shear, may be of equivalent importance, and scale-up in industrial practice is often based on a combination of variables. This is usually done by adjusting the power input per unit volume (P/V), the tip speed (v_s), and the air flow rate. Minor geometric adjustments such as changes in impeller width or diameter are done to achieve the best compromise between conflicting requirements (Wang et al (8)). Holding these key variables constant with scale is not an ideal scale-up technique because the correlations used may not be valid for all combinations of the variables, or a shift in operating domain may have occurred during the scale-up procedure, or the critical parameters may change as the fermentation proceeds. Further, this method generally leads to over-design because full advantage is not taken of the sometimes higher driving forces for mass transfer available in large scale equipment.

METHOD PROPOSED

The method proposed in this work differs in that inhomogeneity is expected in the fermentor. In fact, due to the enormous power requirements for uniform blending, achieving homogeneity may not even be advantageous. Instead, the fermentor is modelled as a number of well-mixed zones, each having a different dissolved oxygen level. Scale-up is based on keeping these local dissolved levels above some predetermined critical value. In this manner the micro-environment of the organism is kept similar on increase in scale.

The next few sections will describe this new technique and illustrate the computations required by means of an example. Critical to the technique is good model of fluid circulation in the fermentor.

The Fluid Circulation Model:

Fluid circulation in the fermentor is estimated by the use of a mathematical circulation model. The model used in this work is shown in Figure 1. Here the fermentor is modeled as a series of well-mixed compartments with recycle and backflow. The model has two parameters: 1) the liquid circulation flow rate f, and 2) the number of stagnant compartments n_c per impeller stage.

The liquid circulation rate f can be computed approximately from the impeller pumping capacity or better yet, it can be measured using the technique presented in an earlier paper (Singh et al (9)). This technique utilized pH transients to estimate the circulation rate in operating fermentors. The transients are generated by acid/base pulses used to control pH in the fermentor. The number of stagnant compartments n_c are a function of the rheological properties of the fermentation broth. Estimation of both these parameters will be discussed later.

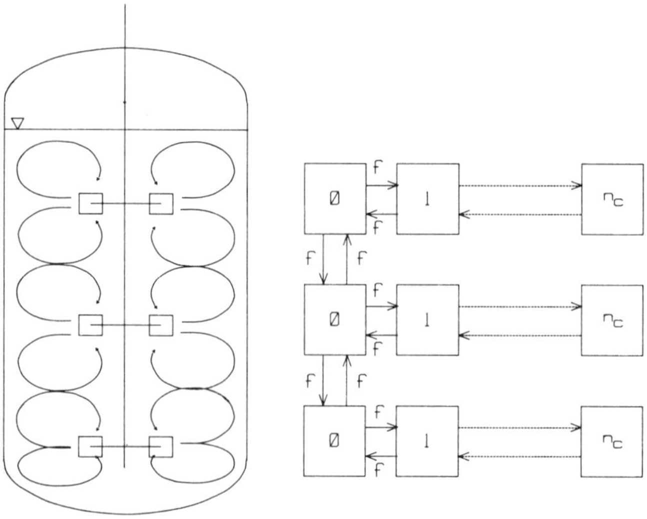

FIGURE 1. Recycle–backmix fermentor circulation model.

This model differs from most models reported in the literature in that it is intended to model multi-impeller systems which are typical in industrial practice. The model is general in applicability and can be adjusted by an appropriate choice of parameter n_c to model low viscosity fluids in turbulent flow, as well as viscous systems with strong recirculation effects.

Oxygen Transfer Correlations:

Oxygen transfer in each compartment is estimated by the use of standard correlations. The use of these correlations is justified because each compartment can be taken as a well-mixed region by proper choice of the parameter n_c.

For the region immediately surrounding the impeller the correlation for non-coalescing media established by van't Riet (2) is used :

$$k_L a = 2 \times 10^{-3} \times \left(\frac{P_g}{V}\right)^{0.7} \times (v_s)^{0.2}$$

This correlation is expected to be valid for well-mixed fluids where $0.5 < P/V < 10$ kW/m^3; $2 < V < 4400$ l. The accuracy quoted is 20 - 40 %.

For regions away from the impeller, oxygen transfer can be assumed to take place from bubbles already formed, and these compartments can be assumed to behave as bubble columns. The following correlation for bubble columns can be used to predict oxygen transfer in these regions:

$$k_L a = k_4 (v_s)^{k_5}$$

typical values are: $k_4 = 0.32$ and $k_5 = 0.7$ (Heijen and Van't Riet (10))

Calculation of Local Dissolved Oxygen Concentrations:

Fluid in the fermentor circulates from regions of high oxygen transfer to regions of poor oxygen transfer. Depending on specific oxygen uptake rate, fluid circulation rate, local mass transfer, and fluid rheology, there may be significant variations in local dissolved oxygen concentrations in different parts of the fermentor. It is possible, by simultaneously solving the compartment oxygen mass balances with local mass transfer correlations, to estimate the dissolved oxygen concentrations in each compartment. Oxygen gradients in the tank can be inferred from these calculated dissolved oxygen values. The technique for such a calculation will be illustrated by means of a simple example:

Figure 2 shows the compartmented model for a fermentor with two impellers. The simplest model with one stagnant compartment per impeller stage (n_c=1) is considered. This model would be applicable to systems that are highly turbulent and exhibit little or no recirculation. This might be the case in typical bacterial or yeast fermentations. Fluid circulates as shown in Figure 2 at a volumetric flowrate f.

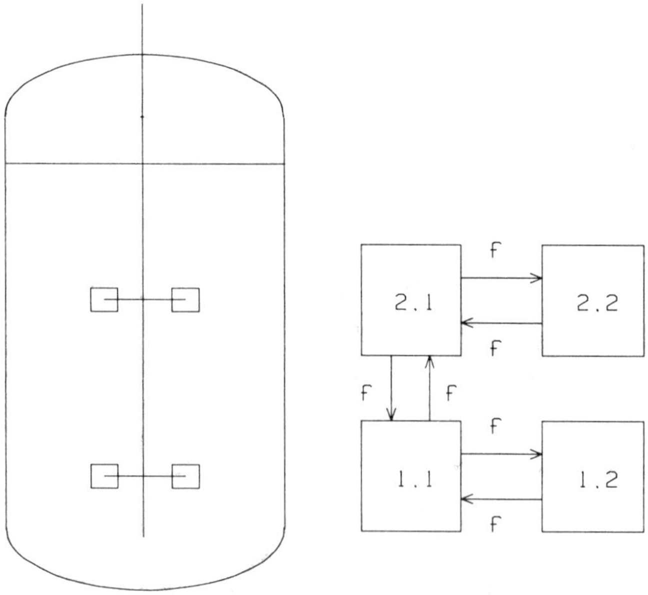

FIGURE 2. Two-impeller system to illustrate technique.

Oxygen balances for each compartment are:

$$\frac{V_c dC_{1,1}}{dt} = fC_{2,1} + fC_{1,2} - 2fC_{1,1} - \text{OUR } V_c + (k_La)_{1,1}(C^*_{1,1} - C_{1,1})V_c \quad (1)$$

$$\frac{V_c dC_{2,1}}{dt} = fC_{1,2} + fC_{2,2} - 2fC_{2,1} - \text{OUR } V_c + (k_La)_{2,1}(C^*_{2,1} - C_{2,1})V_c \quad (2)$$

$$\frac{V_c dC_{1,2}}{dt} = fC_{1,1} - fC_{1,2} - \text{OUR } V_c + (k_La)_{1,2}(C^*_{1,2} - C_{1,2})V_c \quad (3)$$

$$\frac{V_c dC_{2,2}}{dt} = fC_{2,1} - fC_{2,2} - \text{OUR } V_c + (k_La)_{2,2}(C^*_{2,2} - C_{2,2})V_c \quad (4)$$

At steady state the accumulation terms disappear to give :

$$0 = fC_{2,1} + fC_{1,2} - 2fC_{1,1} + (k_La)_{1,1}(C^*_{1,1} - C_{1,1})V_c - \text{OUR } V_c \quad (5)$$

$$0 = fC_{1,2} + fC_{2,2} - 2fC_{2,1} + (k_La)_{2,1}(C^*_{2,1} - C_{2,1})V_c - \text{OUR } V_c \quad (6)$$

$$0 = fC_{1,1} - fC_{1,2} + (k_La)_{1,2}(C^*_{1,2} - C_{1,2})V_c - \text{OUR } V_c \quad (7)$$

$$0 = fC_{2,1} - fC_{2,2} + (k_La)_{2,2}(C^*_{2,2} - C_{2,2})V_c - \text{OUR } V_c \quad (8)$$

For the impeller compartments 1,1 and 2,1 k_La can be estimated from the van't Riet equations:

$$(k_La)_{i,j} = k_1\left(\frac{P_g}{V}\right)^{k_2}(v_s)^{k_3} \quad (9)$$

For the stagnant compartments 1,2 and 2,2 the correlations for bubble columns can be used:

$$(k_La)_{i,j} = k_4(v_s)^{k_5} \quad (10)$$

The flow rate f in these equations must determined by measurement as discussed earlier, or by means of correlations. In the absence of data f can be estimated from the impeller discharge flow rate:

$$f = K_p ND^3 \quad (11)$$

If P_g and v_s are known, then by simultaneously solving equations (5-8), (9),(10) and (11) the dissolved oxygen values

$C_{i,j}$ can be obtained. The values are strongly dependent on the circulation rate. As f increases the value of dissolved oxygen $C_{i,j}$ in the poorly aerated compartments will approach the dissolved oxygen value $C_{i,j}$ in the impeller compartments.

Correlations can be used to estimate P_g in order to relate the dissolved oxygen concentrations in the individual compartments to primary operational parameters such as aeration and agitation rates. This in essence provides the mechanism for scale-up, whereby agitation and aeration rates can be selected to achieve a desired dissolved oxygen profile.

A useful correlation for estimating Pg was reported by Michel and Miller (11):

$$P_g = 0.706 \left(\frac{P_o^2 N D^3}{F^{0.56}}\right)^{0.45} \quad (12)$$

$$\text{where} \quad P_o = \rho N^3 D^5 N_p \quad (13)$$

The power number N_p can be estimated by the following equation by Moo-Young and Blanch (12):

$$Np = 160 \; \frac{LW\,(D-L)}{D^3} \quad (14)$$

This simple example demonstrates the basic mathematics involved in the proposed technique. It also shows that poor circulation can give rise to low dissolved oxygen levels in certain parts of the tank even if local mass transfer rates near the impeller are high. To model an actual fermentor the model needs to be made more complex. Effective methods to select n_c are required in order to properly compartmentalize the tank. Additional equations to account for gas phase mixing, impeller flooding, and hydrostatic effects need to be incorporated.

General Procedure for Fermentor Scale-up:

The general procedure for conducting scale-up by this new technique is summarized in the flowchart in Figure 3. The basic data necessary from small scale experiments are: 1) the critical dissolved oxygen level and, 2) the oxygen uptake rate (OUR), both under well-mixed conditions. The procedure then computes the power input, aeration rate and agitation speed required to keep local dissolved oxygen above critical for a given fermentor configuration.

It is interesting to note that if the procedure on the flowchart is followed then the steady state oxygen mass balances become simply a set of simultaneous linear equations and can be solved easily for all $C_{i,j}$.

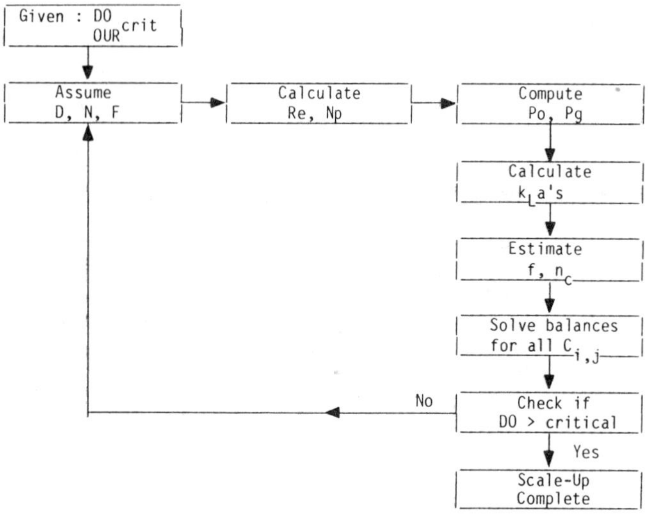

FIGURE 3. Flowchart of proposed technique.

ESTIMATION OF MODEL PARAMETERS

The fluid circulation rate f:

In the absence of operating data, the circulation rate f can be taken as half the impeller pumping capacity. Various correlations for estimating the impeller pumping capacity Q exist. Revill (13) in a review of applicable correlations show that under ungassed conditions:

$$Q = K_p ND^3 \text{ where } K_p = 0.75 \pm 0.15 \quad (15)$$

The effect of impeller geometry can be taken into account by using the equation for K_p reported by Bowen (14):

$$K_p = 6.2 (W/D) (D/T)^{0.3} \quad (16)$$

Data reported by Singh et al (9) show that aeration does not seriously change the circulation rate f provided the system is not operated under flooding conditions.

The number of stagnant compartment n_c per impeller stage:

A key premise of the proposed method is that the model compartments are well-mixed so that correlations obtained in homogenous systems are applicable to each compartment. The parameter n_c provides a means of adjusting the model to actual broth rheology and level of turbulence. Highly viscous systems have a high degree of recirculation (Yuge and O'Shima (15)) and will require a large value of n_c for correct flow modelling. Increasing values of n_c in turn lead to the fermentor being subdivided into more, but smaller well-mixed zones. In this manner viscous systems will be modelled with a large number of small compartments while low-viscosity fluids in turbulent flow may only require a few compartments.

Since the model assumes that all compartments are of equal volume, n_c can be estimated from the volume of the high shear region immediately surrounding the impeller. The volume of this region of impeller influence can be computed from an analysis of the shear field due to the local radial liquid velocity. Costes and Couderc (16) show that the velocity profile near a turbine impeller in turbulent flow can be represented in normalized coordinates by a single curve independant of scale. Their radial velocity profile data for a standard tubine impeller (W=D/5) can be fitted well to the following exponential equation:

$$\frac{v_r}{\pi ND} = 0.7 \exp(-12.16\, z/D) \qquad R^2>0.99 \quad (17)$$

The shear rate in the axial direction is obtained by differentiating this equation with respect to z. The shear rate as a function of velocity is then given by :

$$\frac{dv_r}{dz} = -26.74\, N \exp(-12.16\, z/D) \quad (18)$$

Low-viscosity Systems in Turbulent Flow. It can be shown from equation 18 that the shear field around the impeller decays to around 0 when z/D is about 0.5. This forms the basis for determining the impeller influence zone for low-viscosity fluids in turbulent flow. The volume of the influence zone is calculated by assuming a torus around the impeller in the manner used by Oosterhuis (4) and is shown in Figure 4. For fluids in turbulent flow z/D is taken as 0.5 which corresponds to a torus cross-sectional diameter of five times the impeller blade width W. The number of compartments per impeller stage is then computed by the following equations:

$$V_i = \frac{\Pi^2}{4} D\,(5W)^2 \tag{19}$$

$$n_c = \frac{V}{n_i\,V_i} - 1 \tag{20}$$

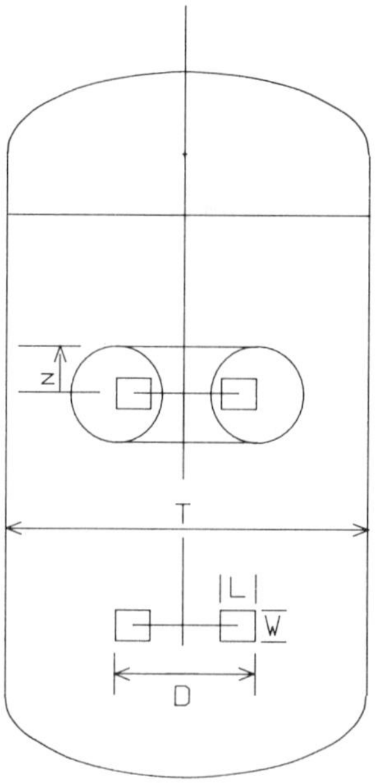

FIGURE 4. Toroidal influence zone around impeller.

Table 1 summarizes the values of n_c calculated in this manner for low viscosity fluids in series of typical fermentors (the fermentor geometry is given in Table 2).

TABLE 1
Calculated Values of n_c for Newtonian Fluids

Fermentor	D(m)	V(m^3)	n_c
F2A	0.13	0.07	3.3
T3A	0.28	0.425	1.6
T2A	0.50	3.8	3.1
T1A	1.12	36.0	2.5

The table shows that for fluids in turbulent flow, the fermentor can be modelled with 1 to 3 stagnant compartments n_c per impeller stage, depending only on tank D/T ratio. The calculations also show that the parameter n_c does not change with scale when geometric similarity is maintained. This implies that scale-up of Newtonian fluids in turbulent flow should be a relatively straight-forward process.

TABLE 2
Representative Fermentor Configuration

Fermentor type	F2A	T3A	T2A	T1A
fermentor volume m^3	0.1	0.5	4.5	40.0
Working volume m^3	0.07	0.425	3.8	36.0
Vessel diameter m	0.35	0.63	1.37	2.89
Impeller diameter m	0.13	0.28	0.5	1.12
D/T ratio	0.37	0.44	0.36	0.39
W/D ratio	0.2	0.2	0.2	0.2
Number of Impellers	3	3	3	3

Non-Newtonian Systems. Most fermentation fluids however have non- Newtonian rheological behavior and it is these cases that cause the greatest difficulty in mixing and mass transfer scale-up. The scale up technique presented here is most useful for dealing with these fluids.

Fermentation broths have been observed to

typically have shear thinning behavior. It has been postulated by a number of authors (Charles (17); Blanch and Bhavaraju (18)) that in these cases the fluid near the impeller is in turbulent flow because of the high shear field; but away from the impeller the shear field decays rapidly and the fluid motion becomes significantly reduced due to increasing apparent viscosity. It is therefore expected that the zone of impeller influence in non-Newtonian fluids is significantly smaller than for a Newtonian fluid in complete turbulence.

A reduction in the zone of influence is reflected by the need for a larger n_c parameter. The zone of influence and the number of compartments can be estimated by a analysis similar to that for the Newtonian case.

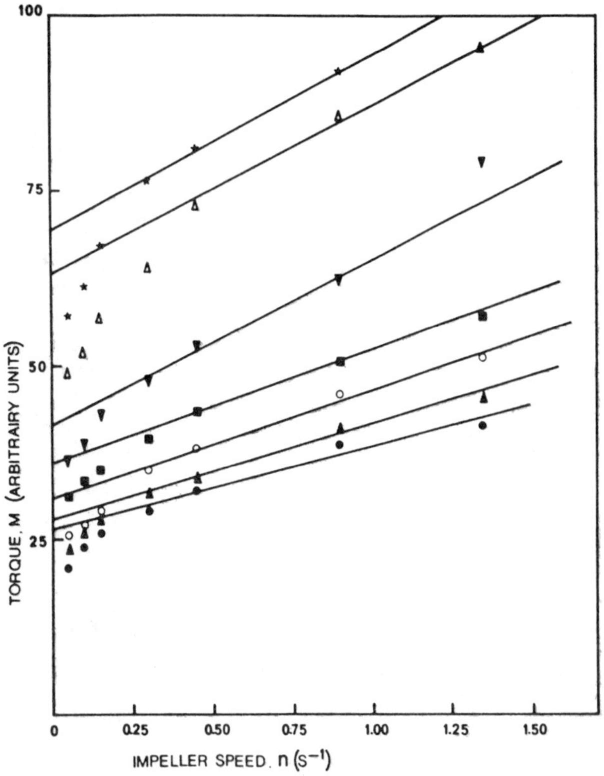

FIGURE 5. Rheogram of *P. chrysogenum*.

The data presented by Roels et al (19) for Penicillium chrysogenum will be used to illustrate the procedure. This broth has a pseudoplastic behavior far from Newtonian (flow consistancy index = 0.22). The rheogram for this broth is reproduced in Figure 5. The original paper reported viscometer speed and not shear rate, however shear rate could be calculated in the manner suggested by Calderbank and Moo-Young (20) with γ = 10 x viscometer speed. The shear rate at which the shear stress - shear rate curve departs from linear (Newtonian) behavior is used as the criteria for estimating the volume of the impeller influence zone. From the rheogram it is apparent that this occurs around a shear rate of 2.5 s^{-1} (n=0.25) for almost all the runs. This is used as the value of $(dv_r/dz)_{min}$ and the radius of the torus of influence z/D can be calculated by rearranging equation (18):

$$\left(\frac{z}{D}\right)_{max} = -\ln\left(\frac{(dv_r/dz)_{min}}{26.74\ N}\right) / 12.16 \qquad (21)$$

The volume of the torus of influence is calculated as before by:

$$V_i = \frac{\pi^2}{4} D\,(2(z/D)_{max}\ 5W)^2 \qquad (22)$$

The number of stagnant compartments is also calculated as before from V_i:

$$n_c = \frac{V}{n_i\ V_i} - 1 \qquad (23)$$

It should be noted that this procedure for non-Newtonian fluids requires fluid specific information from the rheogram, as well as impeller speed, diameter and blade width to estimate the impeller zone of influence.

Table 3 shows the number of compartments required for each of the representative fermentor configurations using P.chrysogenum. The tip speed was kept constant at 6 m/s at all scales for purposes of comparison. It can be seen that as the size of the fermentor increases, the number of compartments increases, reflecting an increasing difficulty in maintaining fermentor homogeneity. This clearly illustrates the problem of scale-up when dealing with non-Newtonian shear thinning fluids.

The scale-up method presented here allows easy analysis of optimum impeller diameter and speed, even in large tanks and with non-Newtonian systems, by providing simultaneous calculation of impeller influence zone, power draw, pumping capacity and dissolved oxygen distribution.

TABLE 3
Calculated Values of n_c for *P. chrysogenum* Broth

Fermentor	$V(m^3)$	D(m)	$N(s^{-1})$	n_c
F2A	0.07	0.13	14.67	5
T3A	0.425	0.28	6.82	4
T2A	3.8	0.50	3.82	10
T1A	36.00	1.12	1.70	14

SIMULATION STUDIES

A computer program was developed to examine the predicted dissolved oxygen profile generated by this technique under widely varying operating conditions. Some plots made from these simulation studies for the 40.0 m^3 tank are presented to illustrate the effect of operating parameters on the dissolved oxygen distributions:

Low Viscosity Turbulent Fermentations ($n_c=1$):

Effect of airflow F. The minimum dissolved oxygen level in the stagnant compartments, and in the impeller compartments were determined at various agitation speeds N. Impeller diameter and OUR were kept constant. The plots of minimum D.O. at various air flow rates in Figure 6 show that airflow has a significant effect on the minimum dissolved oxygen levels. The range of impeller speed is restricted at the high airflow rates due to predicted flooding conditions. Airflow appeared to change both the D.O. (dissolved oxygen) in the impeller, and in the stagnant compartments, by the same relative amount.

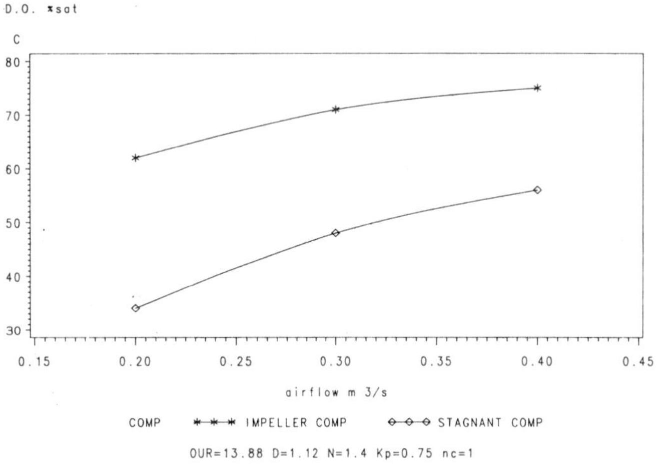

FIGURE 6. Effect of airflow on D.O. distribution.

Effect of D/T ratio. In a similar manner the minimum dissolved oxygen levels were computed for varying D/T ratios. In this case the airflow was held constant. Results presented in Figure 7 show that the D.O. levels in the stagnant compartments are increased greatly by an increase in the impeller D/T ratio. However this is obtained at the cost of increased agitator power input; which for example at N = 1.4 s^{-1}, went from 19,200 W with D/T=0.33, to 180,300 W with D/T=0.5.

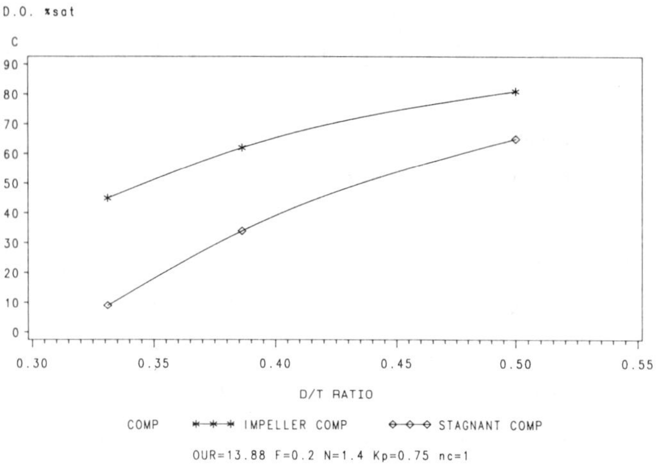

FIGURE 7. Effect of *D/T* ratio on D.O. distribution.

Effect of Impeller Discharge number K_p. The impeller discharge number K_p was varied to test the sensitivity of the dissolved oxygen profile to errors in the prediction of the circulation flow rate. The minimum dissolved oxygen concentrations obtained are presented in Figure 8.

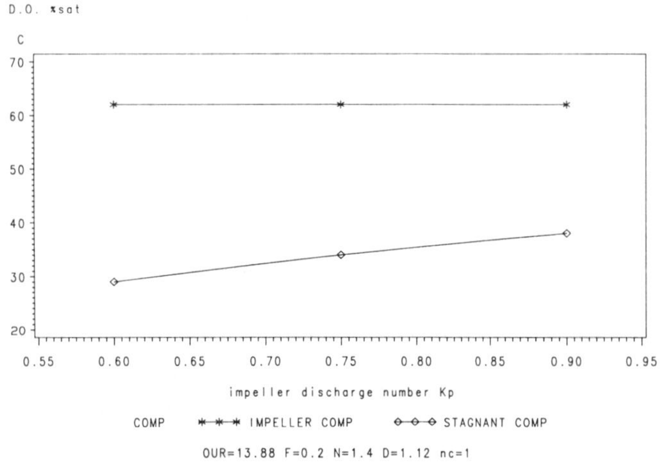

FIGURE 8. Effect of parameter K_p on D.O. distribution.

These show that the D.O. levels in the impeller compartment are not at all effected by changes in K_p and even the D.O. levels in the stagnant compartments only change by about 10%. This indicates that a very accurate estimate of K_p is not essential for a useful prediction of the D.O. profile.

Effect of oxygen demand OUR. Change in oxygen uptake rate can cause steep oxygen gradients in the predicted dissolved oxygen profiles as shown in Figure 9. Here at the high OUR levels only a high impeller speed could provide D.O. concentrations above zero in the most poorly aerated compartments.

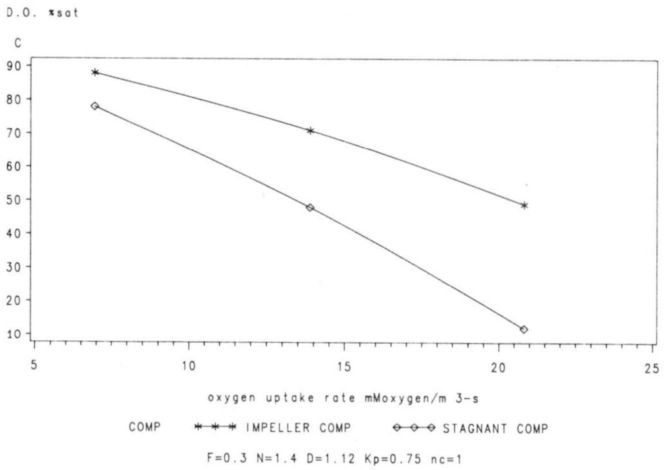

FIGURE 9. Effect of OUR on D.O. distribution.

Effect of number of stagnant compartments n_c:

The effect of poor mixing can be assessed by changing the number of stagnant compartments in the flow model. Figure 10 shows the effect on the minimum dissolved oxygen concentrations as parameter n_c is varied from 2 to 7. The variation in D.O. levels ranges from 79 to 48% when n_c is 2, to a range of 79 to 20% when n_c is 7. It can also been seen that though the D.O. levels in the impeller compartments do not change much with n_c, the relative volume of these compartments compared to the total fermentor volume drops dramatically as n_c is increased.

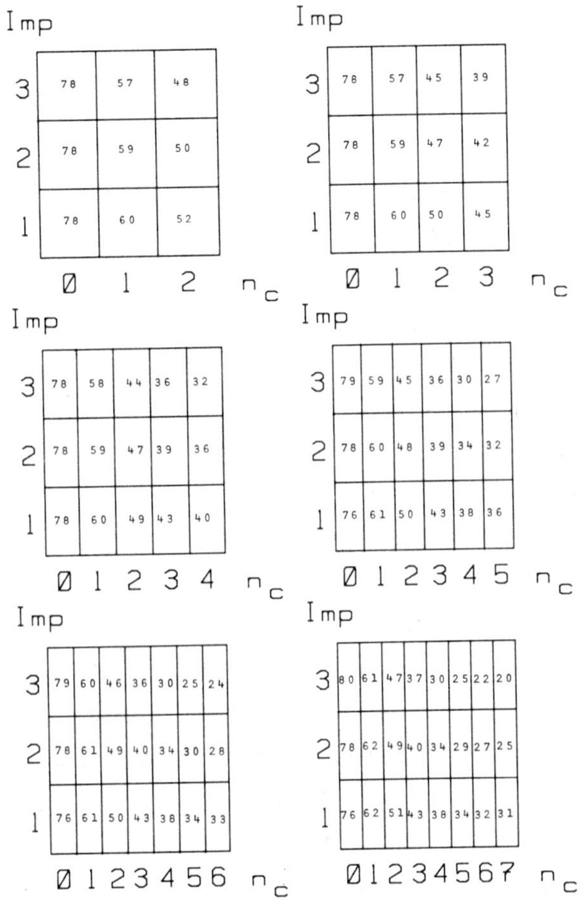

FIGURE 10. Effect of parameter n_c on D.O. distribution.

Effect of scale - P.chrysogenum fermentation:

Dissolved oxygen levels for the non-Newtonian P.chrysogenum fermentation discussed earlier were simulated for both the 0.1 m^3 fermentor (F2A) and the 40.0 m^3 fermentor (T1A). The aeration/agitation rates selected for the simulation are close to the normal operating conditions for these units. The calculated dissolved oxygen (D.O.) values, as percentage of saturation, are presented in the form of 3-D plots, with each impeller position on the x-axis, the compartments on the y-axis, and the corresponding D.O. valves on the z-axis. Figure 11a shows the D.O. profile for the 0.1 m^3 fermentor at a power input of 4.6 k W/m^3. It can be seen that the tank is relatively homogeneous with a range in D.O. from 74% near the impeller to 46% saturation in the most poorly aerated compartment. The 40.0 m^3 fermentor, even at this very high specific power input, shows a much greater variation in D.O. levels ranging from 90% near the impellers to 23% in the most poorly aerated compartment (Figure 11b).

The situation is even worse with a more typical power input of 1.5 kW/m^3 in the 40.0 m^3 tank. As shown in Figure 11c, at this power input, about 40% of the fementor has dissolved oxygen levels below 25% of saturation.

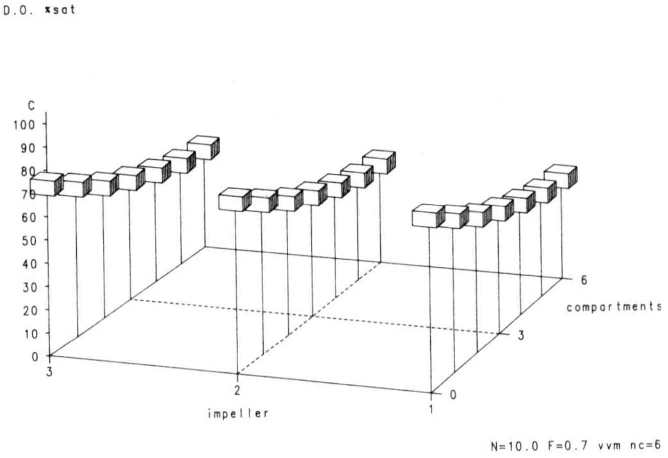

FIGURE 11a. D.O. profiles in 0.1 m^3 fermentor (4.5 kW/m^3 power input).

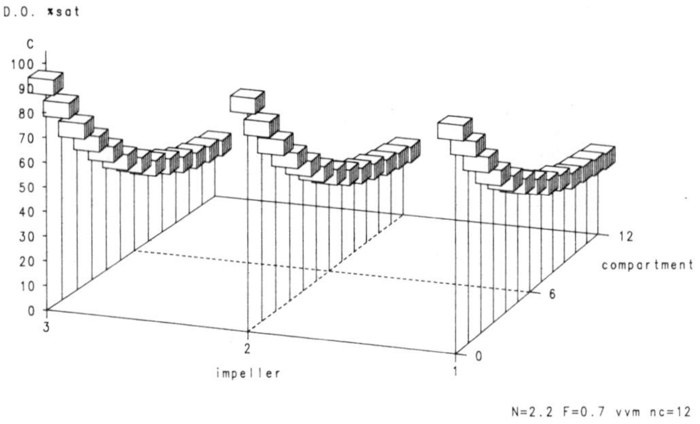

FIGURE 11b. D.O. profiles in 40.0 m^3 fermentor (4.5 kW/m^3 power input).

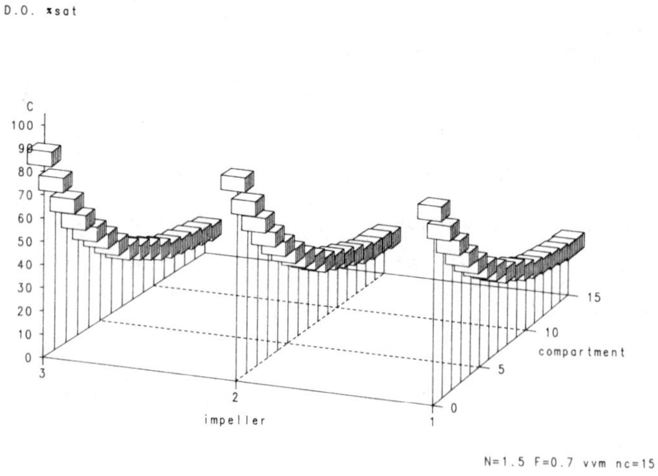

FIGURE 11c. D.O. profiles in 40.0 m^3 fermentor (1.5 kW/m^3 power input).

Difficulty can thus be expected in scaling up from the 0.1 m^3 fermentor to the 40.0 m^3 fermentor with this non-Newtonian broth, unless poorer mixing at large scale is accounted for in the scale up procedure. Similar variations in dissolved oxygen levels are obtained if the fermentation is scaled from the 0.1 m^3 to the 40.0 m^3 fermentor using the conventional equal k_La procedure.

Comments on scale-up criteria:

The technique presented so far enables the estimation of local dissolved oxygen concentrations. However for a practical scale-up method some criteria for determining an acceptable oxygen distribution are required. Some suggested methods are: 1) keeping 90% of the fermentor volume above some critical dissolved oxygen value established in small scale experiments; 2) ensuring all compartments have D.O. levels above this critical value; or 3) keeping the residence time in compartments that have D.O. values below critical less than some predetermined residence time. The choice of appropriate criteria needs to be verified by experiment.

CONCLUSIONS AND FUTURE WORK

The new method presented in this paper provides a structured approach to the scale-up of fermentors where poor liquid mixing is expected. It is computationally simple and can be used to evaluate the effect of a wide variety of fermentor variables. The basic method predicts oxygen distribution in fermentors by considering both liquid circulation and oxygen transfer in the fermentor. This enables the technique, unlike most other methods, to be useful in modelling non-Newtonian pseudoplastic fluids often encountered in antibiotic fermentations.

Besides scale-up, the technique is expected to be useful in optimizing fermentor operation by aiding the selection of appropriate impeller geometry, impeller spacing, and agitation/aeration rates. The technique is being adapted for on-line use in conjunction with an on-line circulation flow rate measurement. This system is expected to continously optimize fermentor power input regardless of changes in rheology or oxygen demand with time.

However much work remains to done with actual fermentations to verify the predictions of this technique. Both Newtonian and non-Newtonian systems are under active investigation and data will be published in a subsequent paper.

Work is also being done to develop better correlations for the prediction of liquid circulation rate in fermentors of varying geometry and impeller configuration. Correlations for oxygen transfer are being tested and refined. Modeling of gas phase flow especially in non-Newtonian fluids is also of great interest.

NOTATION

C	= dissolved oxygen concentration	mMO_2/l or %sat
C^*	= dissolved oxygen concentration at saturation	mMO_2/l or %sat
D	= impeller diameter	m
f	= circulation flow rate parameter	m^3/s
F	= airflow rate	m^3
k_i	= constants	
k_La	= mass transfer coefficient	s^{-1}
K_p	= impeller discharge number	
L	= impeller blade length	m
n_c	= number of stagnant compartments per impeller stage	
n_i	= number of impellers	
N	= impeller speed	s^{-1}
N_p	= power number	
OTR	= oxygen transfer rate	mMO_2/l-h
OUR	= oxygen uptake rate	mMO_2/l-h
P	= power	W
P_g	= gassed power	W
P_o	= ungassed power	W
Q	= impeller discharge flow rate	m^3/s
T	= tank diameter	m
v_s	= superficial gas velocity	m/s
V	= liquid volume in tank	m^3
V_c	= compartment volume	m^3
V_i	= volume of impeller influence zone	m^3
v_r	= radial velocity	m/s
W	= impeller blade width	m
z	= axial distance from impeller center-line	m
ρ	= liquid density	Kg/m^3
γ	= shear rate	s^{-1}

subscripts

i,j	= compartment index i^{th} impeller stage and j^{th} compartment	

LITERATURE CITED

1. Aiba, S. Humphrey, A.E., and Millis, N.F., Biochemical Engineering, Academic Press, New York (1965).

2. Van't Riet, K., "Review of Measuring Methods and Results in Nonviscous Gas-Liquid Mass Transfer in Stirred Vessels", Ind. Eng. Chem. Process Des. Dev.,18,357 (1979).

3. Bartholomew, W.H., "Scale-Up of Submerged Fermentations",in Adv. Appl. Microbiol, ed. W.W. Umbreit, 2,p. 289, Academic Press, New York (1960).

4. Oosterhuis, N.M.G.and Kossen, N.W.F. "Dissolved Oxygen Concentration Profiles in a Production-Scale Bioreactor", Biotech.Bioeng,26,546-550 (1984).

5. Oldshue, J.Y., "The Spectrum of Fluid Shear in a Mixing Vessel", in Chemeca'70, Butterworths (1970).

6. Van Suijdam, J.C. and Metz, B., "Fungal Pellet Breakup as a Function of Shear in a Fermentor", J. Ferment. Technol.,59,4,329-333 (1981).

7. Fox, F.A. and Gex, V.E., AIChE. Journal,2,539 (1956).

8. Wang, D.I.C., Cooney, C.L., Demain, A.L., Dunnill, P., Humphrey, A.E. and Lilly, M.D., Fermentation and Enzyme Technology, J. Wiley & Sons, New York (1979).

9. Singh, V., Hensler, W., Fuchs, R. and Constantinides, A.,"On-Line Determination of Mixing Parameters in Fermentors using pH Transients", International Conference on Biorector Fluid Dynamics, Paper 18, pp. 231-256, BHRA, Cambridge, England (1986).

10. Heijen, J.J. and Van't Riet, K.,"Mass Transfer, Mixing and Heat Transfer Phenomena in Low Viscous Bubble Column Reactors", 4th European Conference on Mixing, BHRA, Fluid Engineering, Paper F1, pp. 195-224, April (1982).

11. Michel, B.J. and Miller, S.A. "Power Requirements of Gas-Liquid Agitated Systems",AIChE.J., 8, 262 (1962).

12. Moo-Young, M. and Blanch, H.W., "Design of Biochemical Reactors. Mass Transfer Criteria for Simple and Complex Systems", Adv. in Biochemical Eng, 19, pp 1-69 (1981).

13. Revill, B.K., " Pumping Capacity of Disc Turbine Agitators - A Literature Review", 4th European Conference on Mixing, BHRA, Fluid Engineering, Paper B1, pp. 11-24, April (1982).

14. Bowen, R.L. "Shear-Sensitive Mixing Systems", Chem. Eng., 11, pp 55-63, June 9, (1986).

15. Yuge, K. and O'Shima, E., "Concentration Responses in High Viscosity Liquid Stirred Tanks", Journal of Chem Eng(Japan),18,2, pp. 151-156 (1975).

16. Costes, J. and Couderc, J.P. "Pumping Capacity and Turbulence Intensity in Baffled Stirred Tanks: Influence of the Size of the Pilot Unit", 4th European Conference on Mixing, BHRA, Fluid Engineering, Paper B2, pp. 25-34, April (1982).

17. Charles, M. "Technical Aspects of the Rheological Properties of Microbial Cultures", Adv. in Biochemical Eng,8,pp. 1-62 (1978).

18. Blanch, H.W. and Bhavaraju, S.M. "Non-Newtonian Fermentation Broths: Rheology and Mass Transfer", Biotech.Bioeng,18,pp. 745-790 (1976).

19. Roels, J.A., Van den Berg, J.,and Voncken, R.M. "The Rheology of Mycelial Broths", Biotech.Bioeng.,16,pp. 181-208 (1974).

20. Calderbank, P.H. and Moo-Young, M. "Prediction of Power Consumption in Agitation of Non-Newtonian Fluids", Trans.Instn Chem Engrs,37,26 (1959).

Scale-up Studies on the Microbial *N*-Dealkylation of Drug Molecules

R. ENGLAND
School of Chemical Engineering
University of Bath
Bath, BA2 7AY, United Kingdom

C.J. SOPER
School of Pharmacy and Pharmacology
University of Bath
Bath, BA2 7AY, United Kingdom

Species of the filamentous fungus *Cunninghamella* have been shown to *N*-demethylate a variety of drugs and drug intermediates in shake flask culture. Studies to scale-up the process were initially carried out in a 12 L stirred tank fermenter using *Cunninghamella bainieri* as the test organism and codeine as substrate. Severe problems were encountered in maintaining and monitoring the fermentation when mycelial pellets from shake flask cultures were used as the inocula, due to aggregation of the fungus and its adhesion to surfaces. Similar difficulties were encountered in a 5.5-L airlift fermenter. Attempts to overcome these problems by using fungal spores as inocula instead of mycelia proved unsuccessful, and yields of *N*-demethylated product were very low. The use of spores immobilised in alginate gel containing proportions of bisoxirane activated and polyethylene-imine cross-linked gel was more successful. In the airlift fermenter, gel immobilised inocula produced the desired pelleted growth and *N*-demethylation of codeine to norcodeine equivalent to that obtained with mycelial inocula in the initial shake flask studies.

Many drug molecules possess N-alkyl functions, usually in saturated cyclic structures or in alkylamine chains. Since the nature of the N-substituent can have a considerable effect on the pharmacological activity of the molecule, there is frequently the requirement in drug synthesis to replace pre-existing N-alkyl functions, usually methyl, with other substituents. This usually involves an N-dealkylation process followed by N-alkylation, N-acylation or N-arylation of the normethyl intermediate to produce the N-substituted derivative (1). Chemical N-dealkylation has traditionally been achieved by the von Braun reaction using cyanogen bromide (2), but the method suffers from the disadvantages of variable product yields and toxicity of the reagent (3). Some improvements in yield have been obtained using chloroformates (1), and diethylazadicarboxylate (4), but these reagents are also highly toxic.

There are obvious potential benefits in effecting N-dealkylation reactions with microbial enzymes rather than the traditional chemical methods (5). A programme of research has therefore been undertaken to screen microorganisms for their ability to N-dealkylate selected drug molecules. Of the many organisms screened, *Cunninghamella sp.* were shown to possess the broadest substrate specificity in their N-dealkylation activity (6). In these studies, a range of drug molecules was used including opiates, 6-7 benzomorphan analgesics, benzodiazepines and phenothiazines to enable the investigation of N-alkyl groups in different chemical environments, including alkylamino chains, saturated cyclic structures, and in amide functions. In subsequent studies to elucidate the mechanism of N-dealkylation and the factors affecting the transformation, a single strain *Cunninghamella bainieri* was used with the opiate codeine as the test substrate, since this drug molecule has the potential for both N-demethylation (to norcodeine) and O-methylation (to normorphine) (7,8). These studies showed that dealkylation occurred selectively at the N-substituent and that maximum N-dealkylation of codeine occurred after complete glucose depletion of the growth medium, when the cells were in the stationary phase of growth. Transformation occurred more readily when the organism exhibited pelleted growth rather than mycelial growth, and under optimal conditions, yields of 12-14% were obtained.

These studies confirmed the potential for using *Cunninghamella bainieri* for the microbial N-dealkylation of drug molecules and encouraged further investigations into scaling up the process. Two lines of investigation were pursued; (1) the use of fermenter systems, and (2) the preparation of gel immobilised cells, and the preliminary findings from these studies are reported in this paper.

MATERIALS AND METHODS

Organism

Cunninghamella bainieri C43 was a gift from the American Cyanomid Company, Lederle Laboratories, New York. USA, and was maintained on Malt Extract Agar slopes at 4 °C.

Test compound and reference standard

Codeine phosphate was purchased from MacFarland Smith Ltd, Edinburgh, UK. Norcodeine base was prepared from codeine by the method of Montska *et al* (9)

Culture media

Malt Extract Agar was obtained from Oxoid Ltd, and was prepared in accordance with the manufacturers' directions. The medium was sterilised by autoclaving at 121 °C for 15 mins.

A chemically defined basal medium was used for all transformation studies. This was prepared from Analar grade reagents (BDH Ltd, Poole, UK) to the following formula (g ℓ^{-1}): KH_2PO_4, 1.36; $NaH_2PO_4.2H_2O$, 5.37; $(NH_4)_2SO_4$, 0.5; Na_2EDTA, 0.202; $MgSO_4.7H_2O$, 0.29; $CaCl_22H_2O$, 0.055; $(NH_4)_6Mo_7O_{24}.4H_2O$, 0.155; $FeSO_4 . 7H_2O$, 0.0058; $ZnSO_4.7H_2O$, 0.00166; $MnSO_4.4H_2O$, 0.00166; $CuSO_4.5H_2O$, 0.00166; $Co(NO_3)_2 . 6H_2O$, 0.00208; $Na_2B_4O_7 . H_2O$, 0.0015; glucose 10; casein hydrolysate 3. The nutrient salts were prepared as a double strength solution and autoclaved at 121 °C for 20 mins. Glucose and casein hydrolysate were prepared as ten-times strength solutions, sterilised by membrane filtration and added aseptically to the nutrient salts solution.

Fermenters

Stirred tank - L H Fermentation Series 2000, nominal capacity 12 ℓ (L H Engineering Co Ltd, Stoke Poges, UK).

Airlift - this was constructed of glass and had a nominal capacity of 5.5 ℓ. The fermenter dimensions were: diameter of fermenter, 10 cm; height of fermenter, 90 cm; diameter of de-entrainment section, 15 cm; height of de-entrainment section, 18 cm; height of draught tube, 70 cm; diameter of draught tube, 6 cm; height of draught tube above bottom, 5 or 10 cm.

HPLC apparatus

This comprised an LDC Constametric III pump, a Rheodyne valve (Model 7125) with a 100 µℓ loop, a Cecil Instruments CE588 Microcomputer scanning spectrophotometer (240 nm) and a Rikadenki recorder. A stainless steel column (150 mm x 4.6 mm i.d.) packed with Partisil 5 (Whatman Ltd, Kent, UK) was used.

Transformation procedures using cell inocula

Shake-flask cultures - the organism was incubated in submerged liquid culture employing a two-stage protocol (7). Sterile growth medium (50 ml) contained in a 250 ml baffled conical flask was inoculated with mycelia from freshly prepared stock slopes of *C. bainieri*. The stage-one cultures were incubated at 27 °C with rotary agitation of 250 rpm for 3-4 days to produce a pelleted growth form. Stage-one culture (10 ml) was inoculated into flasks containing 50 ml growth medium and incubated at 27 °C, 250 rpm. The codeine phosphate drug substrate was added to the stage-two cultures to give a concentration of 1 mM, 24 hours after inoculation with the stage-one culture. Incubation was continued, and duplicate flasks were removed at intervals during the course of the transformation for the determination of microbial growth (as dry cell weight), glucose depletion and codeine N-demethylation.

Stirred tank fermentation - the nutrient salts solution was sterilised *in situ* in the fermenter and the required quantities of glucose and casein hydrolysate solutions added aseptically. Stage-one culture (50 ml) was used to inoculate 8 ℓ of sterile growth medium in the fermenter. Codeine phosphate was added to give a concentration of 1 mM, 24 hours after inoculation. Fermentation was carried out at 27 °C. The pH during fermentation was controlled at 6.5-7.0 by the addition of 5 M phosphoric acid or 5 M sodium hydroxide as required. The oxygen concentration in the medium was maintained by continually sparging sterile air at a rate of up to 10 ℓ min^{-1} into the fermenter below a six bladed turbine impeller operating at 500, 700 or 1000 rpm. 5 ml of Antifoam B (Fisons Ltd, Loughborough, UK) were added at the start of the fermentation. Samples were withdrawn from the fermenter at intervals for analysis.

Airlift fermentation - the glass fermenter was sterilised with steam at 1.4 bar

(109.3 °C) for 4 hours and charged with 5 ℓ of sterile growth medium. Stage-one culture (50 ml) was used to inoculate the fermenter. Codeine phosphate was added to give a concentration of 0.5 g ℓ^{-1}, 24 hours after inoculation. Fermentation was carried out at 27 °C, the temperature being controlled by the use of a water jacket. The pH during the fermentation was controlled at 6.5-7.0 by addition of acid and alkali using a Pye pH controller (Cambridge Instruments, UK), and associated peristaltic pumps. Circulation and oxygen concentration in the growth medium was maintained by sparging in sterile air at a rate of 5 ℓ min^{-1}. Oxygen concentration was measured using a Uniprobe Oxygen Meter (Uniprobe Ltd, Cardiff, UK). Foaming was controlled by the initial addition of 5 ml of either Tween 80 (Sigma Chemical Co, Poole, UK) or Antifoam B (Fisons Ltd, Loughborough, UK). Liquid loss in the exhaust air was minimised by using a refrigerated condenser at -2 °C. Samples were withdrawn from the fermenter at intervals for analysis.

Preparation of spore inoculum

Malt Extract Agar slopes were inoculated with the test organism and incubated for 14 days at 27 °C. Spore production was indicated by a greyish covering of the otherwise white mycelial growth, and was confirmed by microscopic examination. Spores were dislodged into sterile water using a sterile platinum loop and washed by repeated centrifugation (4000 rpm, 5 mins.) and resuspension in sterile water. The spores in suspension were counted in a haemocytometer counting chamber.

When larger volumes of spore suspension were required, *eg* for fermenter inocula or gel immobilisation, the test organism was inoculated into liquid growth medium in large Petri dishes (25 cm) and incubated for 14 days at 27 °C. The mycelial mat was aseptically cut into pieces and added to a conical flask (500 ml) containing 2 mm glass beads (50 g). Sterile water was added and the flask was shaken at 250 rpm for 5 mins. at 25 °C. The spore suspension was decanted from the mycelia and glass beads. The procedure was repeated twice and the spores were concentrated from the combined suspensions by centrifugation (4000 rpm, 5 mins.). The spores were washed by repeated centrifugation and resuspension in sterile water prior to counting.

Determination of codeine and norcodeine

The concentrations of the drug substrate codeine and the N-desmethyl derivative norcodeine in the transformation mixtures was quantitatively estimated by the HPLC procedure described by Gibson (8) using Partisil 5 as stationary phase and a mobile phase comprising 95% ethanol 50%, chloroform 45%, acetonitrile 5% and triethylamine 40 μℓ.100 ml^{-1}. All chemicals were HPLC grade supplied by BDH Ltd, Poole, UK. Column flow rate was 1 ml min^{-1} and the temperature was maintained at 25 °C. An external standard method was employed. 1 ml samples of transformation mixture were centrifuged (3000 rpm. 5 mins.) and the supernatant filtered (3 μ cellulose nitrate membrane). Samples of the filtrate (100 μℓ) and solutions of norcodeine standards in mobile phase were injected alternately, and the concentration of norcodeine in the reaction mixture calculated by comparison of the peak heights.

Determination of dry cell weight

Samples of transformation mixture were filtered through 9 cm diameter Whatman 542 hardened, ashless filter paper with a particle retention of 2.7 μ. The cells were washed with distilled water and the filters and retained cells dried to constant weight at 70 °C and cooled to room temperature in a dessicator over silica gel prior to weighing.

Determination of glucose

The glucose concentration in transformation mixtures was estimated by the method of Raabo and Terkildsen (10), which is based on the oxidation of a colourless indicator O-dianisidine by hydrogen peroxide formed from the oxidation of glucose.

Preparation of gel immobilised spores

1% sodium alginate (100 ml) was mixed with 6N sodium hydroxide (4.5 ml), 1-4-butane-diol diglycidyl ether (20 ml) and sodium borohydride (2 mg ml^{-1}), and stirred for 8 hours at 25 °C. The reaction was terminated by adjusting the pH to 5-6 with concentrated hydrochloric acid. The activated alginate was then dialysed in 10% polyethylene glycol 6000 in 25 mM phosphate buffer, pH 6.5, for 24 hours and adjusted to volume (100 ml). All chemicals were Analar grade supplied by Sigma Ltd, Poole, UK. Mixtures were prepared containing between 20% and 35% activated alginate in 1% normal sodium alginate. An inocolum of spore suspension was added to each mixture and thoroughly dispersed. The alginate/spore mixtures were then added dropwise into 100 ml volumes of 10% calcium chloride solution *via* a 1.1 mm i.d. needle, resulting in the formation of discrete alginate beads approximately 2 mm diameter. The beads were allowed to equilibrate for 24 hours at 25 °C, and the calcium chloride solution was decanted. A 1% solution of polyethylene-imine was added and the beads were left, to allow cross-linking to occur, for a further 24 hours. The beads were then washed three times with sterile water.

Transformation procedures using gel immobilised spore inocula

Shake-flask cultures - 10 g of alginate beads containing spores were weighed into sterile 250 ml baffled conical flasks and 50 ml of growth medium added. The flasks were incubated at 27 °C with shaking at 250 rpm. Codeine was added to a concentration of 0.5 g ℓ^{-1} after 7 days, and incubation was continued. Samples were withdrawn at intervals and assayed.

Airlift fermentation - 500 g of 25% activated alginate beads containing spores were inoculated into the airlift fermenter containing 5.5 ℓ of growth medium. The fermentation was carried out at 27 °C and codeine was added to give a final concentration of 1 g ℓ^{-1}, 48 hours after inoculation. The fermentation was continued as described above for airlift fermentation with cell inocula, with samples being withdrawn at intervals for analysis.

RESULTS AND DISCUSSION

Typical data obtained during the stage-two shake flask culture of *C. bainieri* in chemically defined medium with added codeine are illustrated in Figure 1.

The growth curve is represented by dry cell weight (biomass). Following a short lag period, the organism enters a slow growth phase which continues until the third day of incubation. The short stationary phase which occurs between 3 to 6 days' incubation is followed by a decrease in dry cell weight indicative of cell lysis, resulting in non-retention of cell debris in the filtration process used to determine dry cell weight. Detailed analysis of growth curve data (7) has shown that growth obeys cube root kinetics associated with pelleted growth forms (11) rather than the more usual exponential growth kinetics.

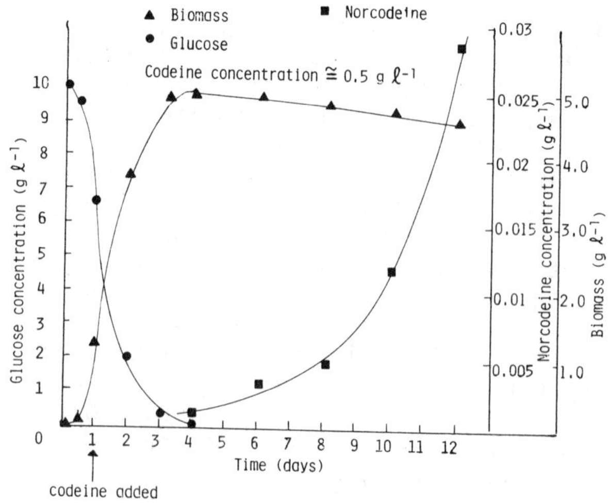

FIGURE 1. Glucose depletion, biomass production and norcodeine production by *C. Bainieri* in shake flasks, 27°C, as a function of time.

Norcodeine, the produce of codeine N-demethylation, is not detected prior to the third day of incubation. Thereafter, the concentration of norcodeine in the medium increases with incubation time, initially (day 3 to 8) at a slow rate, and subsequently at a more rapid rate. Comparison of the growth curve with the data for depletion of glucose from the growth medium suggests that the extent of fungal growth was limited by carbon source exhaustion which occurred between the fourth and fifth day of incubation. Cell growth is seen to parallel depletion of glucose from the medium, a linear relationship being obtained between dry cell weight and glucose utilisation during the active cell growth phase (Figure 2).

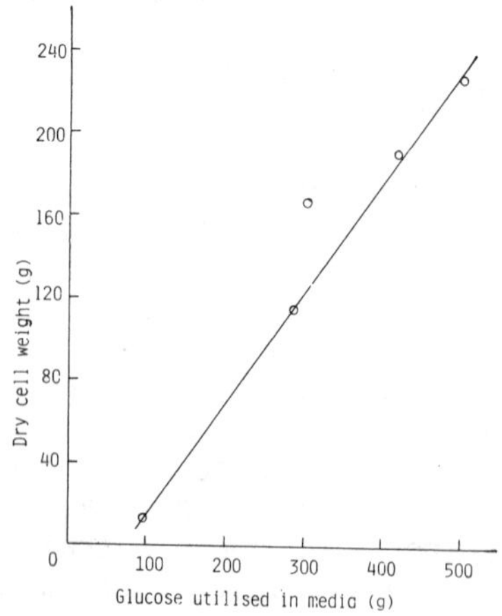

FIGURE 2. Relationship between glucose utilised in media and biomass produced in shake flask culture of *C. bainieri* at 27°C.

It is significant that there was a requirement for extensive, if not total, depletion of glucose from the medium before microbial N-demethylation of codeine was effected, suggesting that a carbon source regulatory mechanism was operating.

Initial scale-up studies were conducted in the stirred tank fermenter using as inoculum the mycelia from the stage-one shake flask culture. Since the shake flask studies had shown that glucose depletion and stationary phase growth were required for N-demethylation the fermenter was operated in batch mode rather than continuous mode. Although good growth of the test organism was achieved, problems of attached growth were severe. It was apparent that attachment of the fungus occurred at all stages of the fermentation. In the early stage of the fermentation, the mycelia attached to small gaps in the fermenter, such as those between seals, and the fermenter walls, and on unwashed surfaces above the liquid level. Later in the fermentation, attachment was observed around the pH and pO_2 probes and the cooling coil of the fermenter (Figure 3).

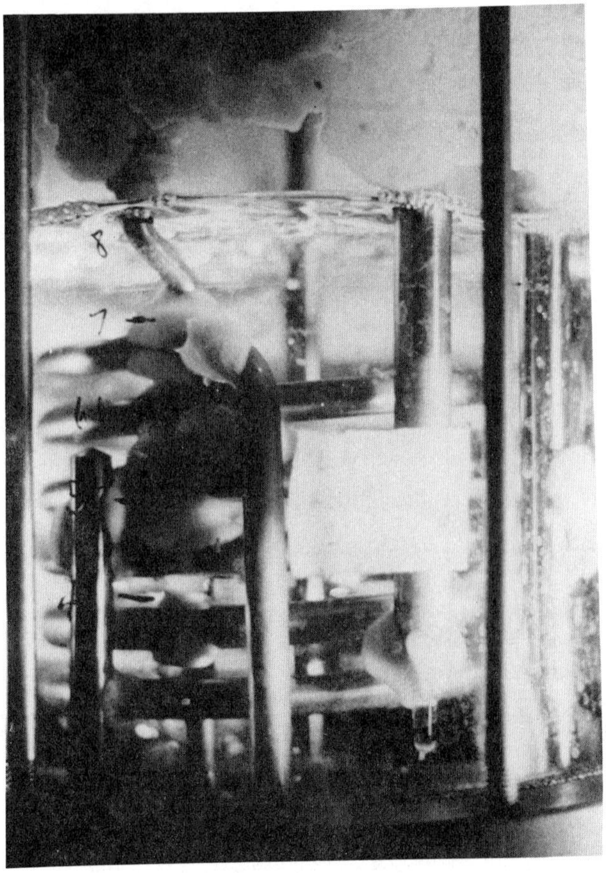

FIGURE 3. Attached growth formed during stirred tank fermentation of *C. bainieri* at 27°C.

The biomass recovered from the fermenter consisted of large pieces of attached fungal growth up to 25 cm^3 (Figure 4).

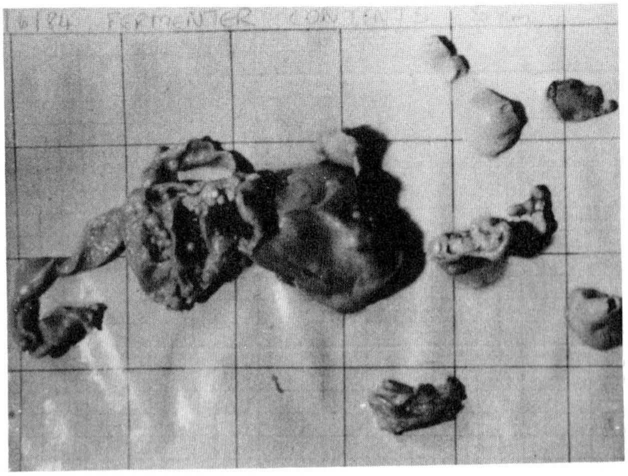

FIGURE 4. Typical large pieces of biomass from stirred tank fermentation of *C. bainieri* at 27°C; grid scale 5 cm.

In most cases, very few free pellets were circulating in the fermenter after 4 to 5 days' fermentation, and no norcodeine was detected even after 10 days' fermentation (Table 1).

The addition of Antifoam B and increasing the stirrer speed overcame the problem of foaming and resulted in the production of some pelleted growth and the detection of norcodeine in the medium after 10 days' fermentation (Table 1). However, the majority of fungal growth was still large aggregates attached to the fermenter surfaces.

It was envisaged that the use of fungal spores rather than mycelia or the inoculum in the fermenter would initiate a more even pelleted growth and minimise fungal aggregation and attachment. In shake flask culture, an inoculum of spores produced similar growth, as determined by glucose depletion from the medium, as an inoculum of mycelia from a stage-one culture (Figure 5).

In the stirred tank fermenter, the use of spore inocula was less satisfactory. Spore inocula usually gave rise initially to pelleted growth rather than mycelia. Although in the early stages of the fermentation there was more uniformly distributed growth in the medium, attachment of mycelia to surfaces was still a problem as the fermentation progressed (Table 1). The conditions of high airflow and fast stirrer speed required to maintain pelleted growth even for short periods, resulted in excessive

TABLE 1
Results Obtained for the Culture of *C. bainieri* in the LH Stirred Tank Fermenter in the Presence of Codeine (0.5 g.L^{-1}) at 27°C.

Run								
Inoculum	5 x 2 day 250 ml shake flask	5 x 2 day 250 ml shake flask	Spores				Pre-germinated spores	
			1.5 x 10^{5}ml^{-1}	3 x 10^{3}ml^{-1}	3 x 10^{3}ml^{-1}	9 x 10^{6}ml^{-1}	4 hrs	24 hrs
Volume ℓ	7	8	8.5	8.5	9.5	9.5	9.5	9.5
Antifoam	none	100 ppm	100 ppm	100 ppm	none	none	none	none
Foaming	yes	no	excessive	excessive	no	no	no	no
Air flow rate ℓ min^{-1}	7	7	10	10	2.5	2.5	2.5	2.5
Stirrer speed, rpm	700	1000	1000	1000	500	500	500	500
Codeine conversion	no	yes ∿ 2% (10 days)	no	yes ∿ 10% (10 days)	no	no	yes ∿ 10% (10 days)	no
Comments	Only attached growth after 3 days	Some pellets mainly attached growth. Norcodeine detected after 10 days	Initially mycelia then pellets. After 2 days attached growth, large liquid loss	Initially large quantity small pellets, then attached growth. Norcodeine detected after 4 days large liquid loss	Mainly attached growth	Run stopped after 4 days Attached growth	Large quantity small pellets	Only attached growth after 3 days

TABLE 2
C. bainieri Culture in Airlift Fermenter in the Presence of Codeine (0.5 g.L^{-1}) at 27°C; Air Flow Rate 5 L-min^{-1} in Each Case.

Run								
Sparger type	Sinter	Single orifice	← 4 pt injection →					
Draught tube gap (cm)	10	10	5	10	10	5	5	5
Inoculum 3x10 ml^{-1}	← spores →				← Pre-germinated spores → 4 hrs	4 hrs	24 hrs	24 hrs
Antifoam Tween 80	5 ml	5 ml	none	none	none	none	none	none
Growth	Initially small pellets then attached	½ cm pellets and attached	Attached growth at base	Pellets and attached at base	Pellets and attached	Pellets and attached	Pellets and some large clumps	Mainly attached
Comments	Pellets settled at base	Very large clumps attached	Flow restricted by attached growth at base	Flow restricted by attached growth at base	Flow restricted by growth at base	∿ 50% of biomass attached	Large liquid loss	Large liquid loss
		Trace amounts only of norcodeine were detected in all cases						

foaming that could not be easily controlled by conventional antifoam treatment. Furthermore, these conditions did not prevent subsequent fungal attachment, and resulted in the additional problem of significant fluid loss from the fermenter. Norcodeine yields with spore inocula were low or nonexistent.

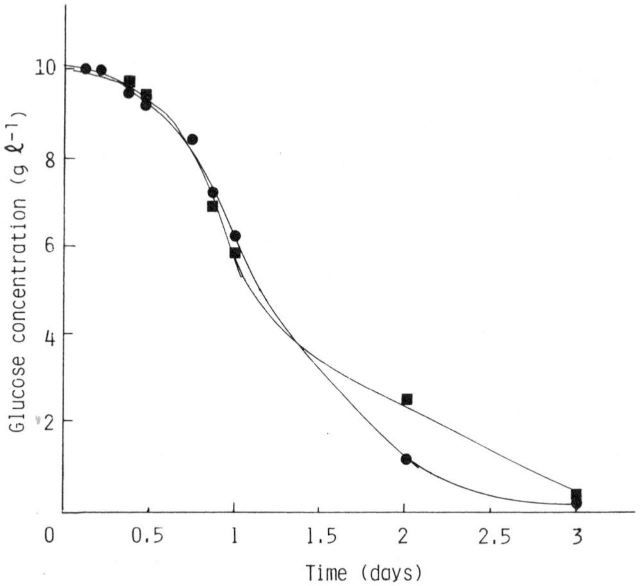

FIGURE 5. Comparison of glucose utilisation by *C. bainieri* spores and mycelia in shake flask cultures versus time at 27°C.

Some improvement in norcodeine yields was obtained by using four hour pre-germinated spores which enabled lower airflow rates and slower stirrer speeds to be employed, thus minimising foaming and fluid loss, and maintaining pelleted growth for longer periods. Increasing the pre-germination time of the spore inoculum to twenty four hours resulted in the same growth and fungal attachment problems that were observed with mycelial inocula.

From the studies with the stirred tank fermenter, it was concluded that the scale-up of microbial N-demethylation using *Cunninghamella bainieri* would require a different bioreactor design. After consideration of the designs available, an airlift fermenter was selected. Inoculation of the airlift fermenter with mycelia from shake flask culture resulted in fungal aggregation into a small number of large clumps, mainly around the sparge tube at the base of the fermenter and the draught tube supports. The use of spore inocula improved the distribution of the biomass throughout the fermenter, but overgrowth of mycelia on the sinter resulting from settling of the pellets was still observed (Table 2). Although foaming occurred, it could be easily controlled by the addition of low concentrations of antifoam agents.

Replacement of the sinter with a four point injection air sparger and redesign of the draught tube resulted in improved circulation of medium in the fermenter, minimal foaming, and growth mainly in the pelleted form (Figure 6).

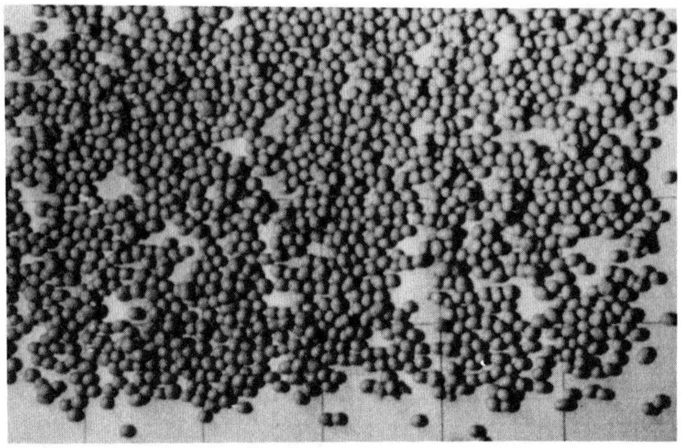

FIGURE 6. Typical pellet growth obtained in airlift fermentation of *C. bainieri* at 27°C; attached growth also occurs; grid scale 5 cm.

The problems of attached growth that were observed with spore inocula in the stirred tank fermenter were greatly reduced in the airlift apparatus. Most attached growth observed in the airlift fermenter was caused by settling of pellets in 'dead areas' in the fermenter or by irregular shaped pellets being trapped in the draught tube supports or around pH, pO_2, temperature and sampling probes (Figure 7).

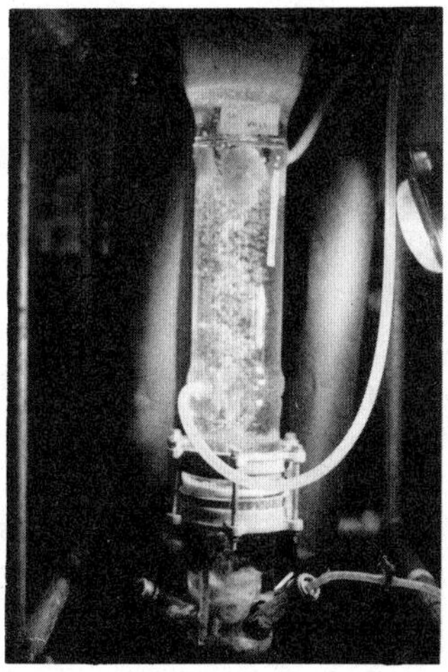

FIGURE 7. Attached growth formed during airlift fermentation of *C. bainieri* at 27°C.

Further improvement in the extent of pelleted growth in the fermenter was obtained by using four hour pre-germinated spores as the inoculum. However, the use of twenty four hour pre-germinated spores produced problems of reduced pelleted growth, large fungal aggregates and attached growth comparable to those observed when identical spores were used to inoculate the stirred tank fermenter (Table 1).

Comparison of performance between the stirred tank and airlift fermenters was not easy because of the nature of the fungal growth and attachment problems in the stirred tank fermenter. However, some comparison can be made using four hour pre-germinated spores as the inoculum since these produced pelleted growth in both the stirred pot fermenter and the airlift fermenter. This comparison is illustrated in Figure 8 which shows the production of biomass and depletion of glucose from the medium as a function of fermentation time for organisms grown in both fermenters, under similar environmental conditions, and from identical pre-germinated spore inocula.

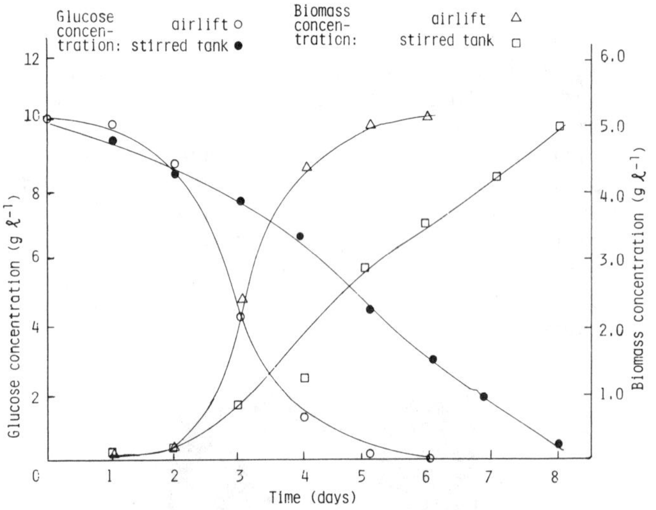

FIGURE 8. Comparison between stirred tank and airlift fermentations of *C. bainieri* using 4 hr pregerminated spores as inoculum versus time at 27°C.

Growth of *Cunninghamella bainieri* in the airlift fermenter can be seen to follow the same pattern as in the stirred tank fermenter. However, growth in the airlift fermenter is faster and the glucose in the medium is totally utilised in 5 to 6 days in contrast to 8 to 9 days in the stirred tank fermenter.

It is significant that only trace amounts of norcodeine were detected in the airlift fermenter when spore inocula were used. Poor yields of norcodeine were also observed with spore inocula in stirred tank fermentation (Table 1). The reasons for this are not clear and experiments are in progress to elucidate this problem.

It was apparent from studies on the characterisation and optimisation of N-demethylation by *Cunninghamella bainieri* that maximal transformation was observed when fungal growth was in the pelleted form (7). In scale-up studies with stirred tank and airlift fermenters it was difficult to obtain the required pelleted growth form consistently with either spore or mycelial inocula, and persistent problems were encountered with fungal aggregation and attachment. In an attempt to overcome these difficulties, the possibility of immobilising *Cunninghamella bainieri* spores in a gel matrix was investigated. It was anticipated that this would produce particles of defined size and shape that would provide nuclei for pelleted growth of the fungus. Germination and outgrowth of spores in the gel particles could be initiated with growth medium and the codeine phosphate substrate could be added when the glucose carbon source had been depleted.

Alginate gel has been widely used as a matrix for cell and enzyme immobilisation (12). The method that has usually been employed is the gelation of alginate by calcium ions. The main drawback of this method is that the calcium alginate gel is unstable in the presence of calcium chelators such as EDTA and phosphate. The medium used for growth of *Cunninghamella bainieri* contains potassium dihydrogen phosphate (1.36 g ℓ^{-1}) and sodium phosphate (5.37 g ℓ^{-1}). The soluble form of the codeine substrate is also the phosphate. The problem of chelation can be overcome by 'hardening' the gel as demonstrated by Birnbaum *et al* (13) whereby alginate is activated and cross-linked, and recently Salleh and England (14) have reported a method of activating the alginate with 1-4-butanediol diglycidyl ether prior to cross-linking with polyethyleneimine. The proportion of activated and cross-linked gel (PEI-activated gel) needed to be controlled within defined limits. Less than 15% PEI-activated gel produced beads that were unstable and greater than 50% resulted in the formation of collapsed disc-like beads. Gel beads containing between 20% and 35% PEI-activated gel were shown to be stable to concentrations of phosphate in excess of that present in the *Cunninghamella* growth medium and to be suitable for the immobilisation of fungal spores (14).

Data obtained when *Cunninghamella bainieri* spores immobilised in 25% PEI-activated gel were incubated with codeine in shake flask culture are shown in Figure 9.

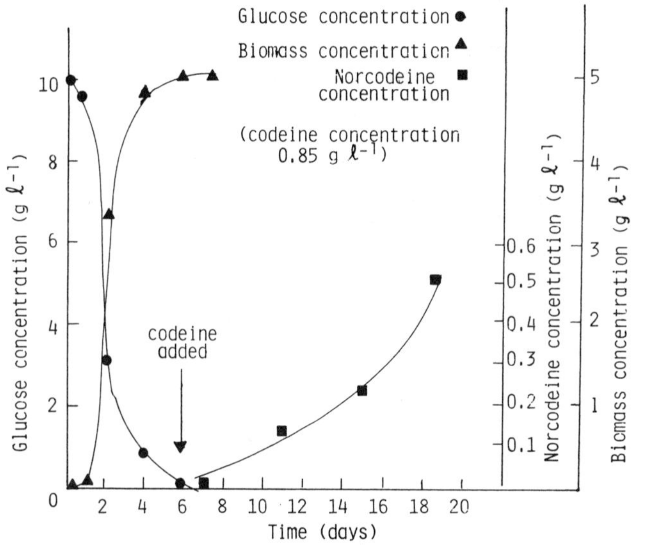

FIGURE 9. Glucose depletion, biomass growth and norcodeine production for shake flask culture of *C. bainieri* spores immobilised in 25% PEI-activated gel.

Biomass production and glucose depletion from the medium with gel immobilised spores was similar to that observed in shake flask cultures inoculated with mycelia or non-immobilised spores. Codeine N-demethylation was also observed with gel immobilised spores after total glucose depletion from the medium. Direct comparison between yields obtained with immobilised and non-immobilised spores is not possible from these data, since different concentrations of codeine substrate were used.

Increasing the proportion of PEI-activated gel used in the immobilisation of the spores from 20% to 35% had no significant effect on the amount of codeine N-demethylation that was observed after 8 days incubation in shake flask culture (Table 3). However, it was observed that beads containing more than 25% PEI-activated gel showed a reduction in diameter of approximately 15% in shake flask culture. The appearance of fragments of gel-like material in the culture medium suggested that the vigorous agitation that occurs in shake flasks caused breakdown of beads already weakened by spore germination and outgrowth within the gel.

Preliminary scale-up studies were undertaken using the airlift fermenter with spores

TABLE 3
Conversion of Codeine to Norcodeine versus % PEI-Activated Alginate in Gel after 4 Days Shake Flask Culture with Codeine (8 Days Fermentation)

% Activated alginate	Conversion codeine to norcodeine %
20	6.6
25	5.8
30	6.6
35	7.3

immobilised in beads containing 25% PEI-activated gel as the inoculum. The rates of biomass production and glucose depletion from the medium indicated that growth in the air-lift fermenter was similar to that observed in shake flask cultures (Figure 10).

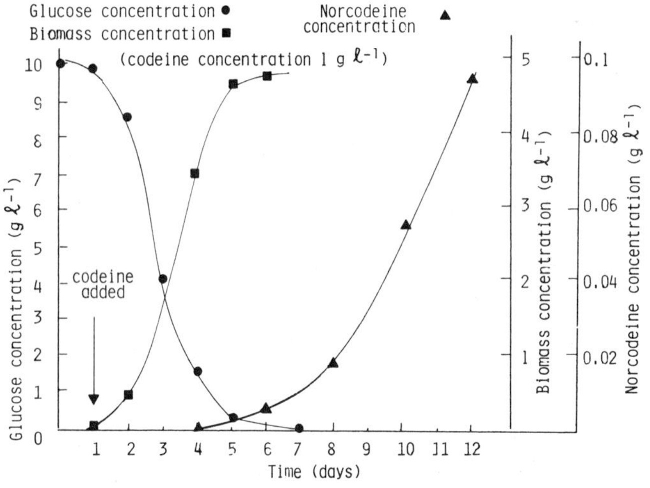

FIGURE 10. Preliminary results for codeine conversion in airlift fermenter using gel immobilised *C. bainieri* spores at 27°C.

In contrast to the shake flask cultures, no gel fragments were observed in the airlift fermenter, the beads remaining intact and discrete, but tending to increase in diameter by up to 20%. No overgrowth occurred on any of the probes and the surfaces of the fermenter remained essentially clear of attached fungal growth. The only areas where fungal growth developed were on the draught tube supports and in the 'dead area' at the base of the sparger tube. Norcodeine was detected after five days of fermentation when glucose depletion from the medium was virtually complete. Continued fermentation resulted in codeine N-demethylation after twelve days equivalent to that obtained with mycelial inocula in shake flask cultures, and very much greater than that obtained with inocula of non-immobilised spores in either stirred tank or airlift fermentations.

These data demonstrate the potential for using gel immobilised spores for large scale microbial N-demethylation with filamentous test organisms such as *Cunninghamellae*. Realisation of this potential will require further investigations to establish methods for large scale production of immobilised spores. The ideal pelleted growth that pertains when gel immobilised spores are used in the airlift fermenter will also permit further studies to improve the yield of the N-desmethyl product. This will require knowledge of the extent to which N-demethylation is regulated by product inhibition and the consequent development of methods for the continuous removal of product during the fermentation.

LITERATURE CITED

1 Palmer, D.C. and Strauss, M.J. *Chem.Rev.* 77, 1 (1977)

2 Hageman, H.A. *Org.React.* 7, 198 (1953)

3 Rice, K.C. *J.Org.Chem.* 40, 1850 (1975)

4 Kanematsu, K., Takeda, M., Jacobson, A.E. and May, E.L. *J.Med.Chem.* 12, 405 (1969)

5 Sewell, G.J., Soper, C.J. and Parfitt, R.T. *J.Pharm.Pharmacol.* 31, 903 (1979)

6 Sewell, G.J., Soper, C.J. and Parfitt, R.T. *Appl.Microbiol.Biotechnol.* 19, 247 (1984)

7 Sewell, G.J. "Studies on the microbial N-dealkylation of drug molecules", PhD Thesis, University of Bath (1982)

8 Gibson, M. "Analytical studies on the N-dealkylation of drug molecules by *Cunninghamella sp.*" PhD Thesis, University of Bath (1984)

9 Montska, T.A., Matiskella, J.D. and Partyka, R.A. *Tetrahedron Lett.* 14, 1325 (1974)

10 Raabo, E. and Terkildsen, T.C. *Scand. J.Clin.Lab.Invest.* 12. 402 (1960)

11 Pirt, S.J. *Proc.R.Soc.London Ser. B.* 166, 369 (1966)

12 Cheatham, P.S.J., Blunt, K.W. and Bucke, C. *Biotechnol.Bioeng.* 21, 2155 (1979)

13 Birnbaum, S., Pendleton, R., Larsson, P-O, and Mosbach, K. *Biotechnol.Lett.* 3, 393 (1981)

14 Salleh, A.B. and England, R. Presentation to Biochem '85, Asia, (1985)

BIOPROCESS APPLICATIONS

Evaluation of a Novel Foam Fermenter in the Production of Xanthan Gum

TUSHAR K. MISRA
STANLEY M. BARNETT
Department of Chemical Engineering and The Biotechnology Center
University of Rhode Island
Kingston, Rhode Island 02881

Fermentation of a culture of *Xanthomonas campestris* NRRL B-1459 was carried out in a foam fermenter. The parameters monitored were growth rate, xanthan gum and glucose concentrations, pH and viscosity of the broth. Flow patterns in the fermenter were studied using dye tracers. Results were a maximum specific growth rate of 0.245 hr^{-1} for 120 hr of fermentation at 28°C, and a final xanthan gum yield of 70% based on initial glucose in the media. The viscosity of the broth at the end of the fermentation was 0.66 PaS. Separate experiments using the "gassing-in" method were conducted to determine the volumetric mass transfer coefficient, k_la, in the foam fermenter under various conditions: in water, in fermentation media, and in media with surfactant. The k_la values showed a direct dependency on air flow rate to the foam generator and were reported to be higher for the case when a surfactant was present in the media. These values decreased with increase in fermentation age and broth viscosity.

INTRODUCTION:

The ultimate goal of present fermentation reactor research is to improve product yield, while minimizing production costs. The research in this area is progressing in two general directions: genetic mutation and manipulation to develop high potential microbial strains, and processing technology improvement, such as the development of fermentation control systems and improved bioreactor designs. This work deals with the latter approach. A critical factor in aerobic cultivation of microorganisms is the transfer of oxygen from the gas phase to the aqueous medium. Traditionally, fermentations have been conducted in stirred tank systems which require considerable power input to attain adequate oxygen transfer. By limiting the degree of oxygenation, restrictions are imposed upon the growth patterns of an organism, often with deleterious results. Hence, both oxygen transfer and mixing characteristics are important in the design and operation of fermentation vessels. It has been stated that as much as 20% of the cost of industrial fermentation could be attributed to aeration and agitation operating expenses[1]. Reduction of these costs is important to the success of future fermentations. Keeping these aims in mind, any new fermenter design must be inexpensive and simple, and be able to achieve oxygen transfer and bulk mixing rates comparable with, if not better than, those achieved in conventional stirred tank fermenters.

Many bioreactors have been designed to give higher oxygen transfer rates and minimal power consumption. For example, tower cycling fermenters[1], hollow fiber membrane reactors[2], tubular loop fermenters and ejector loop fermenters[3,4], concentric and external recirculation air lift bioreactors[5,6] and Koch mixer columns[7] have been reported. A novel fermenter, called a foam fermenter, which utilizes a fine dispersion of air in nutrient broth and has low air and power requirements was developed[8]. It was shown that such parameters as growth rate and product yields compare favorably with traditional stirred tank reactors while the aeration rate to the fermenter was reduced by orders of magnitude and there was a considerable reduction in the power input to the fermenter. In addition to these benefits, ease of separation of product and simultaneous reaction and harvesting, the foaming problems of the traditional fermenters were turned into an advantage.

It was, thus, the objective of this research to evaluate, the performance of the foam fermenter during a viscous fermentation, the production of xanthan gum, a microbial polysaccharide.

Xanthan gum is an exopolysaccharide produced in a pure culture fermentation process by the microorganism Xanthomonas campestris. It is a high molecular weight, polyanionic polysaccharide[9]. Xanthan gum is composed of D-glucose, D-mannose and D-glucuronic acids assembled in pentasaccharide repeating units. Half of the terminal D-mannopyranosyl groups are substituted at position 4 and 6 by pyruvic acid acetal residues, and the D-mannopyranosyl residues are O-acetylated at position 6[10]. Xanthan is extensively used by the petroleum industry, the food industry, and as a suspending and stabilizing agent for herbicides, pesticides, fertilizers and fungicides[9].

Production of microbial polysaccharides present special problems because at the end of the fermentation, the broths attain very high viscosities, amounting to several thousand centipoises[11]. Aeration and agitation are possible because of the high degree of shear thinning exhibited by polysaccharide broths. Even so, out in the bulk fluid of a stirred tank away from the region of high shear near the impeller, stagnant pockets are formed that are starved for oxygen unless adequate shearing action is provided[12]. Hence, for the development of special product areas, such as microbial polysaccharides and other products involving viscous fermentation, efficient fermenters and better means of production are needed.

The present work was undertaken in order to determine the effectiveness of the foam fermenters in the production of xanthan gum. The first phase focuses on the xanthan gum production. Parameters like growth rate, product yield, broth viscosity and pH were measured and compared with corresponding values in the literature. In the second phase, the volumetric mass transfer coefficient, $k_l a$, of an organism-free medium and a microbial culture in the foam fermenter was determined.

MATERIALS AND METHODS:

Culture strains and medium

Viable cultures of X. campestris NRRL B-1459 were used for inoculum build up and fermentation. The medium for inoculum production was composed of (inkg/m^3): Dextrose, 30.0, dipotassium hydrogen phosphate 5.0, magnesium sulphate heptahydrate, 0.1, ammonium nitrate, 4.0 and Tween 20, 300 ppm. A modification in the nitrogen source was made from the original medium composition[13]. Tween 20, a polyoxyethylene derivative of sorbitol fatty acid esters, had been regarded as safe for use as a food additive. It was determined that the surfactant had no effect on the growth of the microorganism. The total period for culture was 120 hours. The parameters were determined every eight hours.

Description of the Foam Fermenter

The fermenter used in the experiments is shown in Fig. 1. Two glass columns, each 0.092m in diameter and 0.66m high, were assembled in such a way that the right hand column was raised above the left hand one. The columns were sealed with rubber stoppers and connected to each other by 0.02m Tygon tubing. The right column worked as a riser while the left one acted as the downcover for the foam. The foam was created by mixing the liquid with a stream of air in a venturi generator. It was then pumped to the bottom of the right hand column. The foam rose to the top from where it cascaded down the left column. A peristaltic pump (248 W) was used to pump the foamed broth back to the riser via the venturi genertor. Air was supplied and regulated through a rotameter. To prevent contamination through the air supply, the air was passed through a 0.23m long sterilization column packed with sterile glass wool, and a 0.2 μm cutoff membrane filter before mixing with the broth. A surfactant was added to the broth to aid in the generation of a stable foam. This recycle of foam was repeated throughout the experiment to produce a dispersion of 50 μm bubbles and to maintain a viable environment for the growth of the microorganisms.

Colloidal Gas Aphrous (CGA)

The method of supplying aeration and growth surfaces involve the use of a foam produced by the CGA method. CGA dispersions are formed in a venturi device, wherein the liquid is injected through a nozzle into a high velocity gas stream. The liquid is then atomized by the formation and subsequent shattering of twisted filaments and thin cuplike film which provide a high degree of turbulence and large interfacial areas for mass transfer[14]. The CGA dispersion can contain up to 65% undissolved gas by volume and produce bubbles from one to one hundred microns in diameter with double surfaces[8]. The CGA used for this work contained, almost exclusively, bubbles of 50 μm in diameter because of repeated recycling of the bubbles through the venturi generator.

Auxiliary Equipment

Cell concentration (in colony forming units/ml) was measured with the help of a Bausch & Lomb Spectronic 21 spectrophotometer. The wavelength for the measurement ws 650 nm. Cell weight was determined by diluting the broth to about 0.07 Pa-S and centrifuged, the precipitate was washed, filtered and dried in an oven at 80°C for 10 hours.

The viscosity of the broth was measured with a Haake Rotovisco RV12, at 25°C and 30 rpm.

For analyzing the glucose concentration, a YSI Model 23A Glucose Analyzer (Fisher Sci.) was used. For the microbiological assays, for determining the cell weight and the xanthan gum yield, the methods prescribed by Rogovin et al, 1961 were followed[13].

The dissolved oxygen (d.o) concentration in the broth was measured with a lead-silver galvanic cell oxygen probe connected to an EXTECH oxygen meter which in turn was connected to a Bausch & Lomb Omniscribe Recorder.

Experimental Procedure

Flow Procedure. In order to determine what kind of flow patterns are produced in the fermenter, a dye was introduced into the system. The dye used was bromophenyl blue (2 Kg/m^3) while the surfactant was ethylhexadecyldimethylammonium bromide (0.3 Kg/m^3). Because of the anionic structure of the dye and the cationic structure of the surfactant, the dye becomes entrained in the foam. Because of this attachment, the dye was used as a tracer and introduced into the bottom of the right hand column using as a $50x10^{-6}m^3$ (50 ml) pipet. The flow pattern observed in the foam fermenter, Figure 2, reveals turbulent agitation in the area about 0.4-0.5from the bottom of the right column. Stagnant liquid appears at the bottom of the column while stagnant foam adheres to the top of the column. Mixing takes place over most of the left column, and the foam quality becomes poorer when it moves from the top of the right column to the left column.

kla Determination

Values of k_1a in water, pure culture medium and medium containing 0.3 Kg/m^3 of the surfactant, Tween 20, were obtained by the gassing-in technique proposed by Bartholomew et al, 1950[15]. Prior to the start of the experiment, the liquid in the fermenter was stripped of oxygen by sparging with nitrogen. When the dissolved oxygen concentration had become sufficiently low (about 0.0008 Kg/m^3) and the liquid was free of gas bubbles, the air supply and the pump were turned on. Increase in dissolved oxygen concentration with time was recorded.

The oxygen balance with respect to a unit volume of liquid is

$$dC/dt = k_1a\,(C_s - C) \qquad (1)$$

where C = conc. of dissolved oxygen in ppm (Kg/m^3)

C_s = value of dissolved oxygen concentration in equilibrium with the average partial pressure of oxygen in the atmosphere.

C_s is the value indicated by the oxygen meter when it has reached a constant after prolonged aeration. Upon integration of Eq. (1),

$$\ln(C_s - C) = -\,k_1at + \text{constant} \qquad (2)$$

Thus plotting the argument of the logarithmic term vs. time allowed the evaluation of k_1a as the slope of the straight line which is obtained by linear regression.

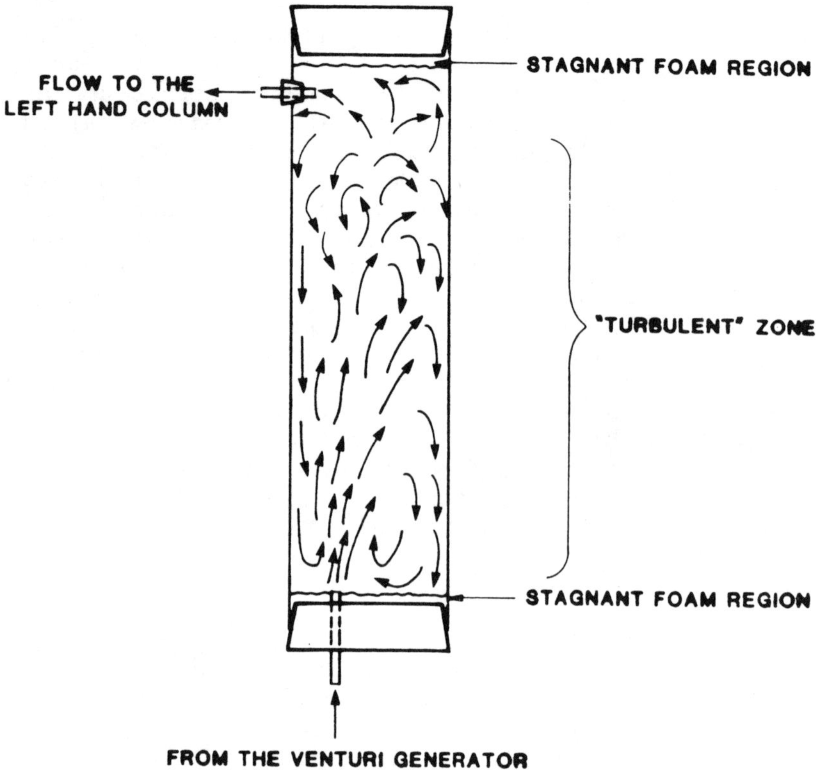

FIGURE 1. The experimental foam fermenter.

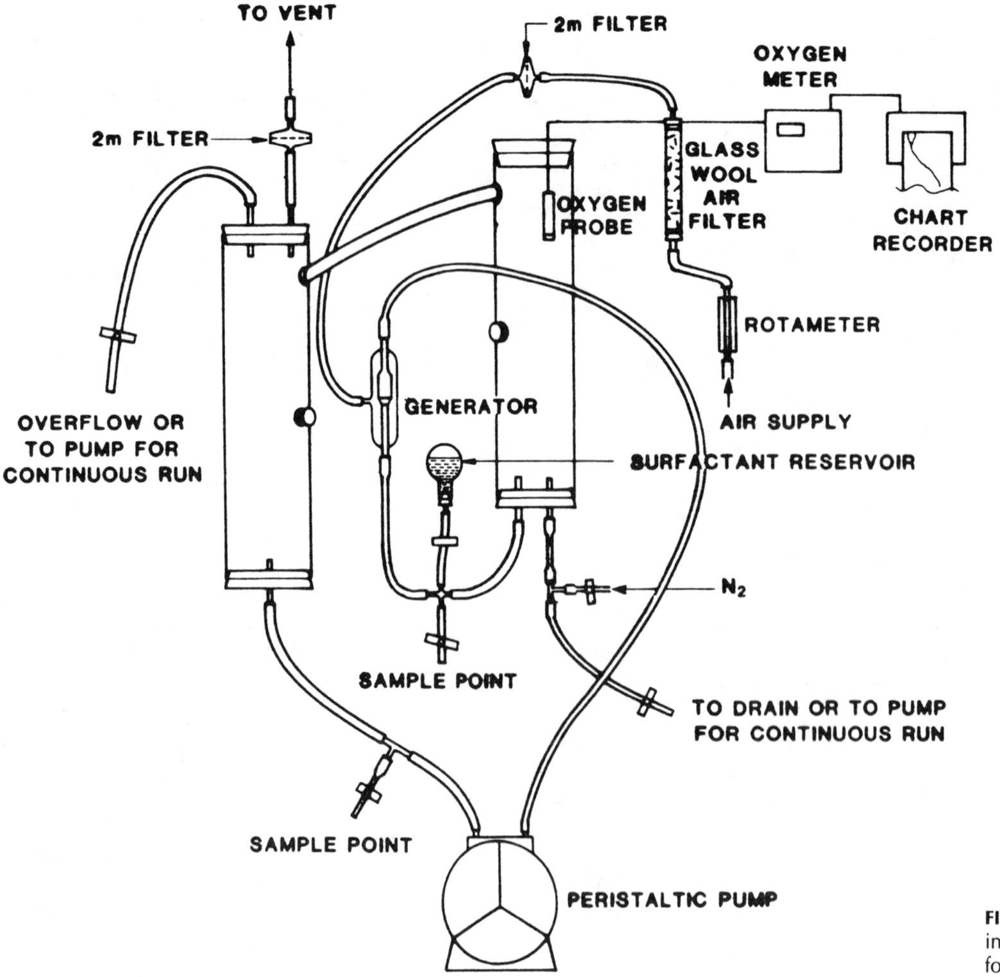

FIGURE 2. Flow patterns observed in the right-hand column of the foam fermenter.

Respiration rate determination

The method of Takahashi and Yoshida[16] (1979) was used for the determination of respiration rate and k_la in the culture. The respiration rate measurement is based on the following oxygen balance for the broth sample,

$$-d\underline{c}/dt = r\underline{x} \qquad (3)$$

In equation (3), $\underline{c}$ and $\underline{x}$ designate c and x in the apparatus for the respiration rate measurement as distinguished from those in the fermenter. Therefore, r can be determined from the negative slope of the chart divided by the cell concentration in the fermenter.

Now, the k_la values in the fermentation broth were determined from the overall oxygen balances which involved the respiration rates measured with a culture present. The k_la for a fermentation broth is given by

$$dC/dt = k_la\ (C_S-C) - rx \qquad (4)$$

Results & Duscussion

Fermentation. Shown in Figure 3 is a composite graph for cell concentration, glucose concentration, pH, viscosity of the broth varying with fermentation time.

Referring to Fig. 3, the cell concentration in colony forming units/ml vs. time shows the traditional lag, exponential and stationary phase of growth. There is a definite evidence of some cell death after 100 hours. From a Lineweaver - Burke plot, the maximum growth rate μ_{max} and the Michaelis constant K_m were obtained as 0.245 hr^{-1} and 0.55 Kg/m^3 respectively. This is comparable to values reported in the literature with Moraine et al[17] reporting a μ_{max} of 0.22 hr^{-1} and a K_m of 0.40 Kg/m^3.

In the first 11 hours when nitrogen was being rapidly metabolized, the pH rose slightly to a maximum of 7.5. Thereafter, the polymer with its free carboxylic groups[18] and other acidic products of sugar metabolism accumulated,the pH decreased slowly to 6.1. During the lag phase, the rate of glucose consumption was low. But after 40 hours, when the exponential growth rate set in the glucose consumption rate increased markedly. At the end of the fermentation, nearly one-third of the sugar was unconsumed. There was no further appreciable utilization of glucose when the fermentation was extended to seven days. Regovin et al[13] had determined that there was a broad optimum pH range around 7.2.

For the first 36 hours, the viscosity was below 0.07 Pa-S due to the lower concentration of the polymer in the broth. Over the next 60 hours, with increased production of the polymer, the viscosity rose steadily to about 0.06 Pa-S. At the end of the fermentation, it had reached a value of 0.66 Pa-S. To negate the effects of high viscosity and to maintain adequate agitation, the aeration rate was turned up. From $1.67\times10^{-6}\ m^3/s$ for the next 32 hours, it was increased to $3\times10^{-6}\ m^3/s$ for the next 48 hours, and to $4.1\times10^{-6}\ m^3/s$ for the final 40 hours.

Xanthan gum formation rate increased with cell concentration, until the stationary phase. Beyond the stationary phase, the gum formation rate still increased, cell growth used little of the glucose, as indicated by the polymer yield not being materially lower during active growth. Energy required for growth comes from assimilation of only a small amount of glucose. Kennedy et al[19] have shown that high nitrogen concentration in the media gives yields of crude product and needs only short fermentation times to achieve maximum product formation. These products, however, have inferior solution rheology compared to those produced from low nitrogen media. This is due partly to high concentration of co-precipitated microbial cells and partly due to difference in tertiary molecular structure. A 3% glucose solution was converted to a 2.1% xanthan solution after 120 hours of fermentation. As shown in Fig. 4, glucose consumption has an approximately linear relationship with polysaccharide production throughout the fermentation. Using a least-squares fit, the equation for the line was found to be

$$P-P_0 = 0.597\ (S_0-S) \qquad (6)$$

where P,P_0 = the product concentration in Kg/m^3 at any time t and at the beginning of the fermentation, respectively, (P_0=0).

S,S_0 = the substrate concentrations in Kg/m^3 at any time t and at the beginning of the fermentation, respectively (S_0=30 Kg/m^3).

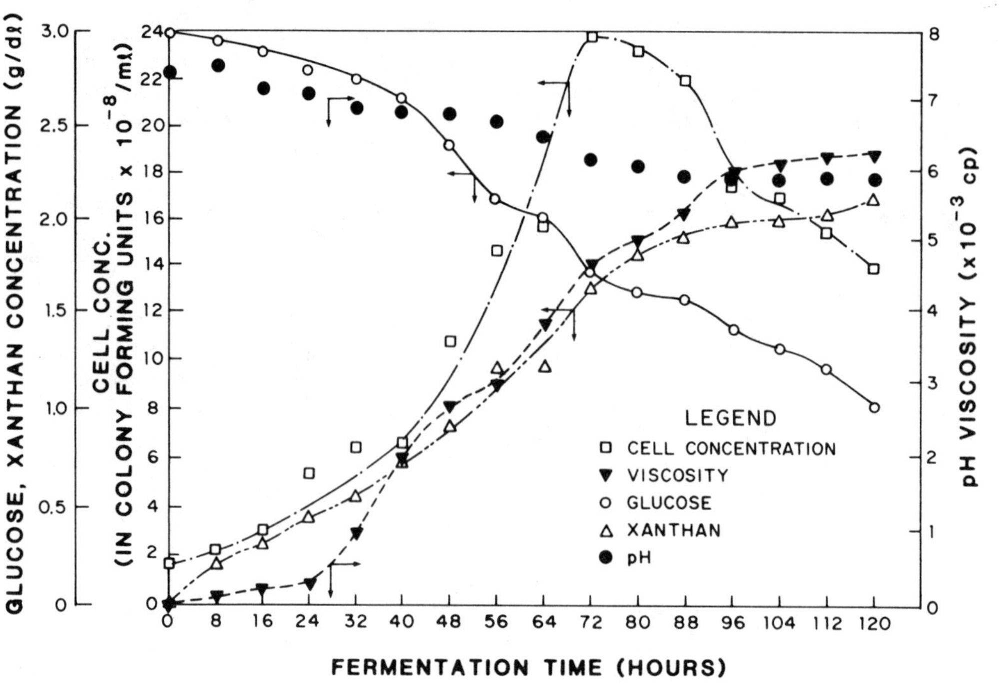

FIGURE 3. Changes in parameters during fermentation of *X. campestris* in the foam fermenter.

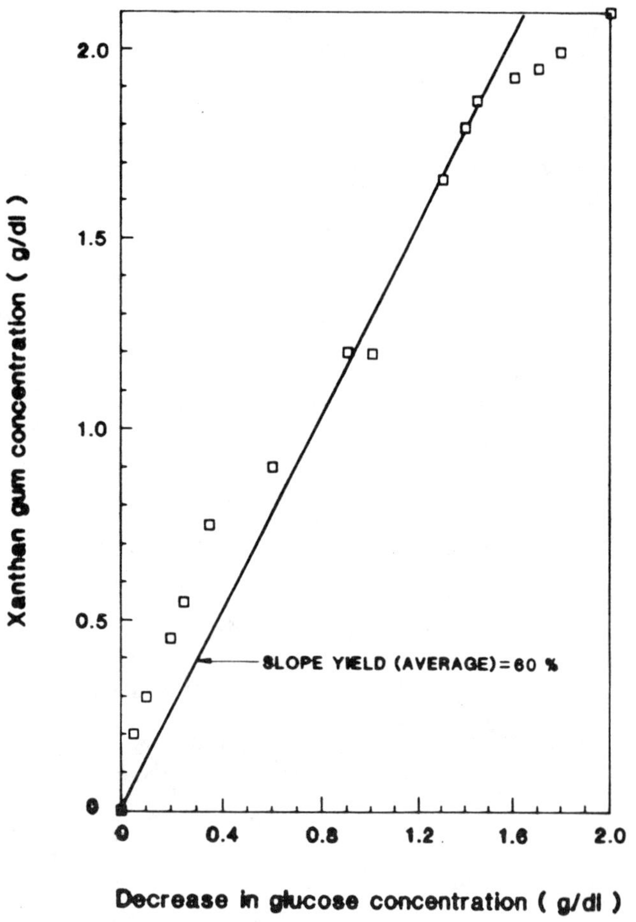

FIGURE 4. The yield of xanthan gum based on the total glucose used in the foam fermenter.

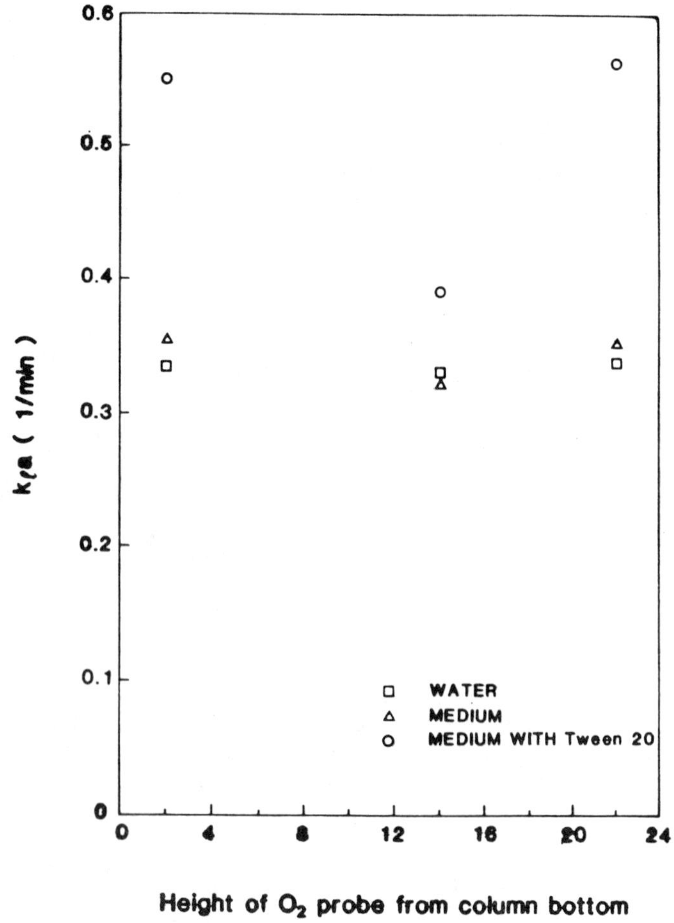

FIGURE 5. Change in k_la along the height of the right-hand column at 100 ml/min aeration rate.

Glucose was converted to polysaccharide at a constant yield of 60% over all phases of the fermentation. And based on the final xanthan content in the solution, the overall xanthan yield based on the original glucose present in the media, was 70%. Values of xanthan gum yield reported in the literature have varied from 60% to 75%.

One of the factors affecting both growth rate and yield is that certain stagnant areas of the fermenter may have poor yield, growth rate and mixing while other areas are well mixed and have correspondingly enhanced growth rate and yield. Thonart et al[20] have shown conversion yields of carbon to product are very high if the fermentation was carried out in two stages. In the first stage, fermentation was carried out at pH 5 and 29°C to promote biomass increase while polymer production was suppressed. At the beginning of the stationary phase, the pH was adjusted to seven, the temperature raised to 33°C over 48 hours. This resulted in yields of 70% to 80% of polymer over glucose while allowing for biomass recycle to the polymer.

Physical oxygen absorption in the foam fermenter

The dissolved oxygen uptake data showed that the dissolved O_2 uptake for water fluctuates during the aeration. Observation through the glass column revealed that the air distribution in the water-air system was not uniform; some large bubbles were mixed with a cluster of fine bubbles and the d.o. probe was frequently hit by large bubbles. The probe measurements, therefore, fluctuated. Votruba et al[21] studied the effect of air bubbles on k_la determination by the dynamic method. They stated that when the large bubbles hit the probe membrane frequently, the probe measures not the bulk concentration but a concentration between the equilibrium and the bulk value. The study of flow patterns in the foam fermenter revealed that the flow is in turbulent agitation throughout the whole reactor column except at the very top and bottom of the column. The data in Fig. 5 and 6 present the values of k_la at different points in the right hand column for air flow rates of $1.67x10^{-6}$ m^3/s and $3x10^{-6}$ m^3/s, respectively. The k_la values are determined in distilled water, pure medium and medium with Tween 20. In Fig. 5, the k_la values are uniform throughout the riser column for distilled water and medium. With the medium-surfactant system, the change in k_la is more pronounced through the length of the riser columns. In Fig. 6, the average k_la values are higher for the medium and the medium with surfactant as compared to those in the distilled water system. At higher aeration rates ($3x10^{-6}$ m^3/s and $4.1x10^{-6}$ m^3/s), there is no overlap of oxygen transfer coefficients and the average value of k_la is higher. This indicates a higher transfer efficiency for the medium surfactant systems.

Shown in Figs. 7, 8 and 9 is the effect of air flow rate on k_la in distilled water, medium and medium with surfactant systems, respectively. The k_la was measured at three different heights to determine the effect of flow patterns and mixing intensities. From observing the three figures, it is seen that the k_la values follow the same trend. The average k_la values for medium-surfactant system is higher than that of the other system. The values of k_la in the upper half of the column are high, which is due to "turbulent" mixing zone at top of the riser which promotes agitation and efficient oxygen transfer.

As shown in Fig. 9, the average values of k_la for the medium-surfactant system are higher than for the air-medium (Fig. 8) or for the air-water system (Fig. 7). This is due to the presence of the surfactant which creates a stable microgas dispersion and improves contact of microorganisms with the air bubbles thus leading to a higher oxygen transfer situation.

It is known that ionic strength can increase interfacial area and reduce the rate of bubble coalescence in aqueous electrolyte solutions. In the k_la measurement for the medium, bubbles in the riser were relatively uniform and small in size. With reduction of hitting frequency on the probe membrane, the oxygen uptake was smoother than the uptake for the case of distilled water. Furthermore owing to the increase of interfacial area in the medium, the mean values of k_la for the medium is greater than that for the water as shown in Figs. 5 and 6. Surfactants are the key factor for production of good CGA dispersions. The surfactant decreases

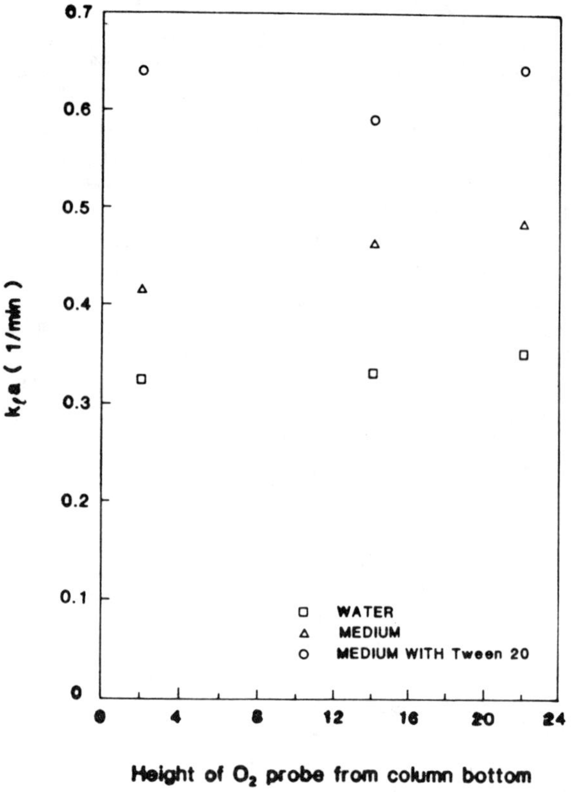

FIGURE 6. Change in $k_l a$ along the height of the right-hand column at 180 ml/min aeration rate.

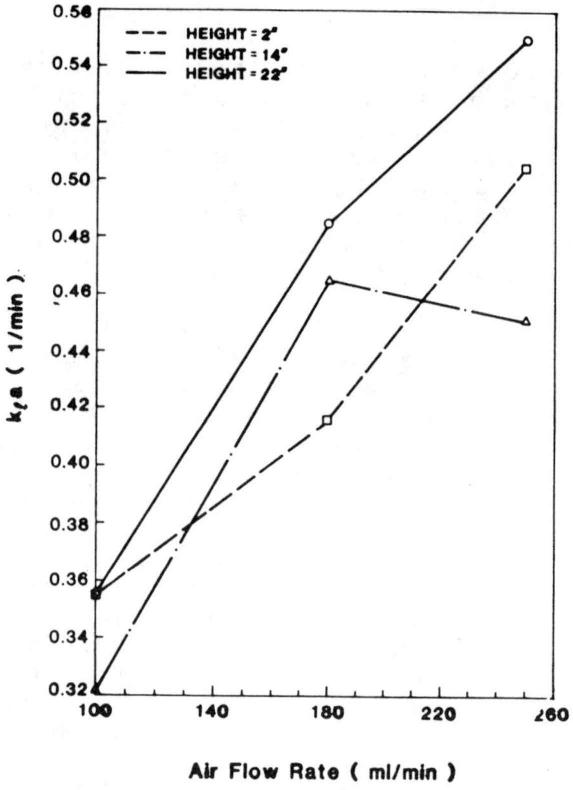

FIGURE 7. Effect of different air flow rates on $k_l a$ in distilled water. HEIGHT refers to the height of the oxygen probe from the bottom of the right-hand column.

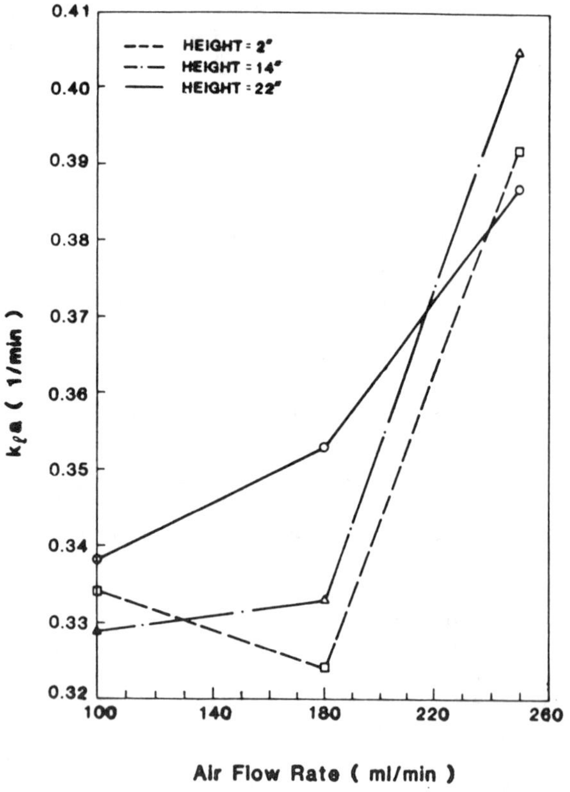

FIGURE 8. Effect of different air-flow rates on $k_l a$ in medium. HEIGHT refers to the height of the oxygen probe from the bottom of the right-hand column.

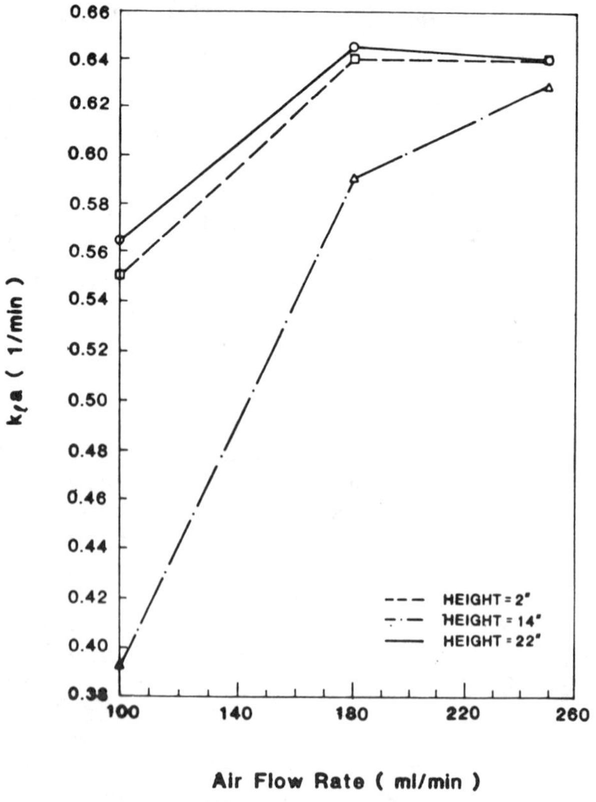

FIGURE 9. Effect of different air-flow rates on $k_l a$ in medium with 300 ppm of surfactant. HEIGHT refers to the height of the oxygen probe from the bottom of the right-hand column.

surface tension in the solution and consequently increases the interfacial area of the foam dispersion. In dealing with an air-medium-surfactant system, pseudo homogeneous flow prevails where the small bubbles have uniform sizes and rise velocity, the dissolved oxygen uptake is smooth and fast. The average value of k_la is greater than those obtained for medium-air, and water-air systems as shown in Figs. 7, 8 and 9. This enhancement of k_la due to the presence of a surfactant was also observed by Carver[21] and seems to imply that variation of k_la with the use of surfactant is caused by the substantial increase in surface area.

Suitable air flow rates in the generator and air pressure are also important to achieve maximum entrainment of gas in the liquid. Very high gas flow rates are not recommended for the foam fermenter, since a homogeneous flow can be changed to heterogenous flow or even slug flow, during which undesirable large bubbles are generated and grow further by coalescence during their rise through the column. This results in a decrease in k_la. On the other hand, if the liquid contains sufficient amount of surfactant to stabilize the gas bubbles, gas flow rates can have a beneficial effect on the quality of CGA dispersions and the value of k_la. The effect of air flow rates on for air-medium Tween 20 system can be seen in Figs. 5-9. At high air flow rates, the turbulent liquid is fully homogeneous and the system reaches steady state rapidly. At low air flow rates, it takes a longer time to reach good foam quality, therefore, k_la is smaller than the k_la at higher gas flow rates.

The values of k_la in the foam fermenter (at a height of 0.56m from the bottom of the column) over the whole course of the fermentation is shown in Fig. 10. The first four points on the plot are kla values that correspond to an aeration rate of 1.67×10^{-6} m^3/s, the next four points are of k_la values evaluated at 3×10^{-6} m^3/s and the final three points correspond to 4.1×10^{-6} m^3/s. For the first 40 hours, the broth viscosity was below 0.1 Pa-S and 1.67×10^{-6} m^3/s air flow was adequate for maintaining proper agitation and foam formation. The increase in k_la during this period was possibly due to the stabilization of the foam. Lee and Tsa[22] have observed that k_la in the broth is slightly enhanced by the increase of respiration rate in the culture media. This extra enhancement may be due to adsorption of organism-produced molecules onto the gas-liquid interface. By the end of 60 hours, the viscosity had risen to about 0.35 Pa-S. Hence, to maintain adequate agitation and bubble formation, the air flow rate was increased to 3×10^{-6} m^3/s. During this period, the k_la decreased slightly due to the steady increase in viscosity with the production of the polymer. At the end of 96 hours, when the viscosity reached 0.6 Pa S, the aeration rate was turned up to 4.1×10^{-6} m^3/s to prevent formation of dead areas or any incomplete mixing zones. Despite the increase in the air flow rate, the k_la gradually decreased to 28.8 sec^{-1}. In the overall process, it was observed that the air flow rate had to be increased to maintain proper agitaiton and mixing of the broth. The increase in air flow rate negated, to a great extent, the large increase in viscosity while maintaining adequate oxygen transfer conditions for the microorganisms to flourish.

Since the k_la values obtained during a culture fluctuates, it is not easy to compare them with those obtained for organism-free media. In comparing Figs. 5-9 with Fig. 10, one might conclude that, for a given flow rate, the mean values of k_la during a culture is slightly smaller than the k_la values for the organism-free medium. This small decrease in k_la during fermentation can be explained by the influence of the microorganisms on the hydrodynamics near the gas-liquid interface. Variations occurring at different growth rates can be attributed to physico-chemical changes of the fermentation broth, such as enzyme and respiration systems and excreted products from the cell.

Table 1 summarizes the results of the foam fermenter. It compares different operative parameters of the foam fermenter with fermentation currently in use in the industry. Though the k_la values obtained for the foam fermenter were lower compared to the other fermenters, the savings in aeration was lowered by 100 times when compared to the tower fermenter and by 50 times for the stirred tank fermenter. Power input was decreased by 10 times when compared to the STF and by 4 times in comparison with the tower fermenter.

TABLE 1
Comparison of Different Operation Parameters of the Foam Fermenter with That of the STF and the TF

Fermenter	k_1a (sec^{-1})	Air Flow (VVM)	Ratio of Power inputs
Foam Fermenter (S. cervisae)	39.0	0.02	0.10
(X. campestris)	36.0	0.02	0.18
Stirred Tank fermenter	160	1.0	1.0 at 500 rpm
Tower fermenter	109.8	2.0-3.0	0.45

CONCLUSION:

The foam fermenter is very effective in the production of xanthan gum. Product yields obtained were high while power input and aeration rates were lower by several orders of magnitude compared to the conventional fermenters. Even though the broth viscosity reached values of up to 0.7 Pa S, there were no substantial dead areas in the fermenter. The constant recirculation and agitation of the foam/media mixture maintained a well mixed environment. It was observed that k_1a values obtained indicate efficient oxygen transfer to the microorganisms. The k_1a values showed a direct dependency on the magnitude of the air flow rate and the surfactant concentrations in the media. Due to its ease of operation, and low power and air consumption, the foam fermenter should be considered for applied production of biomass, ethanol, methanol and pharmaceuticals.

NOMENCLATURE:

- C = dissolved oxygen concentration (Kg/m^3)
- C_s = saturation dissolved oxygen concentration (Kg/m^3)
- $\underline{c}$ = dissolved oxygen concentration in the apparatus for the respiration rate measurement (Kg/m^3)
- k_1a = volumetric mass transfer coefficient, sec^{-1}
- r = respiration rate, sec^{-1}
- x = cell concentration, Kg/m^3 ($CFU/x10^{-6}/m^3$)
- $\underline{x}$ = cell concentration, in the apparatus for the respiration rate measurement, Kg/m^3 ($CFUx10^{-6}/m^3$)
- t = time, hr (sec. or min.)

REFERENCES:

1. H. Lin, B.S. Fang, C.S. Wu, T.Y. Fang, T.F. Kuo, and C.Y. Hu., Biotechnology and Bioengineering, Vol. 18, 1557 (1976).
2. D.S. Inloes, W.J. Smith, Dean P. Taylor, S.H. Cohen, A S. Michaels, C.R. Robertson, Biotech: Bioengg., Vol. 25, 2653 (1983).
3. H. Ziegler, D. Meister, J.J. Dunn, H.W. Blanch and T.W.F. Russel, Biotech. Bioengineeering, Vol. 19, 507 (1977).
4. M. Moresi, G.B. Gianturco, E. Sabastini, Biotech. Bioeng., Vol. 25, 2889 (1985).
5. R.T. Hatch, Single Cell Protein II, (S.R. Tannenbaum, D.I.C. Wang, 454, MIT Press, Cambridge, Mass. (1985).
6. M. Kanazawa in Single Cell Protein II, (S.R. Tannenbaum, D.I.C. Wang), 438, MIT Press, Cambridge, Mass. (1975).
7. K.H. Hsu, L.E. Erickson and L.T. Fan, Biotech. Bioeng., Vol. 19, 247 (1977).
8. K.A. Bradley, S.M. Barnett, Paper presented at the 182nd meeting of the ACS, New York, August (1987).
9. A. Jeanes, J. Polymer Sci., Symposium, No. 45, 209 (1974).
10. P.E. Janson, L. Reeve, B. Lindberg, Carbohydrate Research, Vol. 45, 275 (1975).
11. W.Y. Yen, Ph.D. Thesis, University of Rhode Island (1984).
12. D.I. C. Wang, R.C.J. Fewbes, Developments in Ind. Microbiology, Vol. 18, 39 (1977).
13. S.P. Rogovin, R.F. Anderson, M.C. Cadmus, Biotechnol. Bioeng., Vol. 3, 51 (1961).
14. F. Sebba, J. Colloid Interface Sci., Vol. 35, 643 (1971).

15. W.H. Bartholomew, E.O. Krow, M.R. Sfat, R.H. Wilholm, Ind. Engg. Chem. Vol. 42, 1801 (1950).

16. H. Takahashi, F. Yoshida, J. Ferment. Tech., 'ol. 57, No. 4, 349 (1979).

17. R.A. Moraine and P. Rogovin, Biotechnol. Bioeng., Vol. 8, 511 (1966).

18. J.H. Sloneker, D.G. Orentas, A. Jeanes, Can. J. Chem., Vol. 42, 1261 (1964).

19. J.F. Kennedy, P. Jones and S.A. Barker, Enzyme. Microb. Technol., Vol. 4, 39 (1980).

20. P. Thonart, M. Paquot, L. Hermans, H. Alaoni, P. d'Ippolito., Enzyme Microb. Technol., Vol. 7, 235 (1985).

21. C. Carver, Absorption of Oxygen in Bubble Aeration, in Biological Treatment of Sewage & Industrial Waste, J. McCabe & W.W. Eckenfelder, eds., Reinhold, N.Y. (1956).

22. Y.Y. Lee & G.T. Tsa, Chem. Engg. Sci., Vol. 27, 1601 (1972).

23. M. Nakanoh and F. Yoshida, Ind. Engs. Chem. Proc. Des. Dev., Vol. 19, 190 (1980).

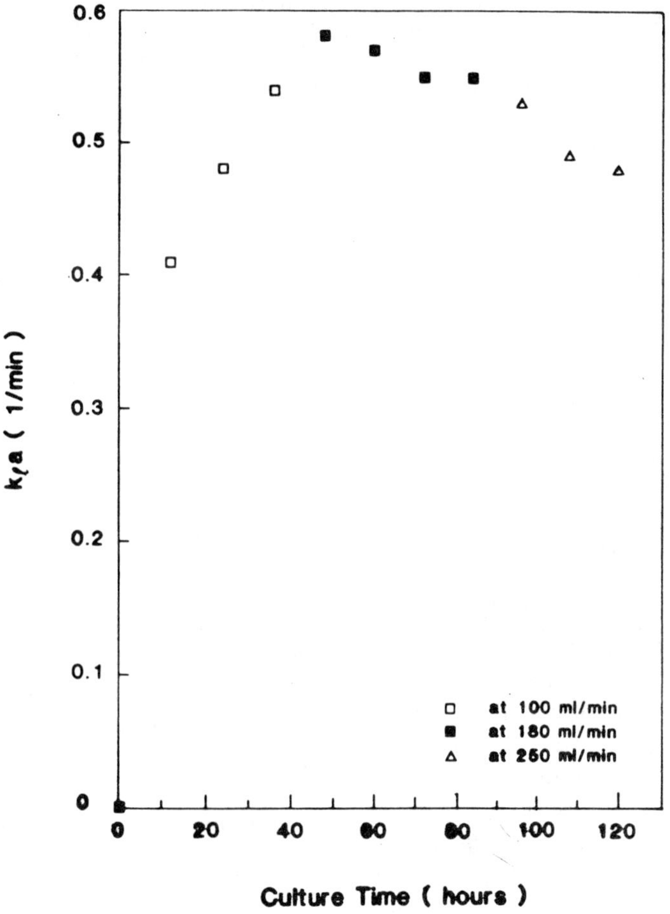

FIGURE 10. Values of $k_{l}a$ in the broth, during fermentation, measured with the oxygen probe located 22 in. from the bottom of the right-hand column.

Modeling the Dynamic Behavior of Immobilized Cell/Enzyme Bioreactors: The Tanks-in-Series Model

THANOS PAPATHANASIOU
NICOLAS KALOGERAKIS
LEO A. BEHIE
Department of Chemical Engineering
The University of Calgary
Calgary, Alberta, Canada T2N 1N4

G. MAURICE GAUCHER
Division of Biochemistry
The University of Calgary
Calgary, Alberta, Canada T2N 1N4

JULES THIBAULT
Department of Chemical Engineering
Laval University
Quebec City, Quebec, Canada G1K 7P4

An analysis of the dynamic response of an immobilized cell/enzyme bioreactor (ICEB) is presented using the tanks-in-series model. The presence of the cell or enzyme support matrix (i.e. beads) means that another term must be considered in the material balance equations to account for bead absorption. Dynamic response curves for a cascade of (N) immobilized cell/enzyme CSTRs in series are presented for typical bead sizes (0.5 to 3.5 mm), reactor dilution rates (0.01 to 100 h^{-1}), and intraparticle diffusivities (1.0×10^{-5} to 1.0×10^{-7} cm^2/s). Furthermore, the effect of key parameters such as bed voidage and partition coefficient on the bioreactor behavior is given. It was found that the presence of polysaccharide immobilization beads alters the response characteristics of the system substantially compared to the traditional series-of-CSTRs or dispersion models as applied to homogeneous reactors. The response curves presented are applicable to two-phase immobilized fixed bed or fluidized bed bioreactors in which the mass transfer from the bulk to the immobilized cells is dominated by intraparticle diffusion. Finally, experimental results are given for the mass diffusivities and partition coefficients of key, low molecular weight substrates (lactose, glucose, ammonium) and typical products (e.g. antibiotics such as patulin and penicillin-G).

INTRODUCTION

Use of continuous fermentations, with the biocatalyst immobilized inside solid, but highly porous, support matrices, is a relatively new application in bioengineering. Through the use of immobilized cells several problems inherent in free cell cultures, such as operation under low dilution rates, poor mass transfer and cell damage due to excessive shear caused by extensive agitation can be overcome. Cell damage is particularly important in the cultivation of fragile mammalian cells. From a chemical engineering point of view, it is often useful to know the residence time distribution characteristics of the bioreactor for modelling and design purposes.

In this work we apply the tanks-in-series model to simulate the transient behavior of a continuous fixed-bed or fluid bed ICEB at high mass Biot numbers. We present a novel method to approximate the series solution of the partial differential equation describing the intraparticle mass transfer, while the ordinary differential equation which gives the mass balance in a CSTR is solved numerically using GEAR's method.

The objectives of our work are:

(a) To obtain the dynamic response curves of the ICEB for a step change in the input concentration of tracer and, to indicate how mass transfer from the bulk liquid to the beads can alter the response characteristics of the reactor, as compared to the response of a simple, one phase, cascade of (N) CSTRs of equal volume.

(b) To show how the fluid-to-particle mass flux and the distribution of the tracer in the intraparticle space change with time and location in the reactor.

In addition, we studied the effect of parameters such as bed voidage (ε) and partition coefficient (β) on the performance of an ICEB. Our model can also be used to determine the axial dispersion in a fixed-bed ICEB at high Biot numbers when the intraparticle mass transfer resistance is dominant.

BACKGROUND

The transient modelling of an immobilized cell bioreactor is essentially the problem of the transient modelling of an isothermal tubular catalytic reactor, which has been the subject of several publications (Kubin, 1965; Kucera, 1965; Schneider and Smith, 1968; and Suzuki and Smith, 1971).

Nevertheless, since these efforts were mainly directed towards parameter estimation in chromatographic columns, no significant advances have been reported in solving the governing differential equations in the continuous time domain. In the above mentioned references the model equations have been solved in the Laplace domain and, utilizing the properties of the Laplace transform, expressions relating the moments of the chromatographic peak with the parameters of the column were obtained. Suzuki and Smith (1971) and Kucera (1968) comment on the complexity and even inadequacy of these procedures.

The absence of a transparent solution in the real time domain hinders a deeper understanding of the operation of a two-phase bioreactor. Given the increasing importance of immobilized bioreactors and having in mind that biocatalysts are very sensitive in nutrients, byproducts or oxygen levels within the immobilization matrix, we consider an easy-to-use solution in the real time domain of primary importance. This has been done recently, for a batch system (Leyva-Ramos et.al, 1985). For the continuous fixed-bed reactor, the only effort to present a solution in the real time domain has been done by Rasmussen et.al (1980,1981,1982). His solution is given in terms of an infinite integral of an oscillatory function, which, in some instances, might be difficult to obtain accurately due to convergence problems. Furthermore, no parametric studies were presented that could support the usefulness of that solution.

The mathematical problem includes the solution of two coupled partial differential equations (PDEs) subject to appropriate boundary conditions. The equation describing the dispersion in the bulk liquid is

$$\frac{\partial C}{\partial \theta} = (\frac{1}{Pe})(\frac{\partial^2 C}{\partial x^2}) - \frac{\partial C}{\partial x} - N_1 - N_2 \qquad (1)$$

and the equation describing the mass transfer inside the bead is

$$D_{eff}\left[\frac{\partial^2 C_b}{\partial r^2} + (\frac{2}{r})\frac{\partial C_b}{\partial r}\right] - N_3 = \varepsilon_p \frac{\partial C_b}{\partial t} \qquad (2)$$

In Eqn. 1, the term $(1/Pe)(\partial^2 C/\partial x^2)$ is the mass flux in the bulk liquid due to axial diffusion, $(\partial C/\partial x)$ is the flux due to convection, N_1 is the mass flux from the bulk liquid to the cell support matrix and N_2 is the mass consumed due to chemical reaction occurring in the bulk liquid. This last term, N_2, will be zero in an ICEB, since the cells are confined in the intraparticle space. In Eqn. 2, N_3 is the mass consumed due to intraparticle chemical reaction.

As an alternative to Eqn. 1, a cascade of a number (N) of continuous stirred tank reactors (CSTRs) of equal volume can be used to simulate the axial dispersion in a tubular reactor. Adopting this approach, Eqn. 1 can be replaced by a set of ordinary differential equations (ODEs) of the form

$$\frac{dC_i}{dt} = (\frac{F}{\varepsilon V})(C_{i-1} - C_i) - N_1 - N_2 \qquad (3)$$

where C_i and C_{i-1} are the effluent concentrations of the ith and $(i-1)^{th}$ CSTR in the cascade respectively.

The computational advantage of using the tanks-in-series model is obvious, since the PDE of the dispersion model is replaced by a set of ODEs. On the other hand, the accuracy of the method in determining the amount of axial dispersion occurring in the reactor is limited because (N) is of necessity an integer. Buffham and Gibilaro (1968) have shown how non-integer values of (N) can be used to fit response data with the tanks-in-series model, for homogeneous transport lines. Unfortunately, their analysis is not directly applicable to a two-phase immobilized cell bioreactor. Nevertheless, the application of the tanks-in-series model is desirable because of the simplicity of the solution procedure which is free of the instabilities, convergence difficulties, and the numerical dispersion problems associated with the numerical solution of the diffusion-convection type equations, especially at high Peclet numbers (Brenner, 1962; Gray and Pinder, 1976).

In this work we present the dynamic response curves of a tubular ICEB for a step change in the input concentration of tracer, including intraparticle diffusion resistance. Our analysis is valid for the case of large Biot number, when fluid-to-particle mass transfer resistance can be considered negligible. In this paper we adopt the tank-in-series approach. Our solution is based on the decomposition of the second order PDE giving the intraparticle mass balance, into a set of first order ODEs. These, along with the ODE for the mass balance in the bulk liquid, are solved numerically using GEAR's method. The obtained solution is simple and computationally efficient, as only the partial derivative $(\partial C_b/\partial r)_{r=R}$ need to be evaluated in each time step. Furthermore, our method is free from problems of inaccuracy and stability associated with the numerical solution of second order PDEs. The numerical results are compared with the analytical solutions obtained for the two limiting cases, that is for infinitely fast and infinitely slow intraparticle mass transfer and the agreement is excellent.

In addition to the response curves for the cascade of CSTRs, we obtained intraparticle concentration profiles for various positions in the reactor and examined the behavior of the mass flux to the beads as function of time. Also, we have shown the importance of the partition coefficient when intraparticle diffusivities of substances are determined using batch absorption data.

MODEL DEVELOPMENT

In general, the mass transfer phenomena that need to be considered in modelling the dynamic response of an ICEB are:

(i) Tracer accumulation in the bulk liquid,
(ii) Mass diffusion of tracer in the stagnant film surrounding the bead (external mass transfer),
(iii) Mass diffusion of tracer within the void space of the bead (pore-volume diffusion),
(iv) Adsorption of tracer on the pore walls according to a linear or non-linear isotherm,
(v) Diffusion of the tracer on the immobilization matrix surface accompanied by chemical reaction.

Here, we assume that the external mass transfer resistance is negligible, and that no adsorption, surface diffusion or chemical reaction takes place. In addition it is assumed that :

(i) The system is isothermal
(ii) Intraparticle diffusivity is independent of time and also constant throughout the bead
(iii) Beads are spherical particles with uniform and constant radius.

At first we consider the transient model for a single CSTR. This is easily extended to give the response of a series of CSTRs simulating the axial dispersion in a packed or fluidized bed immobilized cell/enzyme bioreactor (ICEB). The dynamic mass balance in the bulk liquid is given by

$$\varepsilon\left(\frac{dC_L}{dt}\right) = \left(\frac{F}{V}\right)(C_{in} - C_L) - N_0 \tag{4}$$

In Eqn. 4, C_{in} and C_L are the tracer concentrations in the inlet and outlet streams respectively. The second term on the right-hand side of Eqn. 4 represents the mass flux from the bulk liquid to the beads and is given by

$$N_0 = 3(1-\varepsilon)\left(\frac{D_{eff}}{R}\right)\left(\frac{\partial C_b}{\partial r}\right)_{r=R} \tag{5}$$

The dynamic mass balance for the bead, using a constant effective intraparticle diffusivity D_{eff} gives

$$D_{eff}\left[\frac{\partial^2 C_b}{\partial r^2} + \left(\frac{2}{r}\right)\frac{\partial C_b}{\partial r}\right] = \varepsilon_p \frac{\partial C_b}{\partial t} \tag{6}$$

The initial conditions for Eqns. 4 and 6 are

$$C_L(0) = 0.0 \tag{7}$$

$$C_b(r,0) = 0.0 \tag{8}$$

Those initial conditions can be used even for non-zero (but uniform) initial concentrations C_L and C_b by defining the deviation variables

$$(C_L)^* = C_L(t) - C_L(0)$$

$$(C_b)^* = C_b(r,t) - C_b(r,0)$$

The boundary conditions for Eqn. 6 are

$$C_b(R,t) = C_L(t)/\beta \tag{9}$$

$$C_b(0,t) = finite \tag{10}$$

In Eqn. 9, the partition coefficient, β, is defined as

$$\beta = (C_L)_{eq}/(C_b)_{eq}$$

This coefficient expresses the experimental observation that at equilibrium the concentration of tracer inside the beads can be only a fraction of its concentration in the bulk liquid. Use of Eqn. 9 implies that fluid-to-particle mass transfer resistance is considered negligible.

Eqns. 4 and 6, are coupled through the flux term N_0 given by Eqn. 5. In our approach we apply Duhamel's theorem to obtain the solution of Eqn. 6 for the time-dependent boundary condition (Eqn. 9), using the solution of the same equation for the constant boundary condition (Crank, 1983). Details on the solution procedure can be found in the Appendix. The final set of ODEs simulating the immobilized cell CSTR is as follows

$$\frac{dC_L}{dt_R} = \frac{C_{in} - C_L - 6\varepsilon_p(1-\varepsilon)\, D_R \sum_1^{n_0}\psi_n}{1 + \varepsilon_p(\frac{1-\varepsilon}{\beta\,\varepsilon})\,(1 - \frac{S_n}{S_l})} \tag{11}$$

The functions ψ_n appearing in Eqn. 11 are solutions of the following ODEs

$$\frac{d\psi_n}{dt_R} = -\,\varepsilon\,(n\,\pi)^2 D_R\,\psi_n + (\frac{1}{\beta})\,(\frac{dC_L}{dt_R}) \quad ,$$

$$n = 1,2,3,\ldots,n_0. \tag{12}$$

The behavior of those functions $\psi_n(t)$ is shown in Figure 1. It can be seen that, for a given D_R, the values of the ψ_n approach zero, when n and/or t increases. For $n > n_0$, those functions can be approximated by an expression of the form (quasi steady-state assumption):

$$\psi_n(t) = \gamma_n\,(\frac{dC_L}{dt_R}) \tag{13}$$

The approximation given in Eqn. 13 is the basis for an accurate and efficient approximation of the series solution of Eqn. 6. As explained in the Appendix, the first n_0 terms of this series are evaluated by Eqn. 12, while the rest are approximated using Eqn. 13. As can be concluded by Figure 1, the errors introduced by the above approximation become truly insignificant for large n. A proper value of n_0 is determined indirectly, the choice based on a user specified tolerance. The dimensionless quantities t_R and D_R are defined as

$$t_R = \frac{t}{\varepsilon}\,(\frac{F}{V}) \quad \text{and} \quad D_R = \frac{D_{eff}}{\varepsilon_p R^2\,(F/V)}$$

also

$$S_n = \sum_1^{n_0}(\frac{1}{n})^2 \quad \text{and} \quad S_l = \sum_1^{\infty}(\frac{1}{n})^2$$

Limiting Cases

The following two limiting cases were examined:

Case (I) : No mass flux to the beads.

In this case the flux term N_0 in Eqn. 3 becomes zero, and the equation for one CSTR with active volume $V_a = \varepsilon\,V$ becomes

$$\varepsilon\,V\,(\frac{dC_L}{dt}) = F\,(C_{in} - C_L), \tag{14}$$

or, using the dimensionless time t_R,

$$\frac{dC_L}{dt_R} = C_{in} - C_L \tag{15}$$

The solution of Eqn. 15 is

$$C_L(t_R) = C_{in}\,(1 - e^{-t_R}) \tag{16}$$

Case (II) : Infinitely fast diffusion into the beads.

In this case no concentration gradients develop inside the bead and the equation giving the response of the CSTR is:

$$\frac{dC_L}{dt_R} = (\frac{1}{\gamma})\,(C_{in} - C_L) \tag{17}$$

where $\gamma = 1 + \varepsilon_p(1-\varepsilon)/\varepsilon\beta$

Solution of Eqn. 17 yields

$$C_L(t_R) = C_{in}\,(1 - e^{(-\frac{t_R}{\gamma})}) \tag{18}$$

The accuracy of our numerical method was tested by matching these two limiting cases. Numerical solutions obtained for values of $D_R < 1.0\times10^{-5}$ and $D_R > 100$, matched the analytical solutions corresponding to the two limiting cases with maximum errors less than 0.8 %.

EXPERIMENTAL DETERMINATION OF D

A standard method to measure intraparticle diffusivities of substances in porous particles is described by Tanaka (1984). In this method, the porous particles, initially free of diffusing substance (tracer), are placed in a well stirred solution of finite volume, containing tracer with initial concentration $(C_L)^0$. The system has to be well stirred, so that liquid-to-particle mass transfer resistances are eliminated. Tracer diffuses from the solution to the interior of the particles and the concentration in the liquid phase drops. This drop is recorded and an experimental curve of concentration of tracer in the liquid phase against time similar to that of Figure 2 is obtained.

The problem of absorption of tracer from a solution of finite volume with initial concentration $(C_L)^0$ into spherical porous particles has been solved (Crank, 1983). The analytical solution is as follows

$$C_L(t) = \left(\frac{\alpha\,(C_L)^0}{1+\alpha}\right)\left[1 + \sum_{1}^{\infty}\frac{6\,(1+\alpha)\,e^{-tk_m}}{9+9\,\alpha+(q_n)^2\,\alpha^2}\right] \quad (19)$$

In Eqn. 19, (α) is defined as the ratio of the bulk liquid volume over the volume occupied by the beads:

$$\alpha = \left(\frac{3\,V}{4\,\pi\,R^3\,M}\right)\beta \quad (20)$$

where V is the volume of the bulk liquid, M is the number of beads in the system, R is the bead radius (β) is the partition coefficient, q_n are the non-zero positive roots of the equation

$$\tan(q_n) = \frac{3\,q_n}{3+\alpha\,(q_n)^2} \quad (21)$$

and k_m is defined as

$$k_m = D\,(q_n/R)^2 \quad (22)$$

With the aid of Eqn. 19 one can determine the value of D that best fits, in the least square sense, the experimental curve of liquid concentration versus time.

The introduction of the partition coefficient β, was necessary in order to describe the diffusional behavior of relatively large and/or charged molecular compounds. Tanaka (1984) reported that Eqn. 19, used without partition coefficient, failed to match the batch absorption data for high molecular weight compounds, such as albumin ($MW = 6.9\times10^4$). We experienced the same result when we tried to match batch absorption data for $(NH_4)^+$. Introduction of another adjustable parameter (β) in Eqn. 19 made it possible to match our experimental data for $(NH_4)^+$, as can be seen in Figure 2.

A number of batch absorption experiments with carrageenan beads as porous particles and glucose, lactose, $(NH_4)^+$, patulin and the precursor 6-methyl salicylic acid (6-MSA) as diffusing substances has been conducted. In order to assess the barrier imposed to the diffusion of the substrates by the immobilized cells, three sets of experiments were conducted for each substrate: Diffusion into beads free of cells, diffusion in beads populated by live cells and diffusion in beads populated by dead cells. In the above experiments the intraparticle cell density was 5 g/L, and the mean bead radius 0.16 cm. Patulin and

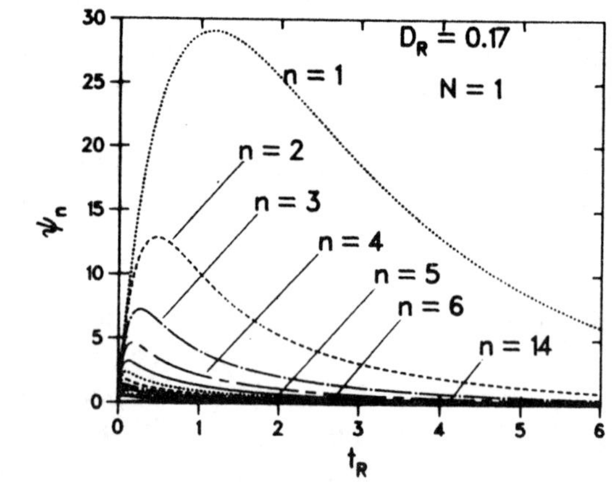

FIGURE 1. Behavior of the functions $\psi_n(t)$ showing the contribution of each term.

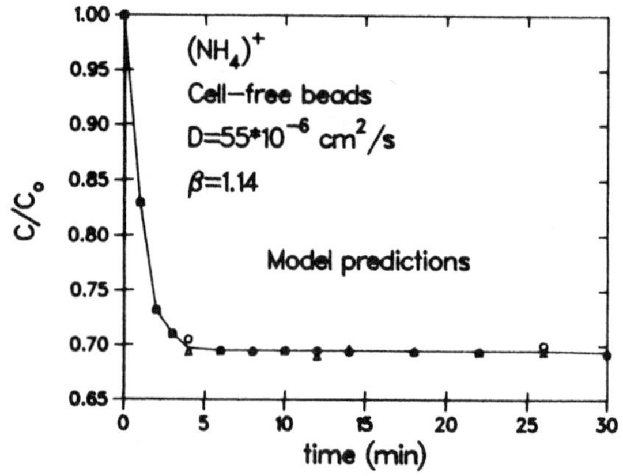

FIGURE 2. Experimental determination of intraparticle diffusivity and partition coefficient for $(NH_4)^+$ in carrageenan beads.

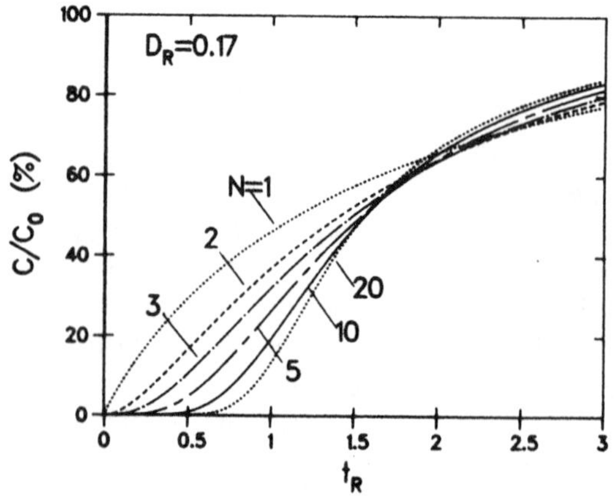

FIGURE 3. Dynamic response curves for ICEB modelled by N CSTRs in series.

6-MSA were used in solutions of concentration 200 mg/L. In the case of glucose and lactose, the effect of the bulk concentration on the determined diffusivity was assessed by carrying out experiments with various bulk concentrations, namely 4.3, 9.2 and 22.5 g/L. It was found that the measured diffusivities were unaffected by the concentration of substrate in the bulk. Dead cells were obtained after placing the beads containing the live cells into a 0.02% solution of toxic sodium azide. Further details on the preparation of cell-free beads and of beads populated by live and dead cells can be found in the literature (Tanaka et al., 1984; Nayar, 1984). The results are summarized in Table 1.

It can be seen that the intraparticle diffusivity of 6-MSA in carrageenan beads is much smaller compared to that of glucose, lactose or patulin. This was attributed to the fact that both 6-MSA and carrageenan immobilization matrix are negatively charged species; this electric charge offers an added resistance to diffusion. In general, the presence of living cells gave rise to higher values for the intraparticle diffusivity for both, glucose and lactose. This is not surprising, since the apparent diffusivity determined when live cells are present, includes a chemical reaction contribution.

The physical barrier imposed by the cells themselves can be assessed by comparing the values of mass diffusivities with and without dead cells. The reduction in D is relatively small for glucose and lactose, but significant in the case of ammonium chloride, patulin and 6-MSA. The effect of the partition coefficient can be observed in the case of $(NH_4)^+$. Figure 2 gives the experimental and predicted values of $C_L(t)$, in the case of cell-free beads, when $(NH_4)^+$ is the diffusing substance in a batch system. The experimental curve is best matched for $D = 55{\times}10^{-6}\ cm^2/s$ and $\beta = 1.14$. This value of β indicates an added resistance for the $(NH_4)^+$ ions that try to enter the bead. The negligible or small decrease in (D) of glucose and lactose in the case of beads containing dead cells was attributed to the small intraparticle cell density used, the relatively small molecular size of these two sugars and their uncharged nature.

SIMULATION RESULTS.

In the numerical experiments presented below, the following parameter values were used:

$1.0{\times}10^{-5} < D < 1.0{\times}10^{-7}$, cm^2/s

$0.01 < F/V < 100,$ h^{-1}

$0.05 < R < 0.35,$ cm

Values for the bed voidage ε and the partition coefficient β were taken as 0.5 and 1.0 respectively in all the runs except of the runs in Figures 7, 8 and 11, where the effect of these parameters was studied. The intraparticle porosity, ε_p, can take a range of values. For example, the porosity of *Celite* 560 was found to be 0.72 (Jones et al., 1986) whereas, the porosity of *carrageenan* beads approaches one. In our simulation studies we used 0.99 for ε_p. In presenting intraparticle concentration profiles, experimentally determined values for the diffusivities and partition coefficient of glucose and ammonium chloride in cell-free polysaccharide beads were used.

Response curves for a number of CSTRs in series: Figure 3 gives the response curves of various cascades of CSTRs in series for a step change in the input concentration of tracer, using the dimensionless time t_R. Those curves correspond to $D_R = 0.17$ and are obviously different from the "classical" response curves of a cascade of homogeneous CSTRs with total volume equal to the total volume of the ICEB, as given by Smith (1981).

The contribution of the mass flux to the beads in the response of the reactor is shown in Figures 4 and 5. In Figure 4 the concentration in the effluent stream C_L, the mass flux to the beads defined as $N_0^* = V/F\ N_0$ and the accumulation in the liquid phase dC_L/dt_R are plotted against time for the case of one CSTR (N=1).

A mass balance in the reactor at any time (t_R) gives

$$C_{in} = C_L(t_R) + N_0^*\ (t_R) + \frac{dC_L}{dt_R} \qquad (23)$$

Eqn. 23 expresses the fact that the mass of tracer entering the reactor at any given time (t_R) has either to exit with the effluent stream, to enter the beads, or to accumulate in the liquid phase.

The amount of tracer accumulating in the liquid phase is not seriously affected by the mass flux to the beads but rather by the bed void fraction. Thus, when fluid-to-particle mass flux is not zero, a decrease in C_L is observed with the new value satisfying Eqn. 23. The higher this flux term is, the lower the corresponding value of C_L and consequently the slower the response of the reactor.

As expected, the mass flux to the solid particles increases as D_R increases. This is shown in Figure 5

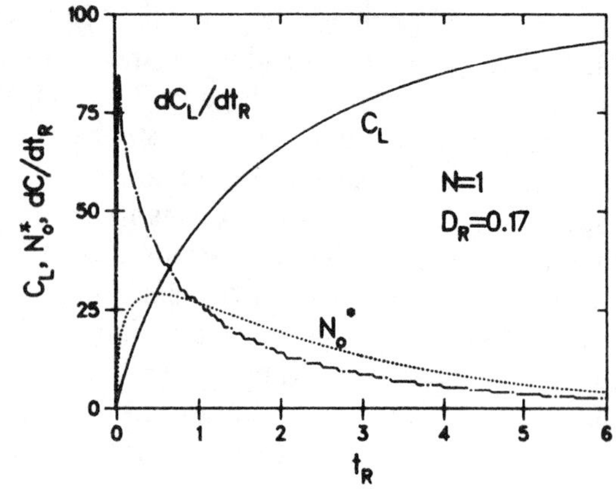

FIGURE 4. Contribution of the various terms in the mass balance equation for a single CSTR.

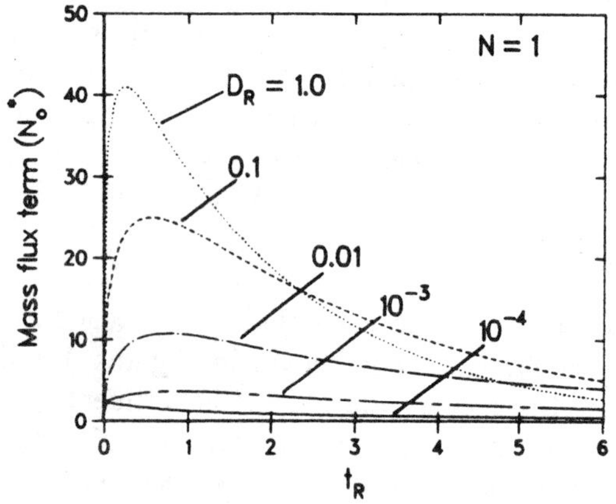

FIGURE 5. Variation of the mass flux to the beads with dimensionless diffusivity D_R.

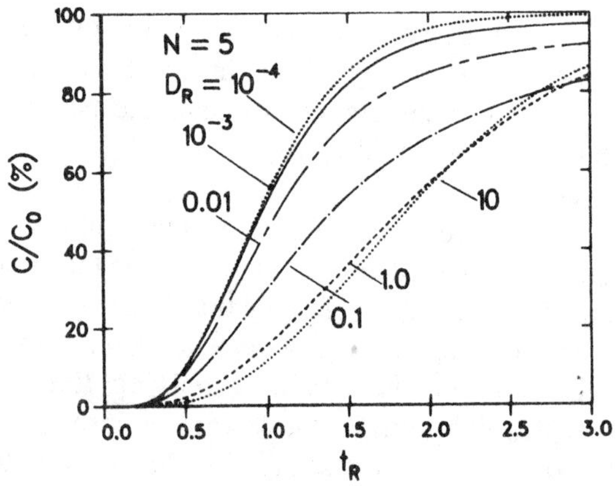

FIGURE 6. Effect of D_R on the response cascade of five CSTRs in series simulating a fixed-bed ICEB.

where N_0^* is plotted against time for various values of the parameter D_R. It can be seen that for large D_R (e.g $D_R = 1$), the mass flux to the beads is large in the beginning and approaches zero relatively fast, while for intermediate values of D_R (e.g $D_R = 0.01$), fluid-to-particle mass flux is more uniform throughout the transient period. Finally, for very small values of D_R (e.g $D_R = 0.0001$), mass transfer to the particles is negligible and the response of the ICEB approaches the response of a homogeneous reactor with active volume $V_a = \varepsilon V$.

Effect of D_R on the response of a cascade of CSTR's:

As outlined above, higher values for D_R result in higher values for the mass flux to the beads and consequently in lower values for the effluent concentration C_L. Figure 6 gives the dynamic response of a cascade of 5 CSTRs for various values of D_R. It was observed that for $D_R > 10$ the response is very much the same as that for zero intraparticle mass transfer resistance. On the other hand, for $D_R < 1.0\times10^{-4}$, response curves tend to coincide with the curve of the limiting case (I). Limiting case (I) is the only case that the response of the immobilized cells bioreactor is the same with the response of an homogeneous tubular reactor of volume εV. For any $D_R > 1.0\times10^{-4}$, fluid-to-particle mass transfer becomes important, the importance increasing with D_R, and the dynamic behavior of the bioreactor deviates from that of the homogeneous system.

Effect of bed voidage (ε) and partition coefficient (β):

Bed voidage and partition coefficient both affect the response of a series of CSTRs very significantly. For a given dimensionless diffusivity (D_R), high (ε) indicates few solid particles in the bed and consequently a small amount of tracer finally transferred to the beads. This is expected to give a fast response that will approach that of a homogeneous reactor as ε approaches 1. This is shown in Figure 7 where the response of a cascade of 5 CSTRs is plotted as function of dimensionless time, with ε taking values from 0.4 to 0.8.

The partition coefficient (β) affects the dynamic response of the reactor in a similar way as presented in Figure 8. As a result of its definition, values of β greater than 1 indicate an added resistance for the tracer to enter the particle and result in smaller mass flux of tracer to the beads. Consequently, higher values of β are expected to yield faster response. For very large values of β this response is expected to coincide with limiting curve (I), indicating a very strong thermodynamic inhibition for the tracer to enter the intraparticle space.

Intraparticle concentration profiles:
The mass diffusivities of frequently used nutrients like glucose, lactose and ammonium chloride in polysaccharide (i.e. carrageenan) beads have been measured as described previously. These values are small and, in general, become smaller when the beads are populated by dead cells. This indicates that the concentration in the intraparticle space will be lower than that in the bulk, and the time required for the entire system to reach steady state will be greater than implied by the reactor effluent concentration.

Figures 9 and 10 give the tracer concentration profiles inside the beads in the first and last CSTR in a cascade of 5 CSTRs when glucose is the diffusing substance. For a dimensionless time $t_R = 2$, the normalized tracer concentration in the bead center is 49% in the first CSTR, whereas it is only 11% in the last CSTR. Figure 11 gives the intraparticle concentration profile in the case of ammonium chloride, a nutrient that partitions between liquid and bead (β=1.14). It can be seen in Figure 11 that intraparticle concentration follows relatively fast the bulk concentration as a result of the high diffusivity of $(NH_4)^+$ into the beads. The effect of the partition coefficient can be observed in that the concentration on the bead surface is always below the concentration in the bulk.

Comparing the simulation results for the penetration of glucose and $(NH_4)^+$ into polysaccharide beads, it can be seen how slow glucose diffuses in the intraparticle space, as compared to $(NH_4)^+$. For the last CSTR in the cascade (N=5) and for a dimensionless time $t_R = 3$, ammonium chloride has almost reached its steady state throughout the bead. For the same CSTR and time, glucose has only reached an average intraparticle concentration of about 60% of its final value, with the concentration near the center being less than 40% of the final value.

CONCLUSIONS AND SIGNIFICANCE

An accurate and efficient analysis of the dynamic response of an immobilized cell/enzyme bioreactor (ICEB) ,in the real time domain, has been presented. The presence of fluid-to-particle mass transfer alters

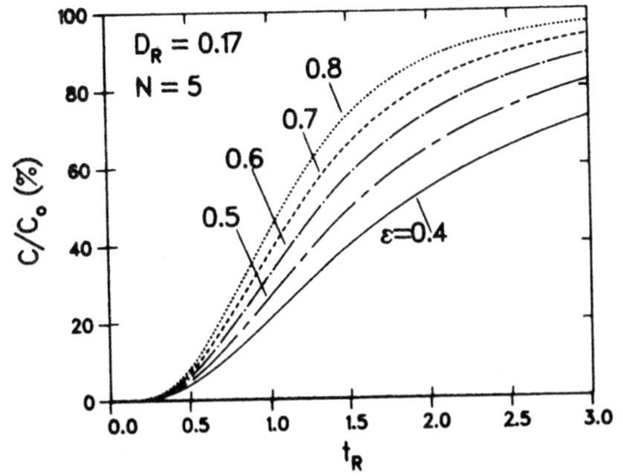

FIGURE 7. Effect of bed voidage (ε) on the dynamic response of an ICEB.

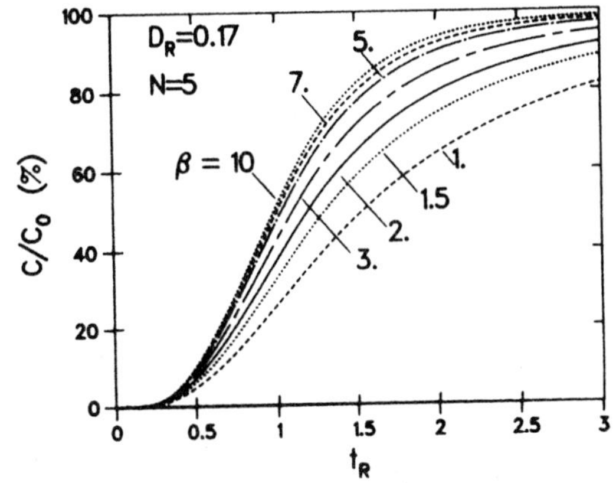

FIGURE 8. Effect of the partition coefficient (β) on the dynamic response of an ICEB.

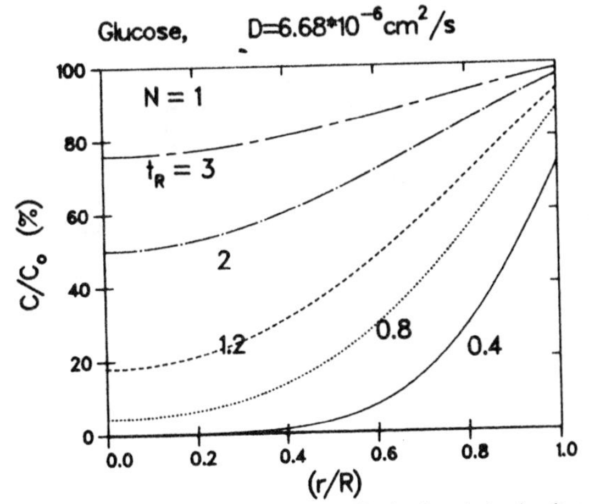

FIGURE 9. Intraparticle concentration profile for beads in the first CSTR in a cascade of five.

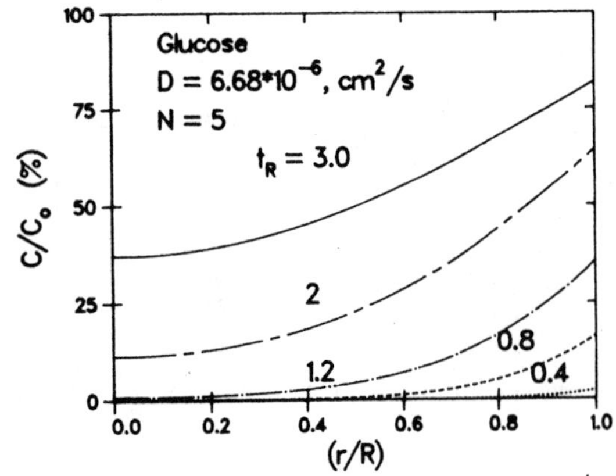

FIGURE 10. Intraparticle concentration profile inside the beads in the last CSTR in a cascade of five.

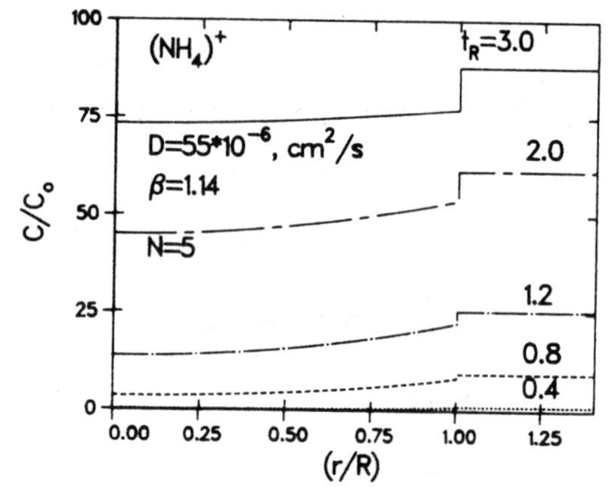

FIGURE 11. Concentration profile inside the beads in the last CSTR in a cascade of five, when $(NH_4)^+$ is the diffusing substance.

the response characteristics of the reactor significantly. Key parameters such as bed voidage and partition coefficient have an important effect on the response of the reactor. Analysis of transient intraparticle concentration profiles for glucose and ammonium chloride, revealed that the time required for the entire system to reach steady state can be significantly larger than indicated by monitoring the effluent stream concentration. Finally, the analysis of batch absorption experimental data for $(NH_4)^+$ diffusing in polysaccharide beads showed the importance of the partition coefficient (β).

NOTATION

C, C_L = Concentration in the bulk liquid

$(C_L)^0$ = Initial concentration

C^* = The solution of the intraparticle dynamic mass balance for constant boundary condition

$(C_L)_{eq}$, $(C_b)_{eq}$ = Concentrations in the bulk and intraparticle liquid at equilibrium

C_{in} = Concentration in the stream entering the reactor

C_b = Concentration in the pore liquid

D = Molecular diffusivity

D_{eff} = Effective intraparticle diffusivity

D_L = Axial diffusivity in the bulk

D_R = Dimensionless diffusivity, $D_R = \dfrac{D_{eff}}{\varepsilon_p R^2 (F/V)}$

F = Volumetric flowrate

k_e = External mass transfer coefficient

k_n = Parameter defined as $k_n = D_{eff}(n\pi/R)^2/\varepsilon_p$

k_m = Parameter defined as $k_m = D(q_n/R)^2$

L = Reactor length

M = Number of beads

6–MSA = 6-Methyl salicylic acid

N = Number of CSTRs

N_0 = Mass flux to the beads, $moles/cm^3s$

N_0^* = $N_0/(F/V)$, in units $moles/cm^3$

Pe = uL/D_L (Peclet number, dimensionless)

q_n = Roots of the equation: $tan(q_n) = \dfrac{3q_n}{3+\alpha(q_n)^2}$

R = Bead radius

r = Radial distance variable (intraparticle)

t = Time variable

t_R = Dimensionless time, $t_R = \dfrac{t}{\varepsilon}(\dfrac{F}{V})$

u = Axial fluid velocity in the reactor

V = Total reactor volume

V_a = Active volume, $V_a = \varepsilon V$

x = Dimensionless axial distance, $x = z/L$

z = Axial distance variable

Greek letters

α = Ratio of the volume of the bulk liquid over the volume occupied by the beads, $\alpha = 3V/(4\pi R^3 \beta M)$

β = Partition coefficient, defined as $\beta = (C_L)_{eq}/(C_b)_{eq}$

ε = Bed voidage

ε_p = particle void fraction

ψ_n = Functions defined as $\psi_n(t) = \int_0^t \frac{\partial\phi}{\partial t} e^{(\tau-t)k_n} d\tau$

$\phi(t)$ = Tracer concentration on the bead surface

CITED LITERATURE

1. Bailey,J.E., and D.F.Ollis, "Biochemical Engineering Fundamentals," 2nd edition, McGraw Hill, 568 (1986)
2. Brenner,H., *Chem.Eng. Sci.*, 17, 229 (1962)
3. Buffham,B.A., and L.G.Gibilaro, *AIChE J.*, 5, 805 (1968)
4. Crank,J., "The Mathematics of Diffusion," 3rd edition, Oxford University Press (1983).
5. Gray,W.G., and G.F.Pinder, *Water Resources Research*, 12,(3), 547 (1976).
6. Jones, A., D.N. Wood, T. Razniewska, G.M. Gaucher and L.A. Behie, *Can. J. Chem. Eng.*, 64, 547(1986)
7. Kubin,B., *Colln.Czech.Chem.Commun.*, 30, 2900(1965)
8. Kucera,E., *J. of Chromatography*, 19, 237(1965)
9. Leyva-Ramos,R., and C.J.Geankoplis, *Chem. Eng. Sci.*, 40, 799 (1985).
10. Nayar,A.K., "Continuous Antibiotic Production," MSc Thesis, U. of Calgary, (1984).
11. Papathanasiou,T., "The Dynamic Modelling of an Immobilized Cell/Enzyme Bioreactor," MSc Thesis, U. of Calgary (in preparation).
12. Rasmuson,A., *Chem. Eng. Sci.*, 37, 411 (1982).
13. Rasmuson,A., and I.Neretkiens, *AIChE J.*, 26,(4), 686 (1980).
14. Rasmuson,A., *AIChE J.*, 27,(4), 1032 (1981).
15. Schneider,P., and J.M.Smith, *AIChE J.*, 14, 762 (1968)
16. Suzuki,M., and J.M.Smith, *Chem. Eng. Sci.*, 26, 221 (1971)
17. Tanaka,H., M.Matsumyra, and I.A.Veliky, *Biotechn. Bioeng.*, 16, 53 (1984)

APPENDIX

Method of solution:

The system of coupled ODE and PDE modelling the dynamic response of an immobilized cells CSTR is as follows:

$$\varepsilon\left(\frac{dC_L}{dt}\right) = \left(\frac{F}{V}\right)(C_{in} - C_L) - N_0 \quad \text{(A-1)}$$

$$N_0 = 3\,(1-\varepsilon)\left(\frac{D}{R}\right)\left(\frac{\partial C_b}{\partial r}\right)_{r=R} \quad \text{(A-2)}$$

$$D_{eff}\left[\frac{\partial^2 C_b}{\partial r^2} + \left(\frac{2}{r}\right)\frac{\partial C_b}{\partial r}\right] = \varepsilon_p \frac{\partial C_b}{\partial t} \quad \text{(A-3)}$$

$$C_L(0) = 0.0 \quad \text{(A-4)}$$

$$C_b(r,0) = 0.0 \quad \text{(A-5)}$$

$$C_b(R,t) = C_L(t)/\beta \quad \text{(A-6)}$$

$$C_b(0,t) = finite \quad \text{(A-7)}$$

The method of solution presented in this work can be described by the following steps :

(i) Find a solution for Eqn. A-3 based on the existing analytical solution for the case of constant surface condition.

(ii) Evaluate N_0, that is, evaluate the partial derivative $(\partial C_b/\partial r)_{r=R}$.

(iii) Substitute in A-1 and solve the resulting ODE using a standard numerical method.

The solution of Eqn. A-3 for constant surface condition, that is for $C_b(R,t) = C_0$, is given by Crank (1983). This solution follows :

$$C^*(r,t) = C_0\left[1 + \left(\frac{2R}{\pi r}\right)\sum_1^\infty \frac{(-1)^n \sin\left(\frac{n\pi r}{R}\right) e^{-tk_n}}{n}\right] \quad \text{(A-8)}$$

where $k_n = D_{eff}(n\pi/R)^2/\varepsilon_p$

Having an expression for $C^*(r,t)$ we can obtain a solution $C_b(r,t)$ of Eqn. A-3 for time dependent boundary condition ($C_b(R,t)=\phi(t)$) using Duhamels theorem which relates C_b to C^*

$$C_b(r,t) = \int_0^t \left(\frac{\partial\phi}{\partial t}\right) C^*(r,t-\tau)\, d\tau \quad \text{(A-9)}$$

Substitute A-8 into A-9, and after performing the integration we derive :

$$C_b(r,t) = \phi(t) + \left(\frac{2R}{\pi r}\right)\sum_1^\infty \frac{(-1)^n \sin\left(\frac{n\pi r}{R}\right)}{n}\psi_n(t) \quad \text{(A-10)}$$

where

$$\psi_n(t) = \int_0^t \frac{\partial\phi}{\partial t} e^{(\tau-t)k_n} d\tau \quad \text{(A-11)}$$

This is an expression for the intraparticle concentration for time-dependent surface condition, as a function of time (t) and distance from the bead centre (r). Differentiating Eqn. (A-11) we obtain the following equation for ψ_n

$$\frac{d\psi_n}{dt} = -k_n\,\psi_n(t) + \frac{d\phi}{dt}$$

or, introducing the partition coefficient

$$\frac{d\psi_n}{dt} = -k_n\,\psi_n(t) + \left(\frac{1}{\beta}\right)\frac{dC_L}{dt} \quad \text{(A-12)}$$

with initial condition:

$$\psi_n(0) = 0.$$

After that, the partial derivative $(\partial C_b/\partial r)_{r=R}$ appearing in the flux term (Eqn. A-2) can be easily evaluated as

TABLE 1
Experimentally Determined Diffusivities of Some Fermentation Substrates and Products ($\times 10^6$ cm^2/s).

Type of experiment	Glucose	Lactose	$NH_4 Cl$	Patulin	6-MSA
Cell-free beads	6.68	6.4	55	6.28	2.92
With dead Cells	6.68	6.0	45	5.03	1.95
With live Cells	7.58	7.2	-	-	-

$$\left(\frac{\partial C_b}{\partial r}\right)_{r=R} = \left(\frac{2}{R}\right) \sum_1^\infty \psi_n(t) \quad \text{(A-13)}$$

Substitution in Eqn. A-1 yields:

$$\varepsilon \frac{dC_L}{dt} = \left(\frac{F}{V}\right)(C_{in} - C_L) - 6(1-\varepsilon)\left(\frac{D_{eff}}{R^2}\right) \sum_1^\infty \psi_n(t) \quad \text{(A-14)}$$

Evaluation of the infinite summation $\sum_1^\infty \psi_n(t)$

Equation A-12 can be written as

$$\left(\frac{1}{k_n}\right)\frac{d\psi_n}{dt} + \psi_n(t) = \left(\frac{1}{k_n}\right)\frac{d\phi}{dt} \quad \text{(A-15)}$$

In Eqn. A-15, the dynamics of the functions ψ_n's depend of the value of the coefficient $\tau = (1/k_n)$; small values of τ indicate that the functions $\psi_n(t)$ follow the forcing function $(1/k_n)(d\phi/dt)$. It is therefore reasonable to assume that for $\tau < \tau_0$, the ψ_n's can be evaluated using the quasi-steady state approximation

$$\psi_n(t) = \left(\frac{1}{k_n}\right)\frac{d\phi}{dt} \quad \text{(A-16)}$$

or, using dimensionless quantities

$$\psi_n(t_R) = \frac{1}{\varepsilon D_R \pi^2 n^2}\left(\frac{d\phi}{dt_R}\right) \quad \text{(A-17)}$$

The value of n_0 is determined as the one corresponding to τ_0:

$$\frac{1}{\varepsilon D_R \pi^2 (n_0^2)} = \tau_0$$

or

$$n_0 = \left(\frac{1}{\pi}\right)\left(\frac{1}{\varepsilon D_R \tau_0}\right)^{0.5}$$

It was found by numerical experimentation, that for values of τ_0 less than 0.01, the resulting response curves, for given D_R were identical to five significant figures. Subsequently, a value of $\tau_0 = 0.01$ was used in all the runs.

After the above considerations, the infinite summation $\sum_1^\infty \psi_n(t)$ is estimated as follows:

$$\sum_1^\infty \psi_n(t_R) = \sum_1^{n_0} \psi_n(t_R) + \sum_{n_0}^\infty \psi_n(t_R)$$

$$= \sum_1^{n_0} \psi_n(t_R) + \frac{1}{\varepsilon D_R \pi^2}\frac{d\phi}{dt_R}\sum_{n_0}^\infty \left(\frac{1}{n}\right)^2$$

and finally

$$\sum_1^\infty \psi_n(t_R) = \sum_1^{n_0} \psi_n(t_R) + \frac{1}{\varepsilon D_R \pi^2}\left(\frac{d\phi}{dt_R}\right)\left[\sum_1^\infty \left(\frac{1}{n}\right)^2 - \sum_1^{n_0}\left(\frac{1}{n}\right)^2\right]$$

Using the property $\sum_1^\infty \left(\frac{1}{n}\right)^2 = \frac{\pi^2}{6}$, and defining:

$$S_n = \sum_1^{n_0}\left(\frac{1}{n}\right)^2 \quad \text{and} \quad S_I = \sum_1^\infty \left(\frac{1}{n}\right)^2 = \frac{\pi^2}{6}$$

we obtain

$$\sum_1^\infty \psi_n(t_R) = \sum_1^{n_0} \psi_n(t_R) + \frac{1}{6\varepsilon D_R}\left(\frac{d\phi}{dt_R}\right)\left(1 - \frac{S_n}{S_I}\right) \quad \text{(A-18)}$$

The concentration at the surface, $\phi(t)$, can be expressed in terms of the bulk concentration $C_L(t)$ using the partition coefficient (β), namely $\phi(t_R) = C_L(t_R)/\beta$

Hence Eqn. A-18 becomes:

$$\sum_1^\infty \psi_n(t_R) = \sum_1^{n_0} \psi_n(t_R) + \frac{1}{6\beta\varepsilon D_R}\left(\frac{dC_L}{dt_R}\right)\left(1 - \frac{S_n}{S_I}\right) \quad \text{(A-19)}$$

Subsequent substitution Eqn. A-19 into A-14 yields:

$$\frac{dC_L}{dt_R} = \frac{C_{in} - C_L - 6\varepsilon_p(1-\varepsilon)D_R \sum_1^{n_0}\psi_n}{1 + \frac{\varepsilon_p(1-\varepsilon)}{\beta\varepsilon}\left(1 - \frac{S_n}{S_I}\right)} \quad \text{(A-20)}$$

The normalized form of Eqn. A-12 is:

$$\frac{d\psi_n}{dt_R} = -\varepsilon(n\pi)^2 D_R \psi_n + \left(\frac{1}{\beta}\right)\left(\frac{dC_L}{dt_R}\right),$$

$$n = 1, 2, 3, ..., n_0. \quad \text{(A-21)}$$

Solution by numerical integration of Equations A-20 and A-21 yields the dynamic response of the reactor.

Dispersal of Insoluble Fatty Acid Precursors in Stirred Reactors as a Mechanism to Control Antibiotic Factor Distribution

FLOYD M. HUBER
RICHARD L. PIEPER
ANTHONY J. TIETZ
Fermentation Technology Department
Eli Lilly and Company
Indianapolis, Indiana 46285

Biosynthesis of factors in the A-21978C antibiotic complex was controlled by addition of appropriate fatty acid precursors. Toxicity associated with higher fatty acids was avoided by continuous addition of the fatty acid at a rate nearly equal to the uptake by the producing organism. Inability of the producing organism to assimilate a solid, insoluble long-chain fatty acid was overcome by dissolution of the acid in another substrate.

Substance A21978C is a complex of antibiotics produced by Streptomyces roseosporus, having a common cyclic polypeptide nucleus and different fatty acid side chains (1) (2) (Figure 1). Separation of the major factors C1, C2, and C3 revealed differences in both in vitro antibiotic activity and toxicology for the different naturally occurring compounds. Deacylation of the alkanoyl side chain, followed by reacylation with a series of fatty acids indicated the n-decanoyl substitution at position R, resulted in the best therapeutic potential (3) (4) (5).

The natural occurrence of the n-decanoyl factor, designated LY146032, was too low to permit isolation in sufficient quantity directly from fermentation broth (Table 1).

In an attempt to produce LY146032 by fermentation, the addition of decanoic acid during the antibiotic production phase of the fermentation was proposed. Initial efforts in shaken cultures were unsuccessful due to either the toxicity or insolubility of the fatty acid. In this paper we will describe efforts to direct the biosynthesis of A21978C factors in continuously stirred reactors operating in a fed-batch mode.

MATERIALS AND METHODS

A mutant strain of Streptomyces roseosporus NRRL 11379 was used to inoculate 50 ml of vegetative medium of the following composition: Trypticase soy broth (Baltimore Biological Laboratories, Baltimore, Maryland), 30 mg/ml; potato dextrin, 25 mg/ml. The inoculated medium was incubated for 48 hours at 30°C in a 250 ml Erlenmeyer flask on a shaker rotating through an arc of two inches in diameter at 250 RPM. One-half ml of the mature vegetative culture was dispensed into multiple containers and stored in the vapor phase of liquid nitrogen. One ml of the stored culture was used to inoculate 800 ml of the vegetative medium described above. The inoculated vegetative medium was incubated in a 2000 ml Erlenmeyer flask at 32°C for 120 hours on a shaker

TABLE 1
Distribution of Naturally Occurring Factors in A21978C Fermentation

A21978C Factor	Concentration µg/ml	% of Total A21978C Complex
C1	77	27.3
C2	113	40.1
C3	72	25.5
C5	trace	-
LY146032	trace	-
	282	

Eli Lilly and Company, Indianapolis, Indiana.

rotating through an arc of two inches in diameter at 250 RPM. The entire contents of the two flasks (approximately 1400 ml after incubation) were used to inoculate 1900 liters of a secondary vegetative stage having the following composition (mg/ml): soybean flour, 5.0; yeast extract (Difco Laboratories, Detroit, Michigan), 5.0; calcium gluconate, 10.0; KCl, 0.2; $MgSO_4 \cdot 7H_2O$, 0.2; $FeSO_4 \cdot 7H_2O$, 0.004; Sag 471 antifoam (Union Carbide, Danbury, Connecticut). The potassium, magnesium, and ferrous salts were prepared separately as follows: 7.6 g $FeSO_4 \cdot 7H_2O$ was dissolved in 76 ml of concentrated HCl. 380 g of $MgSO_4 \cdot 7H_2O$ and 380 g of KCl and deionized water were added to bring the total volume to 3800 ml. The inoculated medium was incubated 24 hours in a stainless steel vessel at 30°C. The vessel was aerated at 0.85 v/v/m and stirred with conventional agitators.

The mature secondary seed (8.33% v/v) was used to inoculate a production medium of the following composition (mg/ml): soybean flour, 22.0; $Fe(NH_4)_2SO_4 \cdot 6H_2O$, 0.66; glucose monohydrate, 8.25; Sag 471, 0.22; potato dextrin, 33.0; and molasses (blackstrap), 2.75.

Two types of stirred reactors were used. The smaller vessel, operated at 120 liters, was agitated with two conventional flat Rushton type impellers at relatively high power input. The larger vessel, operated at 4550 liters, was equipped with impellers having curved paddles and was operated at relatively low power input. Air flow in both reactors was supplied at 0.5 v/v/m by large open tubes which were estimated to contribute very little to the overall mixing. Respiration rates were estimated by difference in inlet and exhaust gas concentration via a Perkin-Elmer mass spectrometer. Distribution of A21978C factors was estimated by high performance liquid chromatography as described previously (2).

Examination of the batch fermentation medium suggested that the growth limiting nutrient was carbon in the form of carbohydrate. It was then hypothesized that in a fed-batch operation a moderately toxic substrate, such as decanoic acid, could be fed continuously to the fermentation if the metabolic consumption rate exceeded the addition rate.

Delivery of decanoic acid to the culture presented a problem. With a melting point of 34°C the compound is a solid at the fermentation temperature of 30°C, and the compound has very low solubility in water. In order to avoid the obvious problems arising in supplying a limiting nutrient as a solid phase, the substrate was dispensed to the stirred reactor as a five percent solution dissolved in a fifty percent ethanol/water mixture. There was an immediate response in oxygen uptake to the onset of the decanoic acid feed, as illustrated in Figure 2. Also, a significant improvement in LY146032 concentration was immediately realized (Table 2).

TABLE 2

Distribution of A21978C Factors with Decanoic Acid Feed[a]

A21978C Factor	Concentration µg/ml	% of Total A21978C Complex
C1	72	19.8
C2	109	29.9
C3	42	11.5
C5	19	5.2
LY146032	122	33.5
	364	

(a) N-decanoic acid/ethanol/water 1:2:2 fed 50 ml per hour to 120 L operating volume.

Material balances suggested that only a small portion of the decanoic acid that was fed could be accounted for by incorporation into the product. Thus, most of the fatty acid was apparently catabolized, presumably by the beta-oxidation pathway. In an attempt to increase the amount of decanoic acid available for the incorporation, the concentration of fatty acid in the

TABLE 3

Distribution of A21978C Factors with Increased Decanoic Acid Feed[a]

A21978C Factor	Concentration µg/ml	% of Total A21978C Complex
C1	131	10.4
C2	189	15.0
C3	107	8.5
C5	52	4.1
LY146032	784	62.1
	1263	

(a) N-decanoic acid/ethanol/water 1:2:2 fed 50 ml per hour to 120 L operating volume.

feed was increased to twenty percent--the solubility limit of decanoic acid in aqueous ethanol. A significant increase in both the LY146032 and total yield was observed (Table 3).

Although the fed-batch fermentation employing the fatty acid/ethanol/water addition proved an effective method of directing the synthesis of the A21978C complex, the presence of the volatile alcohol presented both safety problems and uncertainties in quantitating the carbon balance. It was not known to what extent the producing organism could metabolize ethanol, since it was likely that much of the alcohol was escaping in the exit gases. Methyl oleate was identified in batch shaken cultures as a metabolizable, non-toxic and low volatile solvent. A mixture of equal volumes of methyl oleate and decanoic acid remained liquid at 30°C. The results of feeding the mixture of decanoic acid dissolved in methyl oleate on an equivalent basis to the previously used aqueous ethanol feed resulted in a slightly higher total yield with a similar concentration of the desired LY146032 (Table 4).

TABLE 4
Distribution of A21978C Factors in a Decanoic Acid/Methyl Oleate Fed Fermentation[a]

A21978C Factor	Concentration µg/ml	% of Total A21978C Complex
C1	154	10.6
C2	191	13.1
C3	127	8.7
C5	68	4.7
LY146032	913	62.8
	1453	

(a) N-decanoic acid/methyl 1:1 oleate fed 13 ml per hour to 120 L operating volume

SCALE-UP CONSIDERATIONS

The initial scale-up of the process to a larger 6000 L pilot-scale stirred reactor did not produce equivalent LY146032 factor distribution (Table 5).

The percentage of LY146032 was approximately one-half of that obtained in the smaller equipment. It was hypothesized that the microbial population was oxidizing the decanoic acid preferentially over the methyl oleate, and due to the poorer mixing in the

TABLE 5
Distribution of A21978C Factors with Decanoic Acid/Methyl Oleate Fed to a Larger Reactor[a]

A21978C Factor	Concentration µg/ml	% of Total A21978C Complex
C1	260	17.0
C2	337	22.1
C3	270	17.7
C5	170	11.1
LY146032	489	32.0
	1527	

(a) N-decanoic acid/methyl oleate 1:1 fed 490 ml per hour to 4550 L operating volume

large vessel, the uptake of the fatty acid was not equivalent throughout the population in the reactor (Table 6).

TABLE 6
Mixing Time and Power Input in Pilot-Scale Equipment Used in A21978C Fermentation

Vessel Size (l)	Mixing Time[a] (Seconds)	Power Input (Hp/100 Gal.)
150	7	1.36
6000	43	0.82

(a) Time required for pH to reach equilibrium after addition of a sufficient quantity of 6M NaOH to increase pH by 0.3

The first approach to achieve better incorporation of the fatty acid precursor in the poorly mixed larger vessel was to increase the feed rate of the decanoic acid/methyl oleate mixture. The percentage of LY146032 increased to nearly that achieved in the smaller equipment, but the total yield of A21978C was reduced to about two-thirds of the amount achieved at the lower feed rate (Table 7).

The reduction in overall yield associated with the increased feed of the decanoic acid precursor was believed to be a result of the feed rate of the fatty acid approaching the metabolic rate of consumption. The consequences of introducing the toxic fatty acid

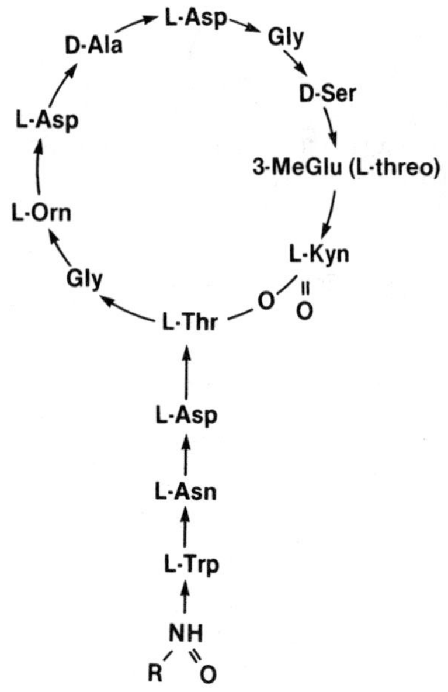

Factor	R
A21978C_0	C_{10} alkanoyl*
A21978C_1	8-methyl decanoyl
A21978C_2	10-methyl undecanoyl
A21978C_3	10-methyl dodecanoyl
A21978C_4	C_{12} alkanoyl †
A21978C_5	C_{13} alkanoyl †

* mixture of two isomers † structure uncertain

FIGURE 1. Naturally occurring A21978C factors.

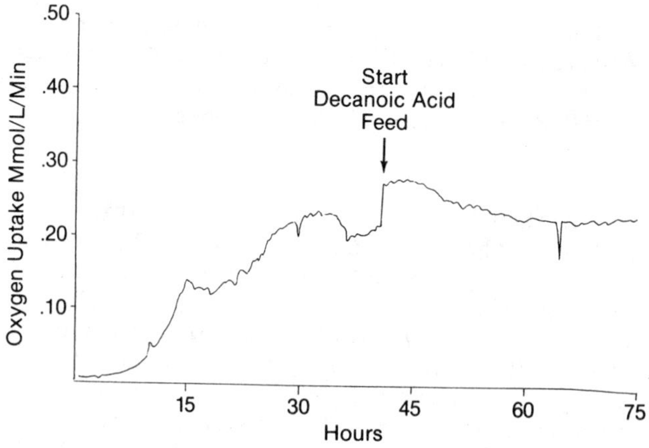

FIGURE 3. The toxic effect of overfeeding decanoic acid.

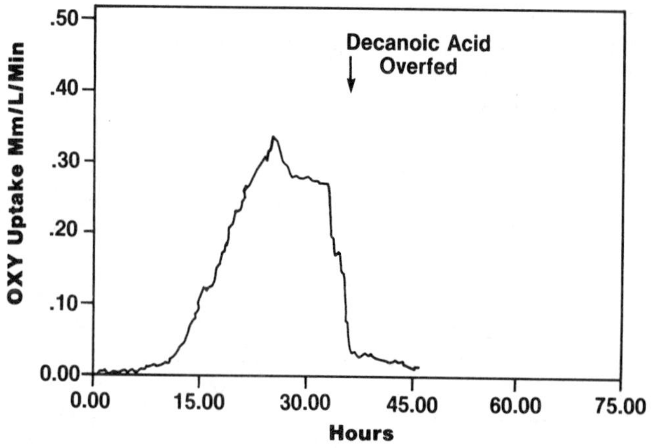

FIGURE 2. The effect of *n*-decanoic acid feed upon oxygen uptake.

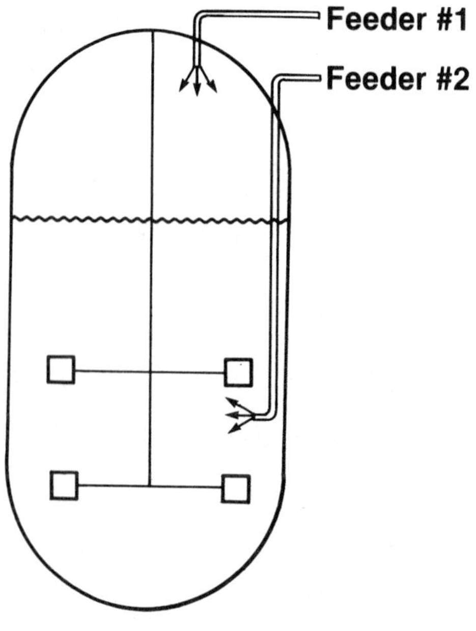

FIGURE 4. Feeding of *n*-decanoic acid/methyl oleate to a large reactor via two entry points.

TABLE 7
Distribution of A21978C Factors with Increased Decanoic Acid/ Methyl Oleate Fed[a] to a Larger Reactor

A21978C Factor	Concentration µg/ml	% of Total A21978C Complex
C1	122	11.9
C2	117	11.4
C3	83	8.1
C5	93	9.1
LY146032	607	59.4
	1022	

(a) N-decanoic acid/methyl oleate 1:1 fed 590 ml per hour to 4550 L operating volume

at a rate greater than the microbial consumption rate are illustrated in Figure 3. In this reactor the feeding system failed, resulting in significant overfeeding of the fatty acid. The resulting accumulation of the fatty acid caused rapid lysis of the culture as evidenced by the rapid cessation of respiration.

A second approach to solve the mixing problem in the large fermenter was to introduce the fatty acid/ester mixture at two widely separated entry points. One entry point was at the top of the vessel and a second was in between the two impellers (Figure 4). The subsequent LY146032 concentration obtained (Table 8) indicated that a greater incorporation of the n-decanoyl side chain had occurred as a result of the dual introduction.

TABLE 8
Distribution of A21978C Factors with Decanoic Acid/Methyl Oleate Fed to a Larger Reactor via Two Entry Points[a]

A21978C Factor	Concentration µg/ml	% of Total A21978C Complex
C1	75	5.3
C2	119	8.5
C3	108	7.7
C5	16	1.1
LY146032	1090	77.4
	1408	

(a) N-decanoic acid/methyl oleate 1:1 fed 212 ml per hour to each of two entry points. 4550 L working volume.

In summary, the following scale-up problems associated with the production of LY146032 have been addressed and solved:

1. A sound strategy was provided to feed the microbial culture a very toxic substance.

2. A mechanism was devised to supply the normally solid substrate to the reactor in a convenient liquid form.

3. Preferential substrate utilization was identified and the necessary feeding hardware was constructed to promote homogeneous utilization of lipoidal materials when the fermentation was scaled up to larger reactors.

LITERATURE CITED

1. Hamill, R.L. and M.M. Hoehn, U.S. Patent 4,208,403 to Eli Lilly and Company (June 17, 1980).

2. Debono, Manuel, U.S. Patent 4,399,067 to Eli Lilly and Company (August 16, 1983).

3. Debono, Manuel, B.J. Abbott, V.M. Krupinski, R.M. Molloy, D.J. Berry, F.T. Counter, L.C. Howard, J.L. Ott and R.L. Hamill, "Synthesis and Structure Activity Relationships of New Analogs of the New Gram Positive Lipopeptide Antibiotic A21978C," Abstract 1077, Interscience Conference on Antimicrobial Agents and Chemotherapy (ICAAC) (October 1984).

4. Fukuda, D.S., B.J. Abbott, D.J. Berry, L.D. Boeck, Manuel Debono, R.L. Hamill, V.M. Krupinski and R.L. Molloy, "Deacylation and Reacylation of A21978C, Acidic Lipopeptide Antibiotic: Preparation of New Analogs," Abstract 1076, Interscience Conference on Antimicrobial Agents and Chemotherapy (ICAAC) (October 1984).

5. Counter, F.T., P.J. Baker, L.D. Boeck, Manuel Debono, P.W. Ensminger, R.L. Hamill, V.M. Krupinski, R.M. Molloy and J.L. Ott, "LY146032 [N-(n-decanoyl) A21978C Nucleus], a New Acidic Lipopeptide Antibiotic: Synthesis and Biological Evaluation," Abstract 1078, Interscience Conference on Antimicrobial Agents and Chemotherapy (ICAAC) (October 1984).

Periodicity in Substrate Concentration in Three-Phase Fluidized-Bed Bioreactors

BRIAN H. DAVISON
TERRENCE L. DONALDSON
Chemical Technology Division
P.O. Box X
*Oak Ridge National Laboratory**
Oak Ridge, Tennessee 37831

Periodic fluctuations in substrate concentrations have been observed within fluidized-bed bioreactor systems for fermentation of glucose to ethanol using biocatalyst beads and for denitrification of wastewater using a mixed biofilm culture attached to coal particles. The periodic fluctuations have been observed as time-variant glucose concentrations at a single location, and as nonmonotonic axial profiles of nitrate concentration. Potential causes of the fluctuations are discussed in terms of hydrodynamic phenomena in the three-phase bioreactors.

Fluidized-bed bioreactors (FBRs) are attracting interest in the research community and are being seriously considered for new process applications because they offer significant advantages relative to suspended-growth systems. These include immobilization of the enzyme or microorganism, which greatly reduces washout and permits higher liquid flow rates; a high concentration of biocatalyst, which creates high volumetric reaction rates; and an approach to plug flow or stagewise operation, which helps to reduce product inhibition effects.

Off-gas is frequently generated in bioreactor systems, or oxygen may be added for aerobic processes. Thus, the FBR is often a three-phase system in which the continuous phase is liquid and the two dispersed phases are solid biocatalyst particles and gas bubbles. The solid particles tend to settle under gravitational force, which is counteracted, in part, by the upward flow of liquid to "fluidize" the bed of particles. The dispersed gas bubbles naturally tend to rise. The overall hydrodynamic flow behavior is obviously complex.

Oak Ridge National Laboratory, Oak Ridge, Tennessee. Operated by Martin Marietta Energy Systems, Inc., under Contract No. DE-AC05-84OR21400 with the U.S. Department of Energy.

At the Oak Ridge National Laboratory, FBRs have been used in a number of process development applications (1-5). In this paper, we describe the nonideal behavior of the substrate concentration in two different FBR systems and report visual observations of liquid flow that appear to be responsible for the behavior of the substrate concentration. One system is the fermentation of glucose to ethanol; the other is the biodenitrification of wastewaters.

BACKGROUND

Plug-flow conditions are approached in the operation of FBRs. In a three-phase system, the deviation from plug flow can be severe due to channeling and entrainment of solid and liquid in the bubble wakes. Virtually all models of three-phase FBRs use dispersion or backmixing to account for these phenomena. In a recent comprehensive review of FBR operations, Muroyame and Fan (6) reported no evidence of periodic behavior in FBRs. However, they recommended additional work to investigate the flow regimes that may appear in such systems. Hydrodynamic studies of systems where gas generation is significant are also lacking.

Backmixing and dispersion models predict that the substrate concentration will decrease monotonically from the inlet to the outlet. This behavior could be observed by sampling at different axial positions along the reactor.

At steady-state operation, the substrate concentration at a fixed axial point would be constant and time invariant.

FERMENTATION OF GLUCOSE TO ETHANOL

In the fermentation system, *Zymomonas mobilis* was immobilized within gel beads and used to convert a feed stream of glucose to equimolar amounts of ethanol and CO_2. The fluidized-bed bioreactor system shown in Figure 1 is 150 cm high and 2.54 cm ID. The biocatalyst beads were 0.18 cm in diameter and composed of 4% κ-carrageenan, 3% Fe_2O_3, and 17 g/L (dry weight) *Z. mobilis*. A feed stream of 100 g/L glucose with 5 g/L yeast extract and 0.1 M KCl was added continuously to the bottom of the column at 21 mL/min.

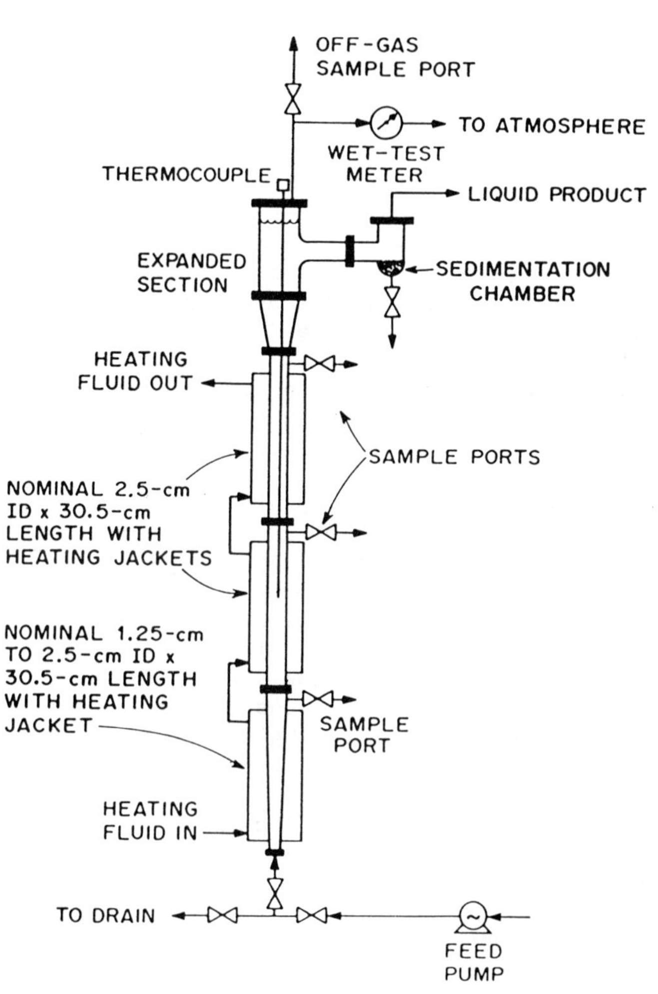

FIGURE 1. Fluidized-bed bioreactor system for fermentation of glucose to ethanol using biocatalyst beads.

This flow rate is approximately the minimum velocity needed to fluidize the beads. The off-gas was measured with a wet-test meter and vented to the atmosphere. The overall conversion of glucose to ethanol was 91%, and 0.5 L/min of CO_2 was produced. Thus, the gas/liquid flow-rate ratio varied from zero at the inlet to ~23 at the top. The bioreactor was operated continuously for ~1 d before collecting the data described below.

The time variation in the concentration of glucose midway along the column after ~1 d of stable operation is shown in Figure 2.

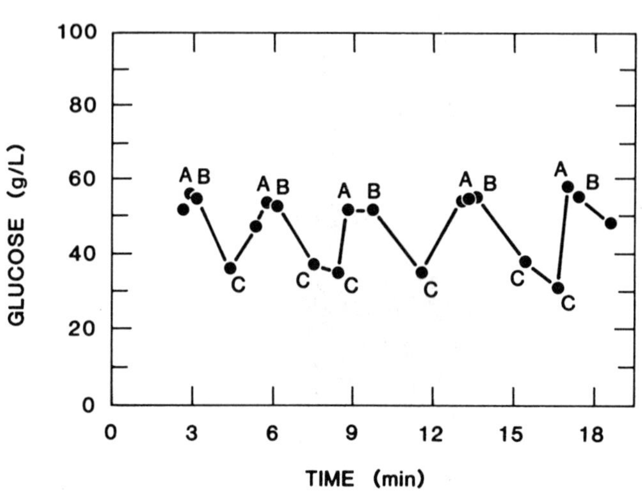

FIGURE 2. Periodic fluctuations in glucose concentration at a fixed point midway up the bioreactor column. The designations A, B, and C refer to phenomena described in Figure 3.

One-milliliter samples were withdrawn, immediately filtered, and analyzed with a YSI glucose analyzer. Concurrent visual observations of the bed conditions at the sample port were made and correlated with the regions denoted A, B, and C in Figure 3. Oscillations of ±10 g/L are readily apparent around the average glucose concentration of 45 g/L. Samples taken at 1-min intervals at the top of the column were stable at 9±1 g/L of glucose. The precision of the glucose assay is ±1 g/L, so these oscillations are not artifacts.

Distinctive three-phase flow patterns were readily visible and are represented schematically in Figure 3; a photograph is shown in Figure 4. This three-phase system is observed to form repeating cell units. The cell appears when a zone of small bubbles forms across the cross section of the column

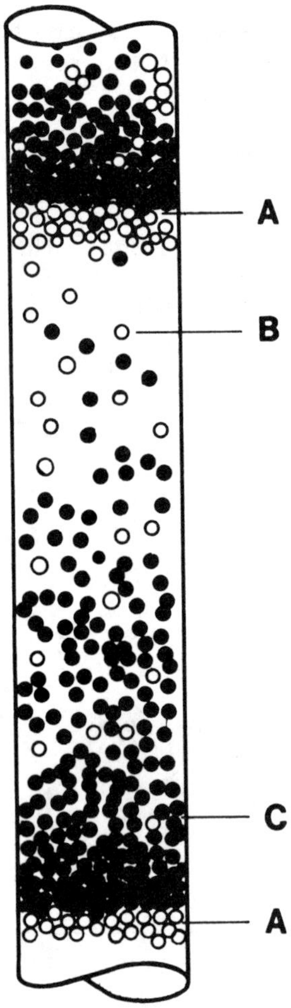

FIGURE 3. Periodic distribution of gas bubbles and biocatalyst in fluidized-bed bioreactor for ethanol fermentation. A = bubble front region; B = free liquid region; C = dense beads region, with gas channeling.

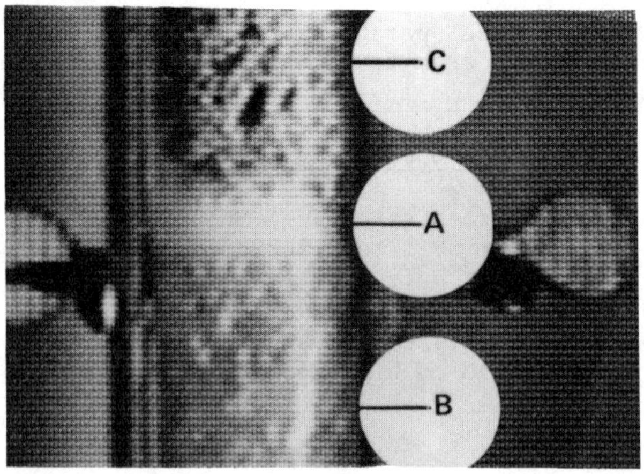

FIGURE 4. Closeup view of fluidized bed showing gas-liquid-solid distribution. A = bubble front region; B = free liquid region; C = dense beads region, with gas channeling.

without coalescing into larger bubbles. The gas continuously escapes into the packed region above and is replenished by the gas generation below. Immediately above this "bubble front," there is a heterogeneous dense-solids region with the gas as the continuous phase instead of the liquid in parts of this region. Below the bubble front there is a region with very little solids until reaching the next cell.

These bubble fronts do not cause gas slugging in the traditional sense of forcing the bed of solids into the disengagement section at the top of the column. In fact, the packed regions are stationary until a front passes. The bubble fronts move slowly and gently upward at the rate of 10 to 30 cm/min, with a separation of 5 to 20 cm. The beads in the bottom of the packed region fall through the bubble front and the liquid region below and settle into the next cell. The front velocity may depend on the superficial liquid velocity, but no clear trend was observed.

This cell structure occurs over a limited, but significant, range of gas and liquid flow rates in this bioreactor. At low gas generation rates (at the bottom of this bioreactor), the gas is able to rise without forming bubble fronts. At high rates (at the top of the bioreactor), the volume of gas is sufficient to create vigorous mixing and to destabilize the fronts. The bubble front behavior has been observed for glucose feed concentrations ranging from 20 to 200 g/L and over a fivefold change in the liquid flow rate, although the cells are less stable at the higher flow rates.

An explanation for the oscillations in the glucose concentration can be deduced from the observed flow pattern. Immediately below the bubble front, there is little biocatalyst and hence little reaction (Figure 3, region B). The substrate concentration will be relatively high in this region. Toward the bottom of the cell (Figure 3, region C), the biocatalyst beads are densely packed, the reaction rate is higher, and the substrate concentration will be relatively low. As these cells progress up the column, a sensor at a fixed location (or samples withdrawn at a fixed location) will display fluctuations as the spatial variation is converted to a time variation, as shown in Figure 2. The A, B, and C notations on the figure indicate samples from the A, B, and C regions, respectively, of the cells, as shown in Figure 3. Data points having no location designator were obtained during the C-to-A transition and exhibit intermediate glucose concentrations.

Similar bubble front behavior was also observed in other experiments with 3.8-cm- and 5.1-cm-ID columns, but the fronts were less stable. While this phenomenon may not be significant in larger bioreactors, it does need to be recognized when bench-scale experiments and data are evaluated for process scale-up.

BIODENITRIFICATION

A pilot-scale fluidized-bed bioreactor 10 cm ID and 12 m high (Figure 5) was operated

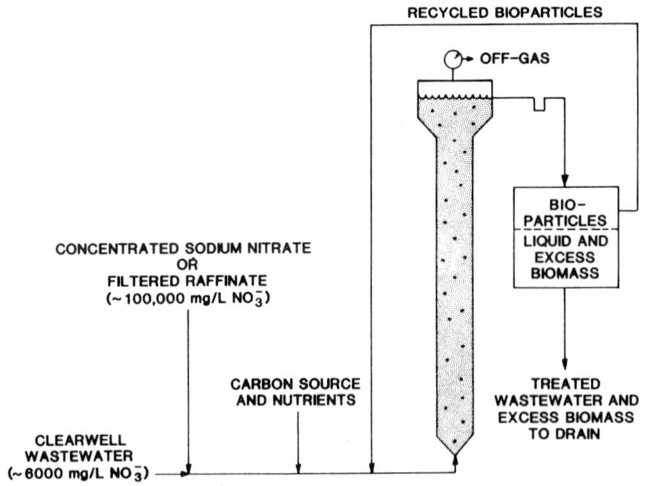

FIGURE 5. Fluidized-bed bioreactor system for biodenitrification of wastewater.

for 3 months during process verification studies for biodenitrification of wastewater obtained from the Feed Materials Production Center at Fernald, Ohio. The bioparticles were 30-60 mesh anthracite coal with attached biofilms of a natural mixed culture. The bioreactor, which was constructed of PVC pipe, was equipped with sample ports located at ~1-m axial spacing. Wastewater containing ~10,000 mg/L of nitrate was fed to the bioreactor at ~4 L/min. The off-gas, composed primarily of nitrogen along with ~5% carbon dioxide, was typically generated at 50 to 100 L/h; therefore, the gas/liquid ratio normally ranged from 0.25 to 0.5.

A representative axial profile of nitrate concentration is shown in Figure 6. The

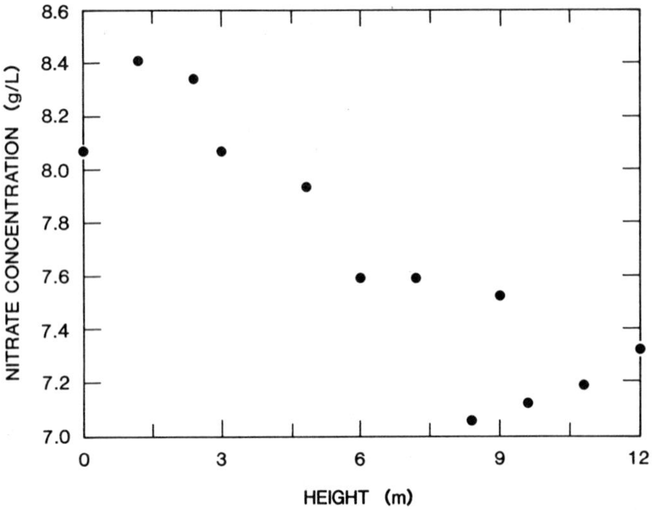

FIGURE 6. Typical axial profile of nitrate concentration in fluidized-bed bioreactor. The zero height location is the bottom of the fluidized bed and does not represent the feed conditions.

nitrate concentration was obtained from assays of 100-mL samples withdrawn from the sample ports over a period of several minutes after a brief flush to remove stagnant liquid. The measured nitrate concentration at zero height is inside the bioreactor at the bottom; it is not the feed concentration.

The assay was the Hach cadmium reduction method (7). Our studies of the precision of the assay indicated an uncertainty of ±5% (~350 mg/L) in this concentration range. A statistical case can be argued, therefore, that the variations seen in Figure 6 result from uncertainties in the assay. However, the pattern was consistent: an initial increase in nitrate concentration, followed by an erratic decline throughout the midsection of the bioreactor, followed by another increase to the final effluent value. This pattern strongly suggests that the uncertainties in the assay are not the sole origin of the trends in the axial profile of nitrate concentration. In contrast to the ethanol fermentation system, nitrate assays of a series of samples over time at a fixed location generally gave a constant value within the precision of the assay.

A potential explanation for the behavior shown in Figure 6 differs from that for the ethanol fermentation system, but it also arises from nonideal flow. No bubble fronts were observed at this lower gas/liquid flow-rate ratio; however, vertical circulation cells were readily visible, as sketched in Figure 7. At a fixed axial position, fluid

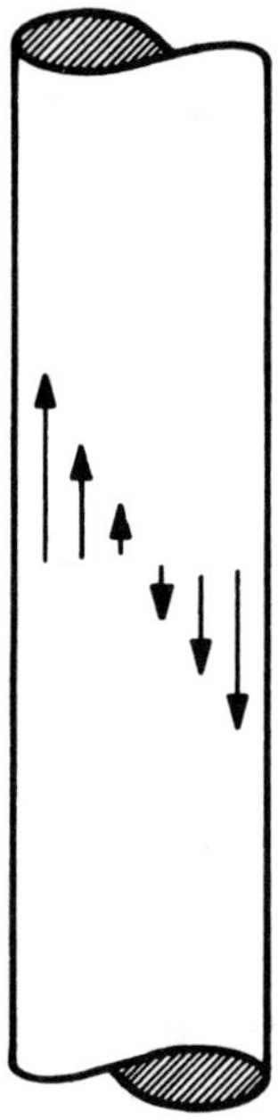

FIGURE 7. Velocity distribution of bioparticles in fluidized-bed bioreactor for biodenitrification.

and particles moved upward on one side of the column and downward on the other side. These flow patterns extended for axial distances of several meters and were relatively stable over observation periods of minutes to hours. This deviation from ideal plug flow will uncouple axial position and liquid residence time and could lead to apparent anomalies in the expected relationship between nitrate concentration and axial position.

The cause of these circulation flows was not investigated; the slight deviation from true vertical orientation of the bioreactor is a possible factor. The bioreactor was constructed from four 10-ft sections of PVC pipe joined with unions. Using a plumb line as a reference, the maximum deviation from vertical was estimated to be about 1 in. over 10 ft, or about 0.5 deg.

SIGNIFICANCE

The nonideal flow patterns in these two fluidized-bed bioreactors have implications for process performance, interpretation of experimental data, and particularly for process control. Effective strategies for dealing with these nonidealities will depend on the amplitude and frequency of the fluctuations and the time constants of other dynamic aspects of the particular process system. These results are presented to document previously unreported nonidealities in three-phase FBRs, as well as to encourage further research in this area.

ACKNOWLEDGMENTS

J. F. Walker, Jr., M. V. Helfrich, and T. A. Cooper assisted in several technical aspects of these studies. This work was supported, in part, by the Energy Conversion and Utilization Technologies Program and the Feed Materials Production Center, U.S. Department of Energy, under Contract No. DE-AC05-84OR21400.

LITERATURE CITED

1. Lee, D. D., C. D. Scott, and C. W. Hancher, J. Water Pollut. Control Fed., 51, 974 (1979).

2. Donaldson, T. L., G. W. Strandberg, J. D. Hewitt, and G. S. Shields, Environ. Prog., 3, 248 (1984).

3. Walker, J. F., C. W. Hancher, R. K. Genung, B. D. Patton, and M. Kowalchuk, Biotech. Bioeng. Symp. No. 11, 415 (1981).

4. Scott, C. D., Biotech. Bioeng., 13, 287 (1983).

5. Davison, B. H., and C. D. Scott, Biotech. Bioeng. Symp. No. 16, 1986, in press.

6. Muroyama, K., and L.-S. Fan, AIChE J., 31(1), 1 (1985).

7. Water Analysis Handbook, p. 2-185, Hach Company, Loveland, Colo. (1985).

Explosion Operation and Biotechnological Application for Effective Utilization of Biomass

TATSURO SAWADA
YOSHITOSHI NAKAMURA
Faculty of Technology
Kanazawa University
Kanazawa 920, Japan

The experiment of explosion for the effective utilization of biomass was carried out under various conditions of steam pressure from 1 to 6 MPa and reaction time from 1 to 20 min. The products were separated into hemicellulose, cellulose, methanol-soluble lignin and Klason lignin. The effects of operating conditions on the characteristics of exploded wood were evaluated from the experimental data of shape and size of particles, amounts of extractive components, enzymatic hydrolysis rate and alcohol fermentation process. As a result, it was found that the explosion with high steam pressure and short reaction time was the most effective operation for enzymatic hydrolysis and alcohol fermentation.

INTRODUCTION

The practical applications of biological conversions such as enzymatic hydrolysis and alcohol fermentation of wood and grass have been expected because of the oil crisis since 1973. However, many technical and economic problems in the pretreatment of biomass have to be overcome for an efficient utilization of biomass by using an industrial low cost energy. Recently, the autohydrolysis and explosion systems which use steam with high temperature and pressure have attracted attention to facilitate the degradation of wood materials. (1,2,3) Since this system has two actions of autohydrolysis as the chemical treatment and explosion as the physical treatment, various macromolecule compounds in wood are converted to the lower molecular weight materials in just a short period by the hydrolysis with steam. Furthermore, their lower molecular weight compounds are separated nearly completely to the three components such as cellulose, hemicellulose and ligin by the simple operations of extraction. (4,5,6) In comparison with other pretreatments (7,8,9,10), this system has several advantages. That is, (1) The activity of enzymatic hydrolysis of cellulose is high. (2) No chemical is necessary. (3) The system is relatively low in energy consumption.

In this work, the trial constructions of explosion system were attempted for development of the most efficient method of pretreatment for energy utilization of woods.

METHODS AND MATERIALS

Figure 1 shows an explosion apparatus which is used in this experiment. This apparatus is constructed of a steam generator, a reactor, a receiver for the exploded materials and a condenser. The system is able to be employed at temperature up to 275 C and maximum steam pressure of 6.0 MPa. Two hundred gram chips of larix leptolepis were put into the reactor. The pressure was rapidly raised to the desired pressure and the chips were treated with saturated steam for desired period. After the steaming, the steam pressure was released instantly to the atmospheric pressure by opening a ball valve. The wood chips were exploded by this treatment.

The extracts were freeze-dried to give a mixture of hemicellulose and water-soluble lignin, and methanol-soluble lignin, respectively. The residual materials in the extraction were heated in 72 % aqueous sulfuric acid for 4 hr to obtain insoluble lignin (Klason lignin) of wood which was weighed after drying method. (11)

The exploded samples was adjusted to pH 5.0 with 0.5 mol phosphate buffer, then are saccharified with Meicelase for 96 hr in 37 C of temperature. The substrate and enzyme concentrations were 2.0% and 0.2%, respectively. The reducing sugars were measured by the Somogyi-Nelson method. (12)

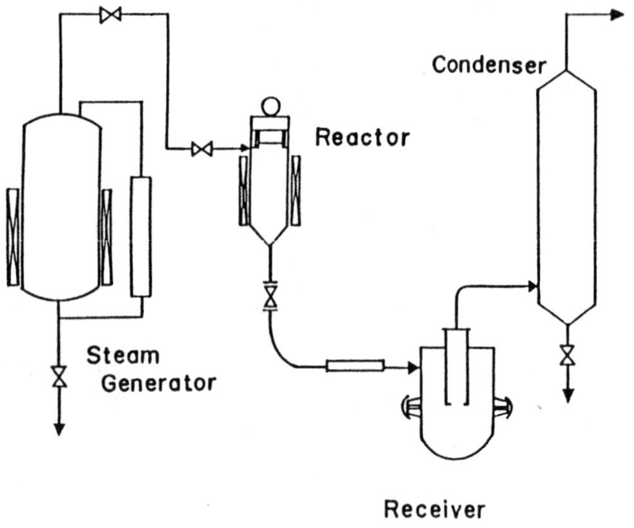

FIGURE 1. Experimental apparatus.

RESULTS AND DISCUSSION

The samples produced from wood chips by the hydrolysis with explosion are the materials of multiphase which consist of the fibers broken into pieces and the broken syrupy liquid with a relatively large amount of water condensated from the vapor water. A part of shredded and fluffy solid causes an aromatic oder which is similar to the plant material burned by heating of steam. The features in shapes and sizes of broken fibers become very remarkable at a higher steam pressure and a longer reaction time.

Figure 2 shows the micro photographs of wood chips at various reaction times under a constant pressure of 4.51 MPa. Although the more progress of fibril was observed most of wood chips remained at a filamental form within one minute of hydrolysis time. However, a rapid fibril of wood chip took place at a longer reaction time. The fibers changed to the broken fibers, and then became the deep brown syrupy liquid. Furthermore, it seemed that the filamental forms were almost similar regardless of the reaction time was over 3 min under a constant pressure at 4.51 MPa.

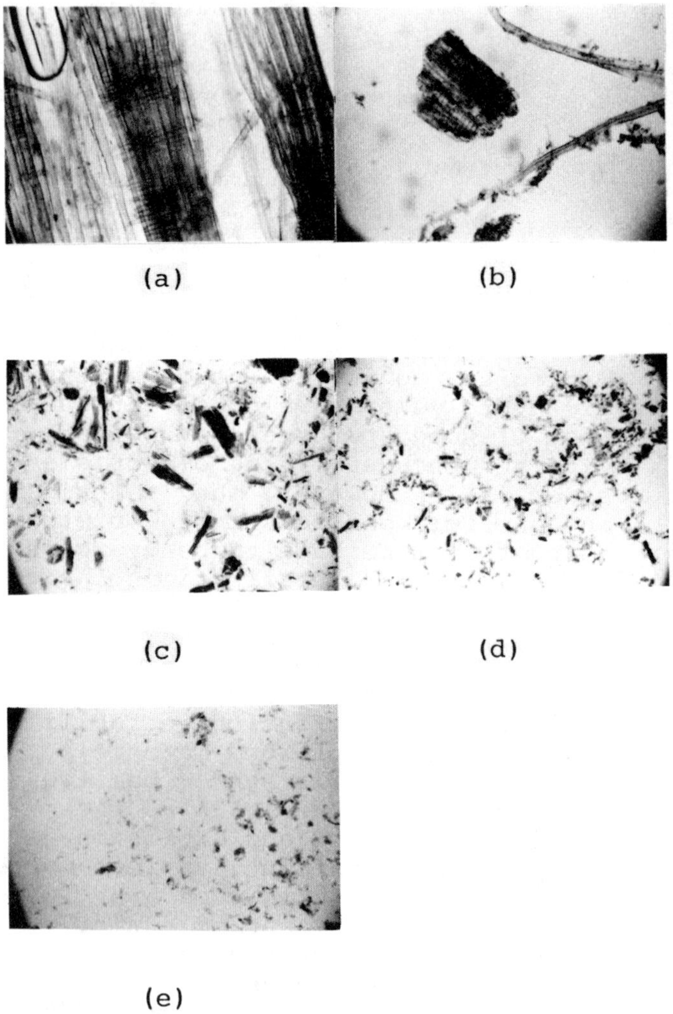

FIGURE 2. Photograph of exploded wood at steam pressure of 4.51 MPa: (a) 0.25 min; (b) 1 min; (c) 5 min; (d) 10 min; (e) 20 min.

Figure 3 shows the frequency distributions for both sizes of length and width in the filamental flakes from 400 to 500 pieces which were measured by naked eye or a microscope. Figure 3 (a) illustrates the distribution of length in the shredded particules of wood chips exploded at a period of hydrolysis of 5 min under various pressure from 1.08 to 4.81 MPa. The raw materials with an almost normal distribution changed to the filamental materials which had the distribution in a wide range from the microscopic size to 20 mm as the steam pressure increases up to 3.53 MPa. The range of distribution in particule lengths by explosion at 4.51 and 4.81 MPa became narrower in comparison with those below 3.53 MPa, and its mean length is below 0.05 mm. From these results, it seemed that the efficiency of explosion on wood chips was very large at pressure over 4.51 MPa. The trends in the changes of distribution curves for the width by pressure as shown in Figure 3 (b) were very similar to those for length. The range of width in distribution curve with increasing the pressure became narrower in spite of the relatively wide range in the raw material. In contrast to the tendency of width, the range of length with increase in pressure elogated in spite of the narrow range in the raw material. It was clear from these figures that steam pressure was one of the most efficient factors on the fibril of fibers.

Figure 4 illustrates a ratio of weight of soluble lignin and that of Klason lignin to weight of exploded dry sample, ξ_S and ξ_K, under various reaction times, t, in the steam pressures of 2.55, 3.04, 3.53 and 4.51 MPa. The amounts of soluble lignin increased as the reaction time and pressure increased. In contrast to the production of soluble lignin, the amount of Klason lignin decreased sharply, then changed to increase as the reaction time getting longer. The timing when the drastic change from the decrease of amount of Klason lignin product to its increase took place for a pressure was in fair agreement with the timing when the amount of soluble lignin became constant

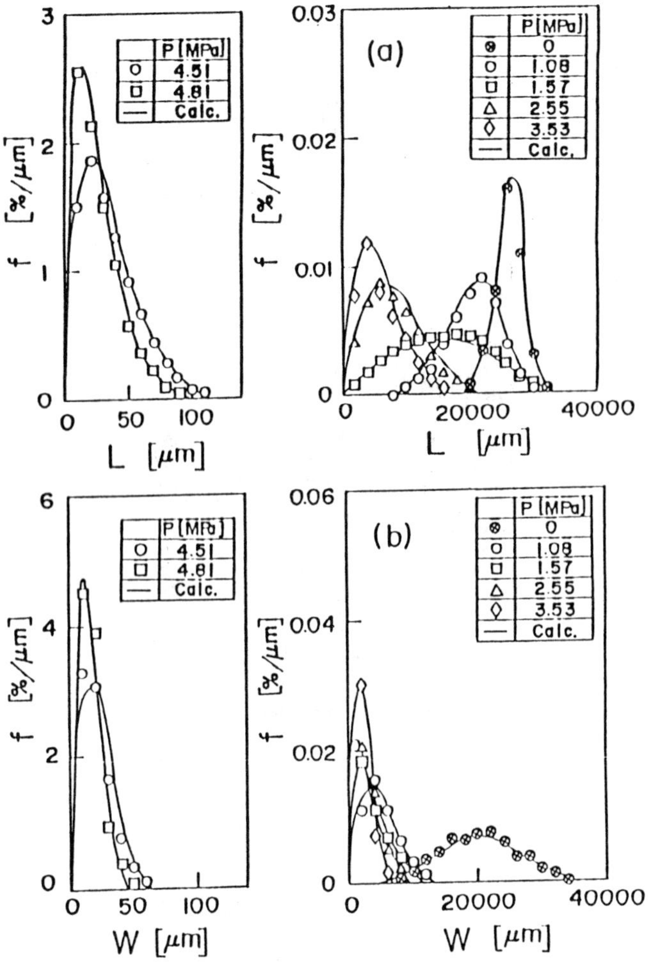

FIGURE 3. Size distribution of exploded wood: (a) length; (b) width.

regardless of reaction time. It seemed that the production of souble lignin could not exceed a certain amount by the hydrolysis, however, the production of Klason lignin proceeded progressive-ly as a results of the condensation with compounds of high reactivity except the soluble lignin or the condensation among the lignins of the lower molecular weight which were produced by the explosion.

Furthermore, the changes in the amounts of cellulose and hemicellulose under several operation conditions are plotted in Figure 5. The amount of cellulose decreased sharply in proportion to the reaction time and pressure. On the other hand, the amount of hemicellulose remained almost constant until the reaction time of about 20 min and the steam pressure of about 4.51 MPa.

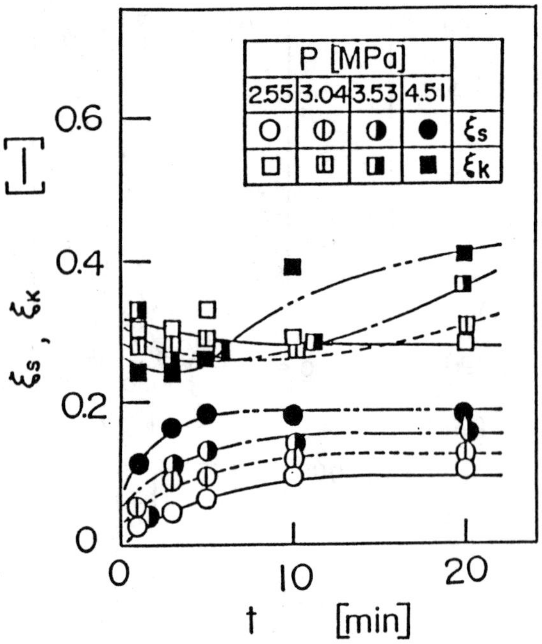

FIGURE 4. Effects of steam pressure and reaction time on ratios of soluble lignin and Klason lignin to dry wood.

Figure 6 (a) and (b) show the relationship between enzymatic saccharification, X, and incubation time, θ, at 2.55 and 4.51 MPa, respectively. Enzymatic saccharification was defined as ratio of weight of reducing sugars produced to that of exploded dry sample. Reducing sugars were measured by Somogyi-Nelson method as glucose. At 2.55 MPa, the small difference in effects of reaction time on enzymatic saccharification was recognized and their conversions were about 0.05 after 100 hr of hydrolysis. On the other hand, at 4.51 MPa, enzymatic saccharification increased remarkably with incubation time and large difference in reducing sugars was recognized. At 5 min of reaction time, the saccharification rate increased sharply and reached about 0.35 after 100 hr of hydrolysis. In the range of t ≧ 5 min, the saccharification rate decreased with reaction time. As a results, it was found that the operation for a few minutes at higher pressure was most effective for enzymatic hydrolysis of exploded wood.

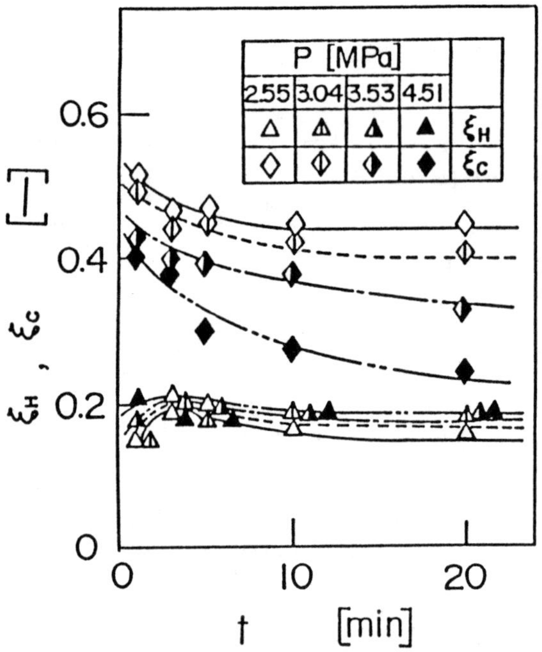

FIGURE 5. Effects of steam pressure and reaction time on ratios of cellulose and hemicellulose to dry weight of exploded wood.

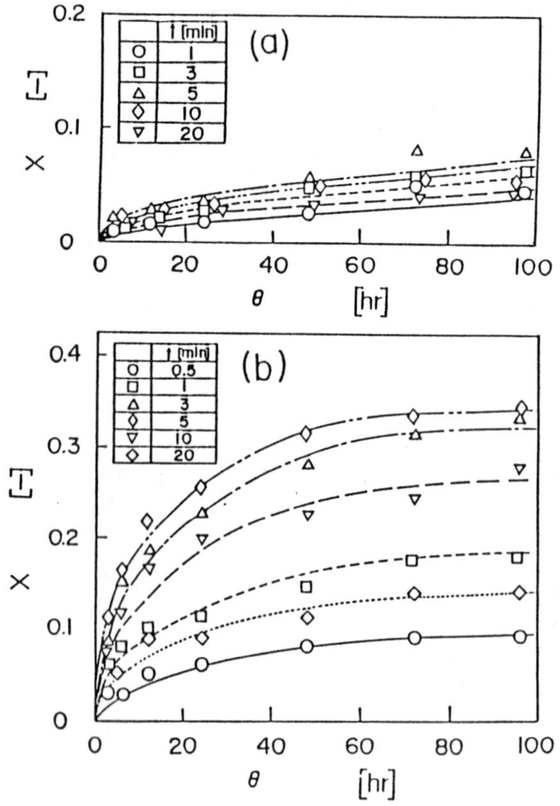

FIGURE 6. Enzymatic hydrolysis of exploded wood: (a) 2.55 MPa; (b) 4.51 MPa.

Equations for the estimation of enzymatic hydrolysis in cellulose were proposed. The general form expressed by Walseth (13) is

$$X = k\ \theta^n \tag{1}$$

where k and n are experimental constant. Taking logarithms of both sides of Eq.(1) is

$$\log X = \log k + n \log \theta \tag{2}$$

The relationship between enzymatic saccharification and incubation time gives essentially straight line on a log scale as shown in Figure 7. The slope was proportional to the 0.34 power of X. Then,

$$n = 0.34 \tag{3}$$

The value of hydrolysis rate constant, k, was obtained as the value of X at $\theta = 1$. Figure 8 shows the value of k under several operational conditions. In the range of $P \leqq 3.53$, the value of k increased with reaction time. On the other hand, in the range $P \geqq 3.53$, the value of k reached a maximum at 5 min and decreased with reaction time. Relationship of k and reaction time, t, was expressed as Eq.(4) using parameters α, β and γ.

$$k = \alpha \ / \ (\beta/t + 1 + t/\gamma) \tag{4}$$

where α, β and γ are maximum hydrolysis rate constant, hydrolysis rate constant at shorter time and hydrolysis rate constant at higher time.

Figure 9 shows values of α, β and γ in applying Eq.(4) to the experimental data. The values of α and β increased with steam pressure and enzymatic saccharification was high regardless of shorter reaction time. On the other hand, the value of γ decreased with steam pressure and suggested a lower enzymatic saccharification at longer reaction time. From these results, it was found that the operation of explosion for a few minutes at higher pressure was most effective for enzymatic hydrolysis of exploded wood.

Figure 10 shows time courses of cell mass concentration, ethanol concentration, reducing sugar concentration and glucose concentration in an alcohol fermentation of exploded wood by saccharromyces cerevisiae. The solid and bloken lines show ethanol production from enzymatic hydrolyzate of exploded wood at 2.55 and 4.51 MPa, respectively. At 4.51 MPa, alcohol concentration attained to 3 g/l after 25 hr of fermentation. On the other hand, at 2.55 MPa, alcohol concentration was only 1.4 g/l after 25 hr of fermentation.

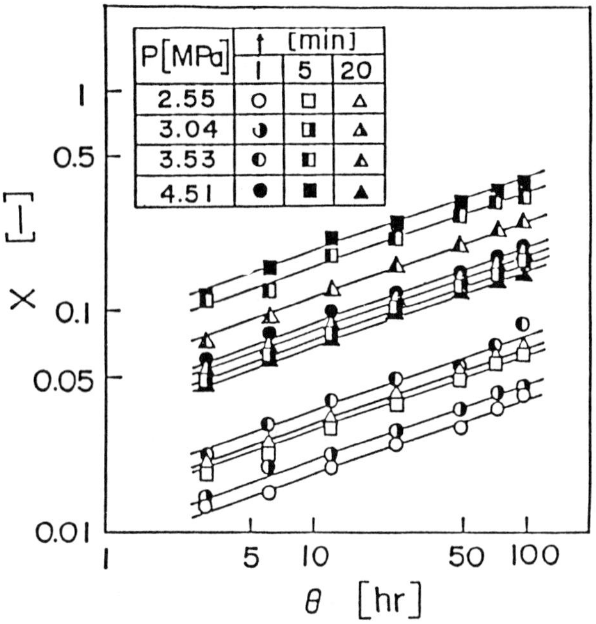

FIGURE 7. Effects of explosion on saccharification of wood.

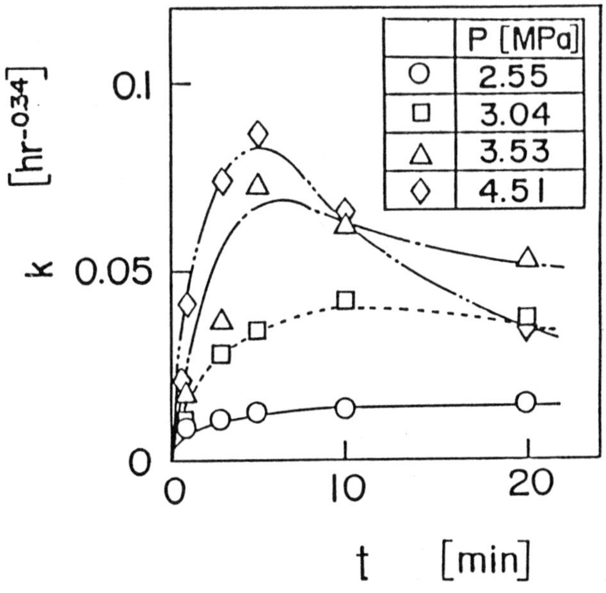

FIGURE 8. Relationship between saccharification rate constant and reaction time.

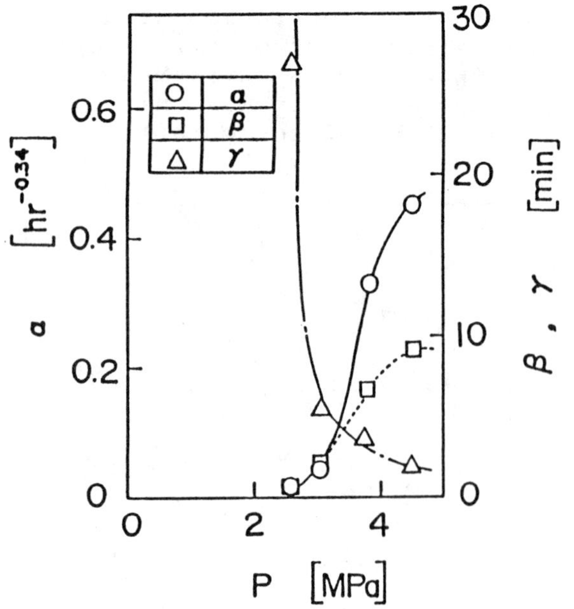

FIGURE 9. Steam pressure dependence on α, β, and γ.

Figure 11 shows effects of steam pressure and reaction time on alcohol production. Alcohol production increased remarkably with reaction time. At 3.53 MPa, alcohol concentration attained to the maximum value at 8 min and decreased with reaction time. Similarly, at 4.51 MPa, it attained to the maximum value at 3 min, then decreased. As a results, it was found that the explosion with high steam pressure and short reaction time was efficient pretreatment of plants materials for enzymatic hydrolysis and alcohol fermentation. Therefore, we expect a practical application for the effective utilization of wood by using the combination system of three operations such as explosion, enzymatic hydrolysis and alcohol fermentation.

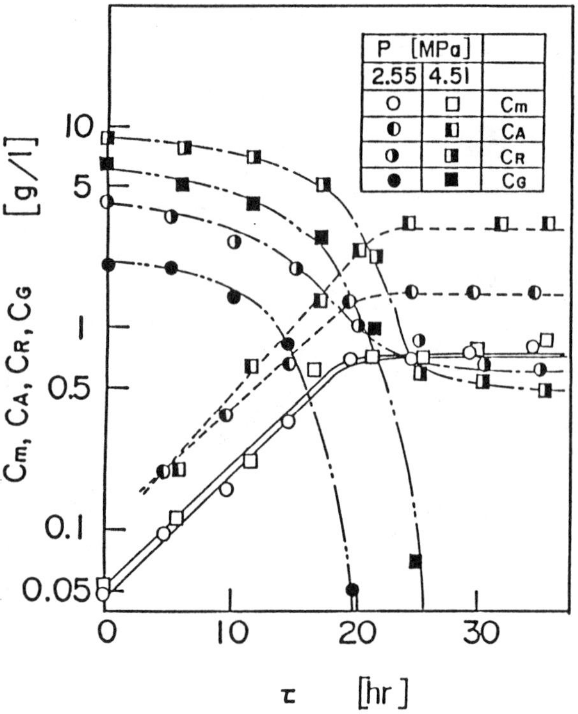

FIGURE 10. Alcohol fermentation of exploded wood.

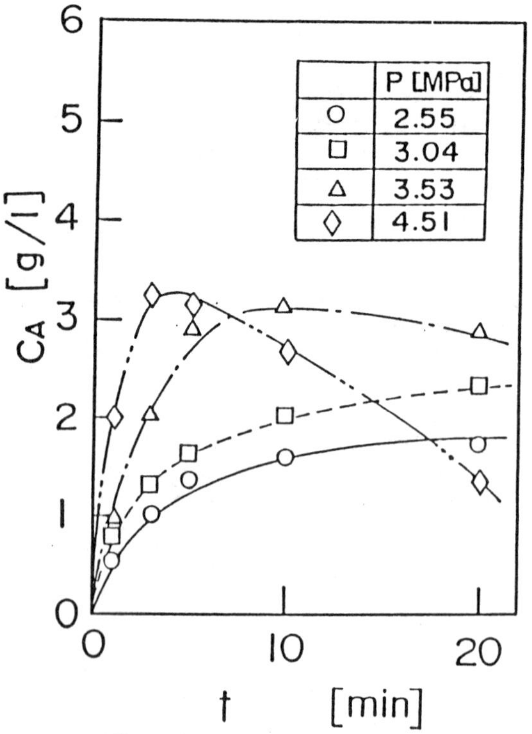

FIGURE 11. Effects of steam pressure and reaction time on alcohol fermentation.

CONCLUSION

The experiments of explosion for the effective utilization of biomass were carried out under various operational conditions. Many of the various macromollecule compounds in wood converted to the lower molecular weight materials in just a short period by the hydrolysis with steam under higher temperature, and their lower molecular weight components were able to be separated relatively easily to the four components such as cellulose, hemicellulose, methanol-soluble lignin and Klason lignin. This investigation will present the basic data required for practical system of separation of wood into main macromolecular components by operations of explosion and extraction, and enzymatic hydrolysis and alcohol fermentation of exploded samples.

NOTATION

C = concentration, g/l
f = particule frequency distribution, %/μm
k = experimental constant defined by Eq.(1), $hr^{-0.34}$
L = length of exploded wood fiber, μm
n = experimental constant defined by Eq.(1), -
P = steam pressure, MPa
t = reaction time, min
W = width of exploded wood fiber, μm
X = saccharification, -
α = experimental constant defined by Eq.(4), $hr^{-0.34}$
β = experimental constant defined by Eq.(4), min
γ = experimental constant defined by Eq.(4), min
θ = enzymatic hydrolysis time, hr
τ = fermentation time, hr
ξ = ratio of extractive component to dry weight of exploded wood, -

(Subscripts)
A = alcohol
C = cellulose
G = glucose
H = hemicellulose
K = Klason lignin
m = cell mass
R = reducing sugar
S = soluble lignin

LITERATURE CITED

1. Maarchessaults, R.H. and J. St-Pierre, Proceedings of Chemrawn Conference, Tront (1978).

2. Morikawa, H., Kagaku to Seibutsu, 19, 286 (1981).

3. Tanahashi, M. and T. Higuchi, Kobunshi Kako, 32, 39 (1983).

4. "PRO-CELL", STAKE Technology LTD. (1980).

5. Sawada T., M. Kuwahara, Y. Nakamura and H. Suda, Kagaku Kogaku Ronbunshu, 12, 1 (1986).

6. Tanahashi, M., S. Takada, T. Aoki, T. Goto, T. Higuchi and S. Hanai, Wood Research, 69, 36 (1983).

7. Ghose, T.K., Biotechnol. Bioeng., 11, 239 (1969).

8. Ishihara, M., Mokuzai Kogyo, 34, 192 (1979).

9. Kumakura, M. and I. Kaetsu, Biotechnol. Bioeng., 20, 1309 (1978).

10. Mandels, M., L. Hontz and J. Nystrom, Biotechnol. Bioeng., 16, 1471 (1974).

11. Wayman, M. and M.G.S. Chua, Can. J. Chem., 57, 1141 (1979).

12. Somogyi, M., J. Biol. Chem., 195, 19 (1952).

13. Walseth, C.S., Tappi, 35, 228 (1952).

Index